STUDY GUIDE
WITH SELECTED SOLUTIONS

TO ACCOMPANY

AN INTRODUCTION TO

CHEMISTRY

MARK BISHOP
MONTEREY PENINSULA COLLEGE

Cover Credit: Tony Asaro, Blakeley Kim, Emiko Rose Koike

ISBN: 0-8053-3211-1

1 2 3 4 5 6 7 8 9 10 -DPC- 04 03 02 01

www.aw.com/bc

Benjamin
Cummings

CONTENTS

CHAPTER 1   An Introduction to Chemistry   1

CHAPTER 1A   A Brief Introduction to Unit Conversions   9

CHAPTER 2   The Structure of Matter and the Chemical Elements   25

CHAPTER 3   Chemical Compounds   35

CHAPTER 4   An Introduction to Chemical Reactions   53

CHAPTER 5   Acids, Bases, and Acid-Base Reactions   65

CHAPTER 6   Oxidation-Reduction Reactions   83

CHAPTER 7   Energy and Chemical Reactions   103

CHAPTER 8   Unit Conversions   119

CHAPTER 9   Chemical Calculations and Chemical Formulas   139

CHAPTER 10   Chemical Calculations and Chemical Equations   159

CHAPTER 11   Modern Atomic Theory   185

CHAPTER 12   Molecular Structure   195

CHAPTER 13   Gases   211

CHAPTER 14   Liquids: Condensation, Evaporation, and Dynamic Equilibrium   233

CHAPTER 15   Solution Dynamics   251

CHAPTER 16   The Process of Chemical Reactions   267

CHAPTER 17   An Introduction to Organic Chemistry, Biochemistry, and Synthetic Polymers   289

CHAPTER 18   Nuclear Chemistry   309

Preface

This is the Student Study Guide for *An Introduction to Chemistry*. It provides many tools that are meant to help you succeed in your chemistry course. There are three necessary ingredients for this success: *time*, *organization*, and *practice*. I cannot help you find more time to study, but I hope this study guide helps you to study more efficiently, so you can learn more in the time you have available. If you learn to use the tools found in this guide, I think it will help you get organized and help make your practice of chemistry-related tasks more efficient.

Let's start with getting organized. It is a good idea to get a skeletal idea of the information in the chapter before you read it carefully. Learning specialists often suggest that readers first skim through the chapter, looking at the section headings and anything else that will give them a general idea of what the chapter contains. The chapters in this study guide make this process easier. Each study guide chapter begins with a list of major and minor section headings for the corresponding text chapter that tell you the topics covered. Next comes a list of the general goals of each section and a brief description of what each section covers. This is followed by what is often called a concept map, which shows how all of the topics in the chapter are interconnected and how information from previous chapters support the present chapter. All of these things provide you with an understanding of the structure of the chapter. When you begin to read the chapter, you already know, in general, what to expect, so it is easier to make sense of the new information you find there.

The Chapter Checklist found in each chapter of this study guide is meant to further support your attempt to get organized. It is a checklist of things you should do to master the topics in each text chapter most efficiently. It might be a good idea to check off each item on this list as you do them. The list includes[1]

- ☐ Read the Review Skills section at the beginning of the text chapter. If there is any skill mentioned that you have not yet mastered, review the material on that topic before reading the present chapter.
- ☐ Read the chapter quickly before the lecture that describes it.
- ☐ Attend class meetings, take notes, and participate in class discussions.
- ☐ Work the Chapter Exercises, perhaps using the Chapter Examples as guides.
- ☐ Study the Chapter Glossary and test yourself on our Web site:
  ### www.chemplace.com/college/
- ☐ Study all of the Chapter Objectives. You might want to write a description of how you will meet each objective. (The most important chapter objectives are often listed.)
- ☐ Memorize (A list of important things to know are listed.)
- ☐ To get a review of the most important topics in the chapter, fill in the blanks in the Key Ideas section.
- ☐ Work all of the selected problems at the end of the chapter, and check your answers with the solutions provided in this chapter of the study guide.
- ☐ Ask for help if you need it.

---

[1] If you have not already done so, you might want to read Chapter 1 of the text. It suggests some important study strategies and describes many of text features with which you want to be familiar before you read more of this preface.

If you have access to a computer connected to the Internet, you might want to take advantage of the study aids provided by our Web site. Each chapter in this study guide lists the Web tools that are related to the text material and that were created at the time of this writing. There will be other tools created with time, so it is a good idea check the Web site regularly to see what else might be available.

www.chemplace.com/college/

Now we turn to another necessary ingredient for success in chemistry—*practice*. You are not going to master the important chemistry-related skills from each chapter without practice, and the selected problems at the end of each chapter in the text provide the opportunity to do that. The numbers for the selected problems are highlighted in the text. This study guide gives you complete solutions to these selected problems. I recommend that you work the problems first without looking at the solution. If you have trouble, it might be best to try to find a similar example in the text chapter for help. When you have either completed the problem or decided that you cannot do it, then check the solution for the problem in this study guide. The solutions not only give you the correct answer, but they often provide you with models for how you can set up problems and how you can reason out their solutions.

If you have the time, if you are organized, and if you practice, you'll be successful as a chemistry student. I hope this study guide helps. Good luck.

Mark Bishop

Monterey Peninsula College

Chapter 1
An Introduction to Chemistry

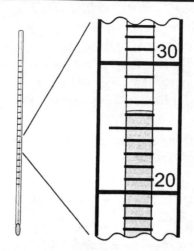

1.1 What Is Chemistry, and What Can Chemistry Do for You?
   Special Topic 1.1: Green Chemistry
1.2 Suggestions for Studying Chemistry
1.3 The Scientific Method
1.4 Measurement and Units
   • The International System of Measurement
   Special Topic 1.2: Wanted: A New Kilogram
   • SI Units Derived from Standards
   • SI Units Derived from Metric Prefixes
   • More About Length Units
   • More About Volume Units
   • Mass and Weight
   • Temperature
1.5 Reporting Values from Measurements
   • Accuracy and Precision
   • Describing Measurements
   • Digital Readouts
♦ Chapter Glossary
   Internet: Glossary Quiz
♦ Chapter Objectives
Key Ideas
Chapter Problems

Section Goals and Introductions

Section 1.1 What Is Chemistry, and What Can Chemistry Do for You?

Goal: To explain a bit about what chemistry is and why it's important.

This section shows you some of the questions that an understanding of chemistry helps to answer and some of the issues of concern to chemists. Throughout the text, you will see Special Topics, such as *Special Topic 1.1: Green Chemistry*, that will reinforce the attitude that chemistry is an important science that expands our understanding of the world around us and helps us to change that world, often in ways that are beneficial to us and to our environment.

This section ends with a commercial message: you cannot get the benefits that an understanding of chemistry brings without first concentrating on the basics, which are not always very interesting and which do not always seem directly related to the real world. Trust me. If you master the basics, you will be explaining the way the physical world works to your friends and family in no time.

Section 1.2 Suggestions for Studying Chemistry

Goals

- *To suggest some study strategies.*
- *To introduce you to some of the unique components of the text.*

It's very important that you understand from the beginning of the course that learning chemistry is a time-consuming task that is best approached in a logical and efficient way. So one goal of this section is to make some suggestions to you about how you might study most efficiently.

Another goal is to be sure that you know the tools that you have available to help you. Be sure you know where to find and how to use each of the following: Review Skills, Examples, Exercises, Glossary, Objectives, and End-of-Chapter Problems.

I want to stress the importance of one of these somewhat unique components, the Objectives. An attempt has been made to write an objective for every skill that you should learn from the text. If you can meet the objectives, you will ace the exams. Be sure to ask your instructor about changes to the list. It's difficult to use someone else's objectives, so your instructor is likely to add objectives and eliminate some of those found in the text.

Section 1.3 The Scientific Method

Goal: To give you an idea of how science is done.

This section describes one way that science is done and shows how this method was applied in the development of an understanding of Parkinson's disease and in the development of treatments for it.

Section 1.4 Measurement and Units

Goal: To introduce units of measurement.

In this section, you will learn that a *value* contains a *number* and a *unit*, and you will learn a lot about the units used in the International System of Measurement (SI). When you are done studying this section, be sure that you know the common SI units used to describe length, volume, mass, and temperature. It is important that you be able to write the relationship between metric units derived from the metric prefixes and the base unit for that same type of measurement. (See Example 1.1.) This section also shows you the relative sizes of English and metric units and explains the difference between mass and weight (two terms that are often confused).

Section 1.5 Reporting Values from Measurements

Goal: To show how scientists report values from measurements.

This section is especially important for students who are taking a course with a laboratory section included. It describes some of the factors that a scientist considers when deciding how to report values derived from measurements.

Chapter Map (Sections 1.4 and 1.5)

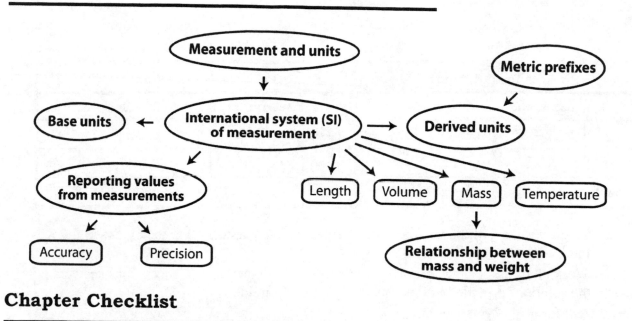

Chapter Checklist

- ☐ Read the chapter quickly before the lecture that describes it.
- ☐ Attend class meetings, take notes, and participate in class discussions.
- ☐ Work the Chapter Exercises, perhaps using the Chapter Examples as guides.
- ☐ Study the Chapter Glossary and test yourself on our Web site:

  www.chemplace.com/college/
- ☐ Study all of the Chapter Objectives. You might want to write a description of how you will meet each objective. (Although it is best to master all of the objectives, the following objectives are especially important because they pertain to skills that you will need while studying other chapters of this text: 3, 4, 5, 6, 7, 10, 11, 14, and 19.)

☐ Memorize the following. (Be sure to check with your instructor to determine how much you are expected to know of the following.)

- *SI base units* (Table 1.1) The table below contains the four base units that you should know now.

Type of Measurement	Base Unit	Abbreviation
Length	meter	m
mass	kilogram	kg
time	second	s
temperature	kelvin	K

- *Metric prefixes* (Table 1.2) The table below contains the most common of these prefixes.

Prefixes for large units			Prefixes for small units		
Prefix	Abbreviation	Value	Prefix	Abbreviation	Value
giga	G	1,000,000,000 or 10^9	centi	c	0.01 or 10^{-2}
mega	M	1,000,000 or 10^6	milli	m	0.001 or 10^{-3}
kilo	k	1000 or 10^3	micro	μ	0.000001 or 10^{-6}

☐ To get a review of the most important topics in the chapter, fill in the blanks in the Key Ideas section.

☐ Work all of the selected problems at the end of the chapter, and check your answers with the solutions provided in this chapter of the study guide.

☐ Ask for help if you need it.

Web Resources   www.chemplace.com/college/

The Web resources that are available for this course require that you have the tools listed below. As of this writing, the following links will take you to Web sites where you can download the latest browsers and the Shockwave and Chime plug-ins.

Netscape Browser: **home.netscape.com/comprod/mirror/index.html**

MS Explorer Browser: **www.microsoft.com/windows/ie/download/windows.htm**

Shockwave: **sdc.shockwave.com/shockwave/download/**

Chime: **www.mdli.com/download/chimedown.html**

There's a glossary quiz in Chapter 1 of our Web site that provides the definitions for each of the glossary terms and asks you to type in the term.

www.chemplace.com/college/

Exercises Key

Exercise 1.1 - Units Derived from Metric Prefixes: Complete the following relationships. Rewrite the relationships using abbreviations for the units. *(Obj #7)*

  a. 1 megagram = **10^6** gram   b. 1 milliliter = **10^{-3}** liter   *Obj #6*

Exercise 1.2 – Uncertainty: If you are given the following values that are derived from measurements, what will you assume is the range of possible values that each represents? *(Obj #19)*

  a. 72 mL   **71 mL to 73 mL**
  b. 8.23 m   **8.22 m to 8.24 m**
  c. 4.55×10^{-5} g   **4.54×10^{-5} g to 4.56×10^{-5} g**

Exercise 1.3 - Uncertainty in Measurement: Let's assume that four members of your class are asked to measure the mass of a dime. The reported values are 2.302 g, 2.294 g, 2.312 g, and 2.296 g. The average of these values is 2.301 g. Considering the values reported and the level of care you expect beginning chemistry students to take with their measurements, how would you report the mass so as to communicate the uncertainty of the measurement? *(Obj #18)*

  2.30 g because the reported values differ by about ±0.01.

Key Ideas Answers

1. Complete this brief description of common steps in the development of scientific ideas: The process begins with **observation** and the collection of **data**. Next, scientists make an initial **hypothesis**. This leads to a more purposeful collection of information in the form of systematic **research or experimentation**. The hypothesis is refined on the basis of the new information, and **research** is designed to test the hypothesis. The results are **published** so that other scientists might repeat the research and confirm or refute the conclusions. If other scientists confirm the results, the hypothesis becomes accepted in the scientific community. The next step of this scientific method is a search for useful **applications** of the new ideas. This often leads to another round of **hypothesizing and testing** in order to refine the applications.

3. The **meter**, which has an abbreviation of **m**, is the accepted SI base unit for length.

5. The **second**, which has an abbreviation of s, is the accepted SI base unit for **time**.

7. Many properties cannot be described directly with one of the seven SI **base units**. Rather than create new definitions for new units, we **derive** units from the units of meter, kilogram, second, kelvin, mole, ampere, and candela.

9. An object's weight on the surface of the earth depends on its **mass** and on the **distance** between it and the center of the earth.

11. The thermometers that scientists use to measure temperature generally provide readings in degrees **Celsius**, but scientists usually convert these values into **Kelvin** values to do calculations.

13. One of the conventions that scientists use for reporting measurements is to report all of the **certain** digits and one **estimated** (and thus uncertain) digit.

PROBLEMS KEY

Section 1.4 Measurement and Units

15. Complete the following table by writing the property being measured (mass, length, volume, or temperature) and either the name of the unit or its abbreviation. *(Objs #3 and 7)*

Unit	Type of measurement	Abbreviation	Unit	Type of measurement	Abbreviation
megagram	**mass**	**Mg**	nanometer	**length**	**nm**
milliliter	**volume**	**mL**	kelvin	**temperature**	**K**

17. Convert the following ordinary numbers to scientific notation. (See Appendix B at the end of the textbook if you need help with this.)

    a. $1,000 = \mathbf{10^3}$

    b. $1,000,000,000 = \mathbf{10^9}$

    c. $0.001 = \mathbf{10^{-3}}$

    d. $0.000000001 = \mathbf{10^{-9}}$

19. Convert the following numbers expressed in scientific notation to ordinary numbers. (See Appendix B at the end of the textbook if you need help with this.)

    a. $10^7 = \mathbf{10,000,000}$

    b. $10^{12} = \mathbf{1,000,000,000,000}$

    c. $10^{-7} = \mathbf{0.0000001}$

    d. $10^{-12} = \mathbf{0.000000000001}$

21. Complete the following relationships between units. *(Objs #6, 10, and 14)*

    a. $\mathbf{10^3}\, m = 1\, km$

    b. $\mathbf{10^{-3}}\, L = 1\, mL$

    c. $\mathbf{10^6}\, g = 1\, Mg$

    d. $1\, cm^3 = 1\, mL$

    e. $\mathbf{10^3}\, kg = 1\, t$ (t = metric ton)

23. Would each of the following distances be closest to a millimeter, a centimeter, a meter, or a kilometer? *(Obj #8)*

    a. the width of a bookcase   **meter**

    b. the length of an ant   **centimeter**

    c. the width of the letter "t" in this phrase   **millimeter**

    d. the length of the Golden Gate Bridge in San Francisco   **kilometer**

25. Which is larger, a centimeter or an inch?

    There are 2.54 centimeters per inch, so an inch is larger.

27. Would the volume of each of the following be closest to a milliliter, a liter, or a cubic meter? *(Obj #9)*

    a. a vitamin tablet   **milliliter**

    b. a kitchen stove and oven   **cubic meter**

    c. this book   **liter**

29. Which is larger, a milliliter or a fluid ounce?

    There are 29.57 milliliters per fluid ounce, so a fluid ounce is larger.

31. Explain the difference between mass and weight. *(Obj #11)*

Mass is usually defined as a measure of the amount of matter in an object. The weight of an object, on Earth, is a measure of the force of gravitational attraction between the object and Earth. The more mass an object has, the greater the gravitational attraction between it and another object. The farther an object gets from Earth, the less that attraction is, and the lower its weight. Unlike the weight of an object, the mass of an object is independent of location. Mass is described with mass units, like grams and kilograms. Weight can be described with force units, like newtons.

32. Which is larger, a gram or an ounce?

There are 28.35 grams per ounce, so an ounce is larger.

36. Which is larger, a degree Celsius or a degree Fahrenheit?

There are 1.8 degrees Fahrenheit per degree Celsius, so a degree Celsius is larger.

38. Which is the smallest increase in temperature: 10 °C (such as from 100 °C to 110 °C), 10 K (such as from 100 K to 110 K), or 10 °F (such as from 100 °F to 110 °F)? *(Obj #16)*

   10 °F

Section 1.5 Reporting Values from Measurements

41. Given the following values that are derived from measurements, what do you assume is the range of possible values that each represents? *(Obj #19)*

We assume ±1 in the last decimal place reported.

   a. 30.5 m (the length of a whale)

   30.5 m means 30.5 ± 0.1 m or 30.4 m to 30.6 m.

   b. 612 g (the mass of a basketball)

   612 g means 612 ± 1 g or 611 g to 613 g.

   c. 1.98 m (Michael Jordan's height)

   1.98 m means 1.98 ± 0.01 m or 1.97 m to 1.99 m.

   d. 9.1096×10^{-28} g (the mass of an electron)

   9.1096×10^{-28} g means $(9.1096 \pm 0.0001) \times 10^{-28}$ g or 9.1095×10^{-28} g to 9.1097×10^{-28} g.

   e. 1.5×10^{18} m³ (the volume of the ocean)

   1.5×10^{18} m³ means $(1.5 \pm 0.1) \times 10^{18}$ m³ or 1.4×10^{18} m³ to 1.6×10^{18} m³.

43. The accompanying drawings show portions of metric rulers on which the numbers
    correspond to centimeters. The dark bars represent the ends of objects being measured.
    (Obj #18)

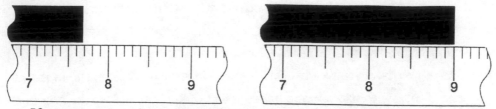

a. If you were not given any specific instructions for reporting your values, what length
   would you record for each of these measurements?

   It's difficult to estimate the hundredth position accurately. For the object on the
   left, we might report 7.67 cm, 7.68 cm, or 7.69. The end of the right object
   seems to be right on the 9 cm mark, so we report 9.00 cm.

b. If you were told that the lines on the ruler are drawn accurately to ±0.1 cm, how
   would you report these two lengths?

   7.7 cm and 9.0 cm

45. At a track meet, three different timers report the times for the winner of a 100-m sprint as
    10.51 s, 10.32 s, and 10.43 s. The average is 10.42 s. How would you report the time of the
    sprinter in a way that reflects the uncertainty of the measurements? *(Obj #18)*

    Our uncertainty is in the tenth position, so we report 10.4 s.

47. The image below represents the digital display on a typical electronic balance. *(Obj #19)*

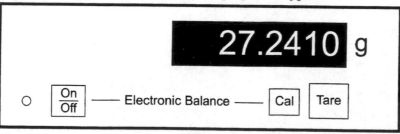

a. If the reading represents the mass of a solid object that you carefully cleaned and
   dried and then handled without contaminating it, how would you report this mass?

   27.2410 g

b. Now assume that the reading is for a more casually handled sample of a liquid and its
   container. Let's assume not only that you were less careful with your procedure this
   time but also that the liquid is evaporating rapidly enough for the reading to be
   continually decreasing. In the amount of time that the container of liquid has been
   sitting on the pan of the balance, the mass reading has decreased by about 0.001 g.
   How would you report the mass?

   Our convention calls for only reporting one uncertain digit in our value. Because
   we are uncertain about the thousandth position, we might report 27.241 g (or
   perhaps even 27.24 g).

Inter-Chapter 1A
A Brief Introduction to Unit Conversions

$$? \, cm = 947 \, mm \left(\frac{10^{-3} \, m}{1 \, mm} \right) \left(\frac{1 \, cm}{10^{-2} \, m} \right) = 94.7 \, cm$$

Desired unit — Given value — Final answer with desired unit

Converts given metric unit into metric base unit

Converts metric base unit into desired unit

♦ Review Skills

1A.1 Common Unit Conversions
  • Metric-Metric Unit Conversions
  • Other Types of Conversions

1A.2 Rounding Off and Significant Figures
  • Rounding Off Answers Derived from Multiplication and Division
  • Rounding Off Answers Derived from Addition and Subtraction
  • Density, Density Calculations, and Rounding
  • A Summary of Dimensional Analysis Steps

♦ Inter-Chapter Glossary
♦ Inter-Chapter Objectives

Key Ideas

Chapter Problems

As the introduction to Inter-Chapter 1A in the text describes, the "inter-chapter" and Chapter 8 both describe the procedures for doing a certain kind of calculation that is very useful in chemistry, but the inter-chapter does so more briefly. Because the inter-chapter allows several possible approaches to learning chemistry's math-related topics, it's important that you check with your instructor to find out which of these options is appropriate for your course.

- Some instructors choose to delay the discussion of the unit conversions until the middle of the course, preferring to present various basic chemical concepts first. These teachers will tell you not to read the inter-chapter.

- Some instructors provide a brief introduction to the unit conversions early in the course, perhaps to prepare you to make simple unit conversions in early laboratory sessions, while postponing the more comprehensive treatment of the subject until later. If your instructor falls into this category, you will read the chapters of this text, including this inter-chapter, in the order that you find them.

- If your instructor wants a broader coverage of the math-related topics early in the course, you will be asked to skip from here to Chapter 8, which describes unit conversions in much more detail than this inter-chapter does. Chapter 8 was written in such a way that you can read it without confusion before reading Chapters 2 through 7.

- Some instructors will ask their students to read this inter-chapter and only one or two of the sections in Chapter 8 before proceeding to Chapter 2.

Section Goals and Introductions

Be sure that you can do the things listed in the Review Skills section before you spend too much time studying this chapter. They are especially important. You might also want to look at Appendices A, B, and C. Appendix A (*Measurement and Units*) provides tables that show units, their abbreviations, and relationships between units that lead to conversion factors. Appendix B (*Scientific Notation*) describes how to convert between regular decimal numbers and numbers expressed in scientific notation, and it shows how calculations using scientific notation are done. Appendix C (*Using a Scientific Calculator*) shows how calculations, such as those found in this chapter, can be done using common scientific calculators.

Section 1A.1 Common Unit Conversions

Goals

- *To introduce a procedure for making unit conversions called dimensional analysis.*
- *To describe a strategy for making metric-metric unit conversions.*
- *To describe how conversion factors can be derived from word problems.*
- *To describe English-metric unit conversions.*

Many chemical calculations include the conversion from a value expressed in one unit to the equivalent value expressed in a different unit. Dimensional analysis, which is described in this section, provides you with an organized format for making these unit conversions and gives you a logical thought process that will help you to reason through such calculations. You may think that you can make the unit conversions in this section more easily without the dimensional analysis technique, but keep in mind that the purpose of this section is to teach you a general technique that allows you to work through many types of problems. You will learn a stepwise procedure that allows you to chip away at a calculation one step at a time. This process allows you to not have to see the whole solution for the problem from the beginning. It is extremely important that you master this technique. You'll be glad you have when you go on to later chapters.

If you have not already done so, there are a few things that you will probably want to memorize from Chapter 1. These include the base units for the International System of Units and their abbreviations, some of the metric prefixes, and some English-metric conversion factors. Be sure to check with your instructor to find out the types of conversions you will be expected to make and to find out which conversion factors you will be expected to know.

One of the most important skills to master for science and math classes is the ability to work *word problems*. One very useful strategy for such problems is to get in the habit of extracting the values (numbers and units) from the problem and writing them down. As you do this, you should be looking for things that can be read as *something per something*. These can be written as ratios and used as conversion factors.

Section 1A.2 Rounding Off and Significant Figures

Goals

- *To describe the procedures for rounding off answers to calculations.*
- *To describe what density is, how it can be used as a conversion factor, and how density can be calculated.*

When you use a calculator to complete your calculations, it's common that most of the numbers you see on the display at the end of the calculation are meaningless. This section shows you simple techniques that you can use to round off your answers. Your instructor might penalize you every time you report an answer that is incorrectly rounded, so mastering the procedures for rounding might be very important for your grade. Be sure that you understand that the guidelines for rounding answers derived from addition and subtraction are different from those used for answers derived from multiplication and division.

Density calculations are common in chemistry. This section describes how densities can be calculated and how they can be used to convert between mass and volume. Density calculations provide more examples that will help you master the dimensional analysis technique and the procedures for rounding.

Chapter Map

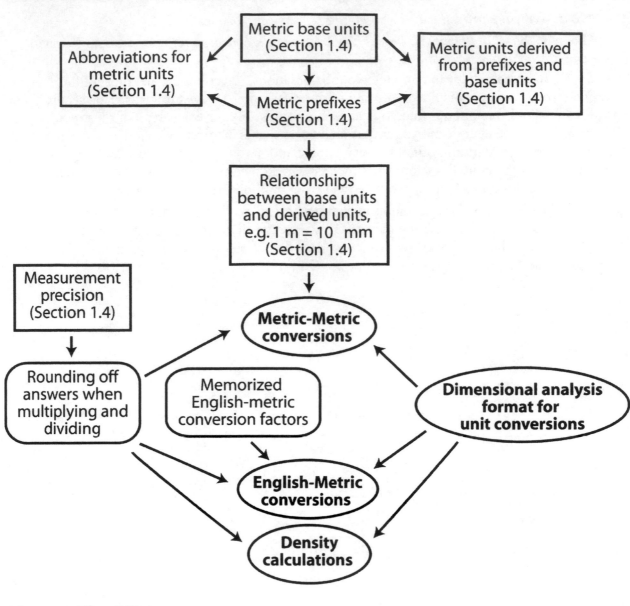

Chapter Checklist

- ☐ Read the chapter quickly before the lecture that describes it.
- ☐ Attend class meetings, take notes, and participate in class discussions.
- ☐ Work the Chapter Exercises, perhaps using the Chapter Examples as guides.
- ☐ Study the Chapter Glossary.
- ☐ Study all of the Chapter Objectives. You might want to write a description of how you will meet each objective. All of these objectives will be important for later chapters.

☐ Memorize at least some of the following English-metric conversion factors. Be sure to check with your instructor to determine how many of these you are expected to know and ask whether there are others that your instructor wants to add.

Type of Measurement	Probably most useful to know	Others useful to know		
Length	$\dfrac{2.54 \, cm}{1 \, in.}$ (exact)	$\dfrac{1.609 \, km}{1 \, mi}$	$\dfrac{39.37 \, in.}{1 \, m}$	$\dfrac{1.094 \, yd}{1 \, m}$
Mass	$\dfrac{453.6 \, g}{1 \, lb}$	$\dfrac{2.205 \, lb}{1 \, kg}$		
Volume	$\dfrac{3.785 \, L}{1 \, gal}$	$\dfrac{1.057 \, qt}{1 \, L}$		

☐ To get a review of the most important topics in the chapter, fill in the blanks in the Key Ideas section.

☐ Work all of the selected problems at the end of the chapter, and check your answers with the solutions provided in this chapter of the study guide.

☐ Ask for help if you need it.

Exercises Key

Exercise 1A.1 – Conversion Factors: Write four conversion factors that relate millimeters and meters. *(Obj #2)*

$$\left(\frac{1 \, mm}{10^{-3} \, m}\right) \left(\frac{10^{-3} \, m}{1 \, mm}\right) \left(\frac{1 \, m}{10^3 \, mm}\right) \left(\frac{10^3 \, mm}{1 \, m}\right)$$

Exercise 1A.2 - Unit Conversions: A basketball has a mass of 612 g. Use the dimensional analysis technique to determine its mass in kilograms. *(Obj #3)*

$$? \, kg = 612 \, g \left(\frac{1 \, kg}{10^3 \, g}\right) = \textbf{0.612 kg}$$

Exercise 1A.3 - Unit Conversions: A particular raindrop has a volume of 0.010 mL. Use the dimensional analysis technique to determine its approximate volume in microliters. *(Obj #3)*

$$? \, \mu L = 0.010 \, mL \left(\frac{1 \, L}{10^3 \, mL}\right) \left(\frac{10^6 \, \mu L}{1 \, L}\right) = \textbf{10 \mu L}$$

✍ Exercise 1A.4 - Rounding Off Answers Derived from Multiplication and Division:

Plants produce carbohydrates through a process known as photosynthesis. In fact, photosynthesizing plants produce an estimated 1.7×10^{11} Mg of carbohydrates per year, which is about 2.8×10^4 kg per person. The dimensional analysis setup that follows shows how to calculate the number of people on earth from this data. Determine the number of significant figures in each inexact value, calculate the answer, and report it to the correct number of significant figures. (You may want to consult Appendix B, "Scientific Notation," and Appendix C, "Using a Scientific Calculator," for help with this.) *(Obj #7)*

$$? \text{ people} = 1.7 \times 10^{11} \text{ Mg carbohydrates} \left(\frac{10^6 \text{ g}}{1 \text{ Mg}}\right) \left(\frac{1 \text{ kg}}{10^3 \text{ g}}\right) \left(\frac{1 \text{ person}}{2.8 \times 10^4 \text{ kg}}\right)$$

Step 1: The numbers 1.7×10^{11} and 2.8×10^4 are not defined, and they cannot be counted so they are not exact. We can assume that they come from a mixture of measurements and calculations. Both 10^6 and 10^3 come from definitions of metric prefixes, so they are exact.

Step 2: Nonzero digits are significant and the exponential term of a number expressed in scientific notation is considered exact, so 1.7×10^{11} and 2.8×10^4 both have two significant figures.

Step 3: When multiplying and dividing, we round our answer off to the same number of significant figures as the value that contains the fewest significant figures, so we report two significant figures in this answer. The calculator reports 6.0714×10^9, which we round off to **6.1×10^4 people**. We need scientific notation to unambiguously report two significant figures.

✍ Exercise 1A.5 – Rounding Off Answers Derived from Addition and Subtraction:

One way to generate electricity is to heat liquid water to steam, which can be used to turn a steam turbine generator. The hot steam is then cooled back to the liquid form in a process that makes use of water comes from natural systems, such as rivers or the ocean. When the water used for cooling is returned to its source, its temperature is 5 K to 10 K higher than before. Thus, water removed at 287.6 K might be returned at 295.7 K, which in some cases is high enough to kill native fish and plants. Convert 295.7 K to °C. (To convert from kelvins to degrees Celsius, subtract 273.15 from the value in K.) *(Obj #8)*

295.7 K − 273.15 = **22.6 °C**

Step 1: The number 295.7 comes from a measurement, so it is not exact. The number 273.15 comes from a definition, so it is exact.

Step 2: The number 295.7 is precise to the tenths position.

Step 3: We round our answer to the tenth position.

⚑ **Exercise 1A.6 - Density and Rounding Off Answers Derived from Multiplication and Division:** High-density polyethylene (HDPE) is a strong, stiff, opaque polymer that is used to make many things, including grocery bags, pipes, and packaging. What is the mass, in kilograms, of 834.6 mL of HDPE that has a density of 0.960 g/mL? (*Objs #7 and 9*)

$$? \, g = 834.6 \text{ mL} \left(\frac{0.960 \, g}{1 \text{ mL}} \right) \left(\frac{1 \, kg}{10^3 \, g} \right) = \textbf{0.801 kg}$$

⚑ **Exercise 1A.7 - Density Calculations:** The following data is collected in a procedure designed to determine the density of a sample of gasoline. First, a graduated cylinder is weighed and found to have a mass of 47.356 g. Next, gasoline is added to the cylinder, and the total mass is found to be 83.836 g. The difference between these two values corresponds to the mass of the gasoline. The volume reading on the graduated cylinder is 45.6 mL. What is the gasoline's density in g/mL? (*Obj #10*)

$$\frac{? \, g}{\text{mL}} = \frac{(83.836 - 47.356) \, g}{45.6 \text{ mL}} = \frac{36.480 \, g}{45.6 \text{ mL}} = \textbf{0.800 g/mL}$$

Key Ideas Answers

1. The **unit-conversion** technique called dimensional analysis is a simple, reliable technique that can be applied to many different types of conversions. It provides a stepwise thought process for solving a great many of the **numerical problems** that arise in chemistry.

3. A conversion factor is a ratio that describes a(n) **relationship between two units**.

5. In converting from one metric unit with a prefix to a related metric unit with a different prefix (from millimeters to centimeters, for example), it is often a good strategy to convert from the given unit to the **base unit** using one conversion factor and then to convert from the base unit to the desired unit with a second conversion factor.

7. The number of significant figures in a value reflects the value's **degree of uncertainty**; a *smaller* number of significant figures indicates a **greater** uncertainty.

9. Numbers in metric-metric unit conversion factors that are derived from the metric prefixes are **exact**.

11. Values that come from measurements are **never** exact.

13. When adding or subtracting, round your answer to the same number of **decimal places** as the inexact value with the **fewest decimal places**.

Problems Key

Section 1a.1 An Introduction to Unit Conversions

14. Convert the following ordinary decimal numbers to scientific notation. (See Appendix B at the end of the textbook if you need help with this.)

    a. $4{,}239 = \mathbf{4.239 \times 10^3}$

    b. $5{,}723{,}845 = \mathbf{5.723845 \times 10^6}$

    c. $0.000415 = \mathbf{4.15 \times 10^{-4}}$

    d. $0.000000001623 = \mathbf{1.623 \times 10^{-9}}$

16. Convert the following numbers expressed in scientific notation to ordinary decimal numbers. (See Appendix B at the end of the textbook if you need help with this.)

    a. $2.76423 \times 10^5 = \mathbf{276{,}423}$ c. $8.8 \times 10^{-7} = \mathbf{0.00000088}$

    b. $6.342 \times 10^2 = \mathbf{634.2}$ d. $5.667 \times 10^{-4} = \mathbf{0.0005667}$

18. Use your calculator to complete the following calculations. (See your calculator's instruction manual or Appendix C at the end of the textbook if you need help with this.)

    a. $27.50 \times 7.000 = \mathbf{192.5}$ c. $27 \times 0.006 \div 8.1 = \mathbf{0.02}$

    b. $7221 \div 14.5 = \mathbf{498}$ d. $(64.240 - 15.568) \div 15.6 = \mathbf{3.12}$

20. Use your calculator to complete the following calculations. (See your calculator's instruction manual or Appendix C at the end of the textbook if you need help with this.)

    a. $10^{12} \times 10^3 = \mathbf{10^{15}}$ d. $10^{12} \times 10^{-16} = \mathbf{10^{-4}}$

    b. $10^6 \div 10^2 = \mathbf{10^4}$ e. $10^9 \div 10^{-2} = \mathbf{10^{11}}$

    c. $10^9 \times 10^3 \div 10^6 = \mathbf{10^6}$ f. $10^{-6} \times 10^3 \div 10^{-2} = \mathbf{10^{-1}}$

22. Use your calculator to complete the following calculations. (See your calculator's instruction manual or Appendix C at the end of the textbook if you need help with this.)

    a. $1.4 \times 10^7 \bullet 6.0 \times 10^4 = \mathbf{8.4 \times 10^{11}}$

    b. $9.39 \times 10^{23} \div 3.00 \times 10^8 = \mathbf{3.13 \times 10^{15}}$

    c. $6.25 \times 10^{12} \times 1.6 \times 10^5 \div 4.0 \times 10^{11} = \mathbf{2.5 \times 10^6}$

    d. $4.2 \times 10^{-5} \bullet 1.5 \times 10^6 = \mathbf{63}$

    e. $3.279 \times 10^{14} \div 10^{-9} = \mathbf{3.279 \times 10^{23}}$

    f. $(8.98 \times 10^{-4} - 4.82 \times 10^{-4}) \div 2.00 \times 10^{-15} = \mathbf{2.08 \times 10^{11}}$

24. Complete each of the following conversion factors by filling in the blank on the top of the ratio.

    a. $\left(\dfrac{\mathbf{10^3}\ g}{1\ kg} \right)$ e. $\left(\dfrac{\mathbf{60}\ min}{1\ h} \right)$

    b. $\left(\dfrac{\mathbf{453.6}\ g}{1\ lb} \right)$ f. $\left(\dfrac{\mathbf{2.54}\ cm}{1\ in.} \right)$

    c. $\left(\dfrac{\mathbf{10^6}\ \mu m}{1\ m} \right)$ g. $\left(\dfrac{\mathbf{12}\ in.}{1\ ft} \right)$

d. $\left(\dfrac{10^{12}\ pm}{1\ m}\right)$ h. $\left(\dfrac{3.785\ L}{1\ gal}\right)$

26. Use the dimensional analysis technique to make the following conversions. *(Objs #2 and 3)*

a. 772 g to kg

$? kg = 772\ g\left(\dfrac{1\ kg}{10^3\ g}\right) = \textbf{0.772 kg}$

b. 0.45 m to cm

$? cm = 0.45\ m\left(\dfrac{1\ cm}{10^{-2}\ m}\right)$

or $? cm = 0.45\ m\left(\dfrac{10^2\ cm}{1\ m}\right) = \textbf{45 cm}$

c. 1.255 kL to L

$? L = 1.255\ kL\left(\dfrac{10^3\ L}{1\ kL}\right) = \textbf{1255 L}$

d. 84 mm to m

$? m = 84\ mm\left(\dfrac{10^{-3}\ m}{1\ mm}\right)$

or $? m = 84\ mm\left(\dfrac{1\ m}{10^3\ mm}\right) = \textbf{0.084 m}$

28. Use the dimensional analysis technique to make the following conversions. *(Objs #2 and 3)*

a. 456.5 kg to Mg

$? Mg = 456.5\ kg\left(\dfrac{10^3\ g}{1\ kg}\right)\left(\dfrac{1\ Mg}{10^6\ g}\right) = \textbf{0.4565 Mg}$

b. 1.549 mm to μm

$? \mu m = 1.549\ mm\left(\dfrac{10^{-3}\ m}{1\ mm}\right)\left(\dfrac{1\ \mu m}{10^{-6}\ m}\right)$

or $? \mu m = 1.549\ mm\left(\dfrac{1\ m}{10^3\ mm}\right)\left(\dfrac{10^6\ \mu m}{1\ m}\right) = \textbf{1549 μm}$

30. An Egyptian pyramid has a mass of about 10^7 Mg. Use the dimensional analysis technique to determine its approximate mass in gigagrams. *(Objs #2 and 3)*

$? Gg = 10^7\ Mg\left(\dfrac{10^6\ g}{1\ Mg}\right)\left(\dfrac{1\ Gg}{10^9\ g}\right) = \textbf{10}^4\ \textbf{Gg}$

32. The accepted SI unit of energy is the joule (J). A 12-inch cheese pizza supplies you with 4940 kJ (kilojoules) of energy. What is this energy value expressed in megajoules? *(Objs #2 and 3)*

$$? \text{ MJ} = 4940 \text{ kJ} \left(\frac{10^3 \text{ J}}{1 \text{ kJ}} \right) \left(\frac{1 \text{ MJ}}{10^6 \text{ J}} \right) = \textbf{4.940 MJ}$$

Section 1a.2 Rounding Off and Significant Figures

34. Assuming that each of the following are not exact, how many significant figures does each number have? *(Obj #7)*

    a. 327.89   **5 significant figures**
    b. 0.214   **3 significant figures**
    c. 1065   **4 significant figures**
    d. 14.002   **5 significant figures**
    e. 782.000   **6 significant figures**

36. Decide whether each of the numbers shown in bold type below is exact or not. If it is not exact, write the number of significant figures in it. *(Objs #6 and 7)*

    a. The estimated mass of the sun, $\textbf{1.989} \times \textbf{10}^{\textbf{30}}$ kg. **Not exact – 4**
    b. A count of **36** natural gas storage tanks at a power plant. **exact**
    c. A measured volume of one of these storage tanks, $\textbf{5.070} \times \textbf{10}^{\textbf{4}} \textbf{ m}^{\textbf{3}}$.
        Not exact – 4
    d. $\dfrac{3 \text{ ft}}{1 \text{ yd}}$ **exact**

    e. $\dfrac{60 \text{ s}}{1 \text{ min}}$ **exact**

    f. $\dfrac{10^3 \text{ mL}}{1 \text{ L}}$ **exact**

    g. $\dfrac{1.609 \text{ km}}{1 \text{ mi}}$ **Not exact – 4**

    h. $\dfrac{2.54 \text{ cm}}{1 \text{ in.}}$ **exact**

    i. The density of magnesite, which is a substance used to make furnace linings, can be calculated from the measured mass of a sample and its measured volume. A portion of magnesite is found to have a mass of **34.100** g **(Not exact – 5)** and a volume of **11.0** mL **(Not exact – 3)**. This yields a density of **3.10** g/mL **(Not exact – 3)**.

38. Report the answers to the following calculations to the correct number of decimal positions. Assume that each number is ±1 in the last decimal position reported. *(Obj #9)*

    f. 37 – 35.4 = **2**   b. 0.9905 + 999.06 = **1000.05**

Because the ability to make unit conversions using the dimensional analysis format is an extremely important skill, be sure to set up each of the following calculations using the dimensional analysis format, even if you see another way to work the problem, and even if another technique seems easier.

40. Amoebas are single-celled animals that move by extending their cellular material outward to form false feet. A typical amoeba is 2.3 mm. What is this length in meters? in centimeters? *(Objs #3 & 8)*

$$? \, m = 2.3 \, mm \left(\frac{1 \, m}{10^3 \, mm} \right) = \mathbf{0.0023 \ m}$$

$$? \, cm = 2.3 \, mm \left(\frac{1 \, m}{10^3 \, mm} \right) \left(\frac{10^2 \, cm}{1 \, m} \right) = \mathbf{0.23 \ cm}$$

42. In 1866, a child found a "pebble" in the Orange River in South Africa that turned out to be a 21-carat diamond. This discovery led to the development of the greatest diamond fields in the world. Convert 21 carats to grams. (There are exactly 5 carats per gram.) *(Obj #8)*

$$? \, g = 21 \, carat \left(\frac{1 \, g}{5 \, carat} \right) = \mathbf{4.2 \ g}$$

44. A chlorophyll molecule called P680 absorbs light best that has a wavelength of 680 nanometers. What is this length in meters? in picometers? *(Objs #3 and 8)*

$$? \, m = 680 \, nm \left(\frac{1 \, m}{10^9 \, nm} \right) = \mathbf{6.80 \times 10^{-7} \ m}$$

$$? \, pm = 680 \, nm \left(\frac{1 \, m}{10^9 \, nm} \right) \left(\frac{10^{12} \, pm}{1 \, m} \right) = \mathbf{6.80 \times 10^{5} \ pm}$$

46. The estimated mass of the sun is 1.989×10^{30} kilograms. What is this mass in grams…in gigagrams? in pounds? *(Objs #3, 4, and 8)*

$$? \, g = 1.989 \times 10^{30} \, kg \left(\frac{10^3 \, g}{1 \, kg} \right) = \mathbf{1.989 \times 10^{33} \ g}$$

$$? \, Gg = 1.989 \times 10^{30} \, kg \left(\frac{10^3 \, g}{1 \, kg} \right) \left(\frac{1 \, Gg}{10^9 \, g} \right) = \mathbf{1.989 \times 10^{24} \ Gg}$$

$$? \, lb = 1.989 \times 10^{30} \, kg \left(\frac{10^3 \, g}{1 \, kg} \right) \left(\frac{1 \, lb}{453.6 \, g} \right) = \mathbf{4.385 \times 10^{30} \ lb}$$

48. The rainfall in tropical rain forests is over 200 cm/yr. Convert 214 cm into meters, into inches, and into feet. *(Objs #3, 4, and 8)*

$$? \, m = 214 \, cm \left(\frac{1 \, m}{10^2 \, cm} \right) = \mathbf{2.14 \ m}$$

$$? \, in. = 214 \, cm \left(\frac{1 \, in.}{2.54 \, cm} \right) = \mathbf{84.3 \ in.}$$

$$? \, ft = 214 \, cm \left(\frac{1 \, in.}{2.54 \, cm} \right) \left(\frac{1 \, ft}{12 \, in.} \right) = \mathbf{7.02 \ ft}$$

50. About 2.4×10^7 Mg of asphalt are produced in the United States per year. What is this mass in grams?   in kilograms?   in pounds?   in English short tons? (There are 2000 pounds per English short ton.) *(Objs #3, 4, and 8)*

$$? g = 2.4 \times 10^7 \text{ Mg} \left(\frac{10^6 \text{ g}}{1 \text{ Mg}} \right) = \textbf{2.4} \times \textbf{10}^{\textbf{13}} \textbf{ g}$$

$$? \text{kg} = 2.4 \times 10^7 \text{ Mg} \left(\frac{10^6 \text{ g}}{1 \text{ Mg}} \right) \left(\frac{1 \text{ kg}}{10^3 \text{ g}} \right) = \textbf{2.4} \times \textbf{10}^{\textbf{10}} \textbf{ kg}$$

$$? \text{lb} = 2.4 \times 10^7 \text{ Mg} \left(\frac{10^6 \text{ g}}{1 \text{ Mg}} \right) \left(\frac{1 \text{ lb}}{453.6 \text{ g}} \right) = \textbf{5.3} \times \textbf{10}^{\textbf{10}} \textbf{ lb}$$

$$? \text{ton} = 2.4 \times 10^7 \text{ Mg} \left(\frac{10^6 \text{ g}}{1 \text{ Mg}} \right) \left(\frac{1 \text{ lb}}{453.6 \text{ g}} \right) \left(\frac{1 \text{ ton}}{2000 \text{ lb}} \right) = \textbf{2.6} \times \textbf{10}^{\textbf{7}} \textbf{ ton}$$

52. The largest diamond ever found, called the Cullinan diamond, was discovered in the Premier Mine in South Africa in 1905. It was 3106 carats before cutting and was cut to form 128 gems, the largest being the Star of Africa. At 530.2 carats, this is the largest cut diamond in the world. What is the mass of the Star of Africa in grams?   in kilograms?   in pounds? (There are exactly 5 carats per gram.) *(Objs #3, 4, and 8)*

$$? g = 530.2 \text{ carat} \left(\frac{1 \text{ g}}{5 \text{ carat}} \right) = \textbf{106.0 g}$$

$$? \text{kg} = 530.2 \text{ carat} \left(\frac{1 \text{ g}}{5 \text{ carat}} \right) \left(\frac{1 \text{ kg}}{10^3 \text{ g}} \right) = \textbf{0.1060 kg}$$

$$? \text{lb} = 530.2 \text{ carat} \left(\frac{1 \text{ g}}{5 \text{ carat}} \right) \left(\frac{1 \text{ lb}}{453.6 \text{ g}} \right) = \textbf{0.2338 lb}$$

54. The longest recorded flight for a tagged Monarch butterfly is 2.9×10^3 km. What is this length in meters?   in miles? *(Objs #3, 4, and 8)*

$$? \text{m} = 2.9 \times 10^3 \text{ km} \left(\frac{10^3 \text{ m}}{1 \text{ km}} \right) = \textbf{2.9} \times \textbf{10}^{\textbf{6}} \textbf{ m}$$

$$? \text{mi} = 2.9 \times 10^3 \text{ km} \left(\frac{10^3 \text{ m}}{1 \text{ km}} \right) \left(\frac{10^2 \text{ cm}}{1 \text{ m}} \right) \left(\frac{1 \text{ in.}}{2.54 \text{ cm}} \right) \left(\frac{1 \text{ ft}}{12 \text{ in.}} \right) \left(\frac{1 \text{ mi}}{5280 \text{ ft}} \right)$$

$$= \textbf{1.8} \times \textbf{10}^{\textbf{3}} \textbf{ mi}$$

$$\text{or} \quad ? \text{mi} = 2.9 \times 10^3 \text{ km} \left(\frac{1 \text{ mi}}{1.609 \text{ km}} \right) = \textbf{1.8} \times \textbf{10}^{\textbf{3}} \textbf{ mi}$$

56. Because of their low density, plastics have replaced steel in many common objects. To get a sense of why, calculate the mass, in kilograms, of 2.2055 L of steel with a density of 7.852 g/mL and the mass, in kilograms, of 2.2055 L of a plastic with a density of 2.010 g/mL. (*Objs #3, 8, and 10*)

$$? \text{ kg} = 2.2055 \text{ L} \left(\frac{10^3 \text{ mL}}{1 \text{ L}} \right) \left(\frac{7.852 \text{ g}}{1 \text{ mL}} \right) \left(\frac{1 \text{ kg}}{10^3 \text{ g}} \right) = \textbf{17.32 kg}$$

$$? \text{ kg} = 2.2055 \text{ L} \left(\frac{10^3 \text{ mL}}{1 \text{ L}} \right) \left(\frac{2.010 \text{ g}}{1 \text{ mL}} \right) \left(\frac{1 \text{ kg}}{10^3 \text{ g}} \right) = \textbf{4.433 kg}$$

58. Because of the large concentration of salts dissolved in the water of the Dead Sea, its density is significantly higher than pure water. Pure water has a density of 1.0 g/mL, and water in the Dead Sea has a density of 1.8 g/mL. What is the volume, in liters, of 5.000×10^4 Mg of Dead Sea water? (*Objs #3, 8, and 10*)

$$? \text{ L} = 5.000 \times 10^4 \text{ Mg} \left(\frac{10^6 \text{ g}}{1 \text{ Mg}} \right) \left(\frac{1 \text{ mL}}{1.8 \text{ g}} \right) \left(\frac{1 \text{ L}}{10^3 \text{ mL}} \right) = \textbf{2.8} \times \textbf{10}^{\textbf{7}} \textbf{ L}$$

60. Albite, also called soda spar, is a mineral used by the ceramics industry. Moonstone is an opalescent variety of this mineral. A 2.456-carat moonstone is found to have a volume of 0.19 mL. What is its density in g/mL? (There are exactly 5 carats per gram.) (*Objs #8 and 11*)

$$\frac{? \text{ g}}{\text{mL}} = \frac{2.456 \text{ carat}}{0.19 \text{ mL}} \left(\frac{1 \text{ g}}{5 \text{ carat}} \right) = \textbf{2.6 g/mL}$$

62. To avoid potentially serious problems, scuba divers are taught to ascend at a rate of about 60 feet per minute. How long will it take a diver to return 245 feet to the surface at 55 feet per minute? (*Objs #5 and 8*)

$$? \text{ min} = 245 \text{ ft} \left(\frac{1 \text{ min}}{55 \text{ ft}} \right) = \textbf{4.5 min}$$

64. Albacore can swim for long distances at over 25 kilometers per hour and can exhibit short bursts of almost 90 km/h. How far will an albacore travel in 15 minutes at a speed of 27 mi/h? (*Objs #5 and 8*)

$$? \text{ mi} = 15 \text{ min} \left(\frac{1 \text{ h}}{60 \text{ min}} \right) \left(\frac{27 \text{ mi}}{1 \text{ h}} \right) = \textbf{6.8 mi}$$

66. There are about 7×10^6 Mg of gold in the ocean. If it were possible to extract the entire amount (it is not), what would its value be at $15 per gram? (*Objs #5 and 8*)

$$? \$ = 7 \times 10^6 \text{ Mg gold} \left(\frac{10^6 \text{ g}}{1 \text{ Mg}} \right) \left(\frac{15 \$}{1 \text{ g gold}} \right) = \textbf{1} \times \textbf{10}^{\textbf{14}} \textbf{ \$}$$

68. Deepwater currents move from 1 to 2 meters per day. If you were floating in a deepwater current moving at 1.237 m/day, how many years would it take you to travel 625 km? (*Objs #5 and 8*)

$$? \text{ day} = 625 \text{ km} \left(\frac{10^3 \text{ m}}{1 \text{ km}} \right) \left(\frac{1 \text{ day}}{1.237 \text{ m}} \right) \left(\frac{1 \text{ yr}}{365 \text{ day}} \right) = \textbf{1.38} \times \textbf{10}^{\textbf{3}} \textbf{ yr}$$

70. It is estimated that 7×10^8 Mg of hydrogen are converted into helium per second in the sun. How many gigagrams of hydrogen are converted into helium in 1 hour? *(Objs #5 and 8)*

$$? \, Gg = 1 \, h \left(\frac{60 \, min}{1 \, h} \right) \left(\frac{60 \, s}{1 \, min} \right) \left(\frac{7 \times 10^8 \, Mg \, H}{1 \, s} \right) \left(\frac{10^6 \, g}{1 \, Mg} \right) \left(\frac{1 \, Gg}{10^9 \, g} \right) = \mathbf{3 \times 10^9 \, Gg \, H}$$

72. In 1999, it was estimated that 40% of the 7.9×10^7 vehicles in the United States had air bags. (This means that for every 100 vehicles, 40 of them have air bags.) Use this data to calculate the approximate number of cars in the United States with air bags. *(Objs #5 and 8)*

$$? \, vehicles \, with \, air \, bags = 7.9 \times 10^7 \, vehicles \, total \left(\frac{40 \, vehicles \, with \, air \, bags}{100 \, vehicles \, total} \right)$$

$$= \mathbf{3.2 \times 10^7 \, vehicles \, with \, air \, bags}$$

74. In 1998, the average car was made of 12% plastic (for every 100 Mg of car, there are 12 Mg of plastic) for an average mass of plastic of 136 kg. What was the average mass of a car in 1998 in megagrams? *(Objs #5 and 8)*

$$? \, Mg \, car = 136 \, kg \, plastic \left(\frac{10^3 \, g}{1 \, kg} \right) \left(\frac{1 \, Mg}{10^6 \, g} \right) \left(\frac{100 \, Mg \, car}{12 \, Mg \, plastic} \right) = \mathbf{1.1 \, Mg \, car}$$

76. Atmospheric pressure decreases with increasing distance from the center of the earth. The average pressure at sea level is about 101 kPa, the average pressure in Denver, Colorado (which is at an elevation of about 1 mile), is 83 kPa, and the average pressure at 10,700 m (the altitude for normal jet flights) is 24 kPa. Convert these pressures to millimeters of mercury (mmHg). (There are 101.325 kPa per atmosphere, and there are 760 mmHg per atmosphere.) *(Objs #5 and 8)*

$$? \, mmHg = 101 \, kPa \left(\frac{1 \, atm}{101.325 \, kPa} \right) \left(\frac{760 \, mmHg}{1 \, atm} \right) = \mathbf{758 \, mmHg}$$

$$? \, mmHg = 83 \, kPa \left(\frac{1 \, atm}{101.325 \, kPa} \right) \left(\frac{760 \, mmHg}{1 \, atm} \right) = \mathbf{6.2 \times 10^2 \, mmHg}$$

$$? \, mmHg = 24 \, kPa \left(\frac{1 \, atm}{101.325 \, kPa} \right) \left(\frac{760 \, mmHg}{1 \, atm} \right) = \mathbf{1.8 \times 10^2 \, mmHg}$$

78. The portion of the sun where visible light is created is called the photosphere. The approximate temperature of the photosphere is 5.51×10^3 °C. Convert 5.51×10^3 °C to kelvins. (To convert from degree Celsius to kelvin, add 273.15 to the value in °C.) *(Obj #8)*

$$5.51 \times 10^3 \, °C + 273.15 = \mathbf{5.78 \times 10^3 \, K}$$

Additional Problems

80. Fireflies, which are actually beetles, produce a chemical called luciferin. When this chemical reacts with oxygen, light is produced. What is the length, in centimeters, of a firefly that is 0.55 inches long?

$$? \, cm = 0.55 \, in. \left(\frac{2.54 \, cm}{1 \, in.} \right) = \mathbf{1.4 \, cm}$$

82. When the human digestive tract is stretched out, it has a length of from 20 ft to 30 ft. What is the length of a 24.6-foot digestive tract in centimeters?   in meters?

$$? \text{ cm} = 24.6 \text{ ft} \left(\frac{12 \text{ in.}}{1 \text{ ft}}\right) \left(\frac{2.54 \text{ cm}}{1 \text{ in.}}\right) = \textbf{750 cm or } \mathbf{7.50 \times 10^2} \textbf{ cm}$$

$$? \text{ m} = 24.6 \text{ ft} \left(\frac{12 \text{ in.}}{1 \text{ ft}}\right) \left(\frac{2.54 \text{ cm}}{1 \text{ in.}}\right) \left(\frac{1 \text{ m}}{10^2 \text{ cm}}\right) = \textbf{7.50 m}$$

84. An electron microscope can show details as small as 0.2 nm. What is this length in meters? in micrometers?

$$? \text{ m} = 0.2 \text{ nm} \left(\frac{1 \text{ m}}{10^9 \text{ nm}}\right) = \mathbf{2 \times 10^{-10}} \textbf{ m}$$

$$? \text{ μm} = 0.2 \text{ nm} \left(\frac{1 \text{ m}}{10^9 \text{ nm}}\right) \left(\frac{10^6 \text{ μm}}{1 \text{ m}}\right) = \mathbf{2 \times 10^{-4}} \textbf{ μm}$$

86. Albatross, which are birds that have been nicknamed "gooneys," have a wingspan of up to 3.4 meters. What is this length in centimeters?   in inches?   in feet?

$$? \text{ cm} = 3.4 \text{ m} \left(\frac{10^2 \text{ cm}}{1 \text{ m}}\right) = \mathbf{3.4 \times 10^2} \textbf{ cm}$$

$$? \text{ in.} = 3.4 \text{ m} \left(\frac{10^2 \text{ cm}}{1 \text{ m}}\right) \left(\frac{1 \text{ in.}}{2.54 \text{ cm}}\right) = \mathbf{1.3 \times 10^2} \textbf{ in.}$$

$$? \text{ ft} = 3.4 \text{ m} \left(\frac{10^2 \text{ cm}}{1 \text{ m}}\right) \left(\frac{1 \text{ in.}}{2.54 \text{ cm}}\right) \left(\frac{1 \text{ ft}}{12 \text{ in.}}\right) = \textbf{11 ft}$$

88. The largest of the anaconda are about 10 meters long. What is the length of an anaconda that is 9.6 meters long in centimeters?   in inches?   in feet?

$$? \text{ cm} = 9.6 \text{ m} \left(\frac{10^2 \text{ cm}}{1 \text{ m}}\right) = \mathbf{9.6 \times 10^2} \textbf{ cm}$$

$$? \text{ in.} = 9.6 \text{ m} \left(\frac{10^2 \text{ cm}}{1 \text{ m}}\right) \left(\frac{1 \text{ in.}}{2.54 \text{ cm}}\right) = \mathbf{3.8 \times 10^2} \textbf{ in.}$$

$$? \text{ ft} = 9.6 \text{ m} \left(\frac{10^2 \text{ cm}}{1 \text{ m}}\right) \left(\frac{1 \text{ in.}}{2.54 \text{ cm}}\right) \left(\frac{1 \text{ ft}}{12 \text{ in.}}\right) = \textbf{31 ft}$$

90. The planet Pluto orbits the sun at an average distance of 39.44 astronomical units (AU). An astronomical unit is a unit of length equivalent to the average distance between the earth and the sun, or 1.496×10^8 km. Convert 39.44 AU into the equivalent lengths in kilometers and miles.

$$? \text{ km} = 39.44 \text{ AU} \left(\frac{1.496 \times 10^8 \text{ km}}{1 \text{ AU}}\right) = \mathbf{5.900 \times 10^9} \textbf{ km}$$

$$? \text{ mi} = 39.44 \text{ AU} \left(\frac{1.496 \times 10^8 \text{ km}}{1 \text{ AU}}\right) \left(\frac{1 \text{ mi}}{1.609 \text{ km}}\right) = \mathbf{3.667 \times 10^9} \textbf{ mi}$$

92. The shortest distance between the earth and the sun is 1.471×10^8 km. What is this length in miles?

$$? \text{ mi} = 1.471 \times 10^8 \text{ km} \left(\frac{10^3 \text{ m}}{1 \text{ km}} \right) \left(\frac{10^2 \text{ cm}}{1 \text{ m}} \right) \left(\frac{1 \text{ in.}}{2.54 \text{ cm}} \right) \left(\frac{1 \text{ ft}}{12 \text{ in.}} \right) \left(\frac{1 \text{ mi}}{5280 \text{ ft}} \right)$$

$$= \mathbf{9.142 \times 10^7 \ mi}$$

$$or \quad ? \text{ mi} = 1.471 \times 10^8 \text{ km} \left(\frac{1 \text{ mi}}{1.609 \text{ km}} \right) = \mathbf{9.142 \times 10^7 \ mi}$$

94. The Great Pyramid in Egypt is made from about 2.3×10^6 blocks that have an average mass of 2.5 Mg. What is the total mass of the Great Pyramid?

$$? \text{ Mg} = 2.3 \times 10^6 \text{ blocks} \left(\frac{2.5 \text{ Mg}}{1 \text{ block}} \right) = \mathbf{5.8 \times 10^6 \ Mg}$$

96. The air traffic control system in the United States is responsible for 7.3×10^7 flight operations (takeoffs and landings) per year. What is the average numbers of flight operations in 1 day?

$$? \text{ flight operations} = 1 \text{ day} \left(\frac{1 \text{ yr}}{365 \text{ day}} \right) \left(\frac{7.3 \times 10^7 \text{ flight operations}}{1 \text{ yr}} \right)$$

$$= \mathbf{2.0 \times 10^5 \ flight \ operations}$$

98. We hear because our ears can detect vibrations of the air. We can hear sound waves that vibrate at a rate as low as 25 vibrations per second. How many times does the air vibrate for such a sound that lasts for 2.5 minutes?

$$? \text{ min} = 2.5 \text{ min} \left(\frac{60 \text{ s}}{1 \text{ min}} \right) \left(\frac{25 \text{ vibrations}}{1 \text{ s}} \right) = \mathbf{3.8 \times 10^3 \ vibrations}$$

100. The sun moves around the center of our galaxy at 2.2×10^2 kilometers per second. It takes 2.5×10^8 years to complete 1 orbit. How far will it have traveled in this time? The sun is thought to be 4.6×10^9 years old. How many orbits of the galaxy has it made?

$$? \text{ km} = 2.5 \times 10^8 \text{ yr} \left(\frac{365 \text{ d}}{1 \text{ yr}} \right) \left(\frac{24 \text{ h}}{1 \text{ d}} \right) \left(\frac{60 \text{ min}}{1 \text{ h}} \right) \left(\frac{60 \text{ s}}{1 \text{ min}} \right) \left(\frac{2.2 \times 10^2 \text{ km}}{1 \text{ s}} \right)$$

$$= \mathbf{1.7 \times 10^{18} \ km}$$

$$? \text{ orbits} = 4.6 \times 10^9 \text{ yr} \left(\frac{1 \text{ orbit}}{2.5 \times 10^8 \text{ yr}} \right) = \mathbf{18 \ orbits}$$

Chapter 2
The Structure of Matter and the Chemical Elements

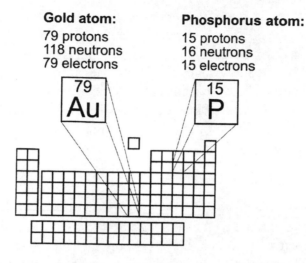

Gold atom:
79 protons
118 neutrons
79 electrons

Phosphorus atom:
15 protons
16 neutrons
15 electrons

♦ Review Skills

2.1 Solids, Liquids, and Gases
  • Solids
  • Liquids
  • Gases
  Internet: Kinetic Molecular Theory

2.2 The Chemical Elements
  Internet: Element Names and Symbols

2.3 The Periodic Table of the Elements

2.4 The Structure of the Elements
  • The Atom
  • The Nucleus
  • The Electron
  • Ions
  • Isotopes
  • Atomic Number and Mass Number

Special Topic 2.1: Why Create New Elements?

Internet: Isotope Notation

2.5 Common Elements
  • Gas, Liquid, and Solid Elements
  Internet: Element Properties
  • Metallic Elements

♦ Chapter Glossary
  Internet: Glossary Quiz

♦ Chapter Objectives

Review Questions

Key Ideas

Chapter Problems

Section Goals and Introductions

Section 2.1 Liquids, Solids, and Gases

Goals

- *To describe a model that allows you to visualize the particle nature of matter.*

- *To describe the similarities and differences among solids, liquids, and gases in terms of this model.*

This is a very important section because it presents a model that you will use throughout your chemistry education and beyond to visualize matter at the submicroscopic level. Be sure you take the time and try to actually visualize the interactions among particles and visualize the movement of these particles. It will be time well spent. The animation found in Chapter 2 of our Web site will help you develop your ability to visualize the particle nature of matter.

     www.chemplace.com/college/

Section 2.2 The Chemical Elements

Goal: To describe the chemical elements, which are the building blocks of matter.

This section introduces the chemical elements. It is best to memorize all of the element names and symbols for the elements found in Table 2.1. Many instructors will consider this excessive, but I think it really pays off in saved time later. Be sure to ask your instructor which names and symbols you are expected to learn for exams. The tutorial in Chapter 2 of our Web site will help you practice converting between names and symbols of elements.

     www.chemplace.com/college/

Section 2.3 The Periodic Table of the Elements

Goal: To describe the periodic table of the elements and show you how you can use it.

The periodic table shown in this section is one of the most important tools of the chemist. It organizes the chemical elements in a way that allows you to quickly obtain a lot of information about them. Be sure that when you are done studying this section, you know (1) how the columns and rows on the periodic table are numbered; (2) how to classify an element as a metal, nonmetal, or metalloid; (3) how to classify an element as a representative (or main-group) element, transition metal, or inner transition metal; (4) how to identify the number for the period in which an element is found; and (5) how to identify an element as a gas, liquid, or solid at room temperature. You should also be able to identify the elements that are alkali metals, alkaline earth metals, halogens, and noble gases.

Section 2.4 The Structure of the Elements

Goal: To describe the structure of the atoms that provide the structure of the elements.

This section introduces atoms for the first time. You will learn about the protons, neutrons, and electrons that form atoms, and you will get an introduction to how these particles are arranged in the atom. Knowledge of the structure of the atom allows us to understand why each element is different from the others. You will discover that electrons can be lost or gained by atoms to form ions, and you will discover why all atoms of an element are not necessarily the same. Different species of atoms of the same element are called isotopes. Chapter 2 of our Web site contains information on the notation used to describe isotopes.

     www.chemplace.com/college/

Section 2.5 Common Elements

Goal: To apply the information described in the first four sections of this chapter to the description of some common elements.

This section brings the chapter full circle to the particle nature of solids, liquids, and gases; but that when you read this section, you will know more about the particles that compose solid, liquid, and gaseous elements. The section helps you to visualize the particle nature of the elements instead of relating to them as just symbols on the page.

Chapter 2 of our Web site contains an animation will help you visualize the elements mentioned in this section.

www.chemplace.com/college/

Chapter 2 Map

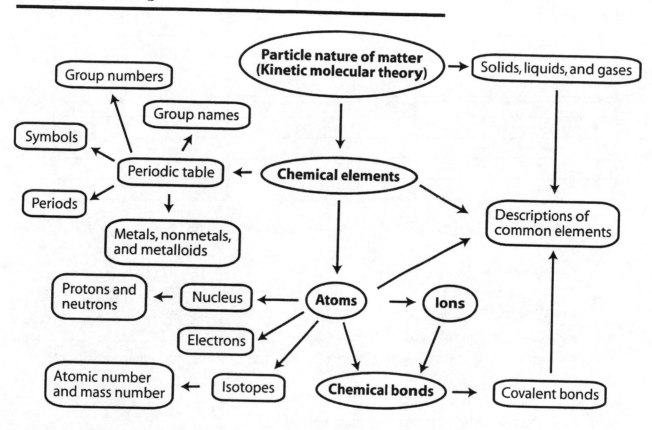

Chapter Checklist

- ☐ Read the Review Skills section. If there is any skill mentioned that you have not yet mastered, review the material on that topic before reading this chapter.
- ☐ Read the chapter quickly before the lecture that describes it.
- ☐ Attend class meetings, take notes, and participate in class discussions.
- ☐ Work the Chapter Exercises, perhaps using the Chapter Examples as guides.
- ☐ Study the Chapter Glossary and test yourself on our Web site:

    www.chemplace.com/college/

☐ Study all of the Chapter Objectives. You might want to write a description of how you will meet each objective. (Although it is best to master all of the objectives, the following objectives are especially important because they pertain to skills that you will need while studying other chapters of this text: 2, 11, 12, 13, 15, 16, 17, 20, 22, and 24.)

☐ Memorize the names and symbols of the elements on the following table. Be sure to check with your instructor to determine how many of these you are expected to know and ask whether your instructor wants to add any others.

Element	Symbol	Element	Symbol	Element	Symbol
aluminum	Al	gold	Au	oxygen	O
argon	Ar	helium	He	phosphorus	P
barium	Ba	hydrogen	H	platinum	Pt
beryllium	Be	iodine	I	potassium	K
boron	B	iron	Fe	silicon	Si
bromine	Br	lead	Pb	silver	Ag
cadmium	Cd	lithium	Li	sodium	Na
calcium	Ca	magnesium	Mg	strontium	Sr
carbon	C	manganese	Mn	sulfur	S
chlorine	Cl	mercury	Hg	tin	Sn
chromium	Cr	neon	Ne	uranium	U
copper	Cu	nickel	Ni	xenon	Xe
fluorine	F	nitrogen	N	zinc	Zn

☐ Learn how to use the periodic table to classify the elements with respect to the following categories:

- Groups 1 to 18
- Groups 1A to 8A
- Alkali metals, alkaline earth metals, halogens, and noble gases
- Metals, nonmetals, and metalloids
- Representative (main-group) elements, transition metals, and inner transition metals
- Periods 1 to 7
- Solids, liquids, or gases at room temperature

☐ To get a review of the most important topics in the chapter, fill in the blanks in the Key Ideas section.

☐ Work all of the selected problems at the end of the chapter, and check your answers with the solutions provided in this chapter of the study guide.

☐ Ask for help if you need it.

Web Resources www.chemplace.com/college/

Kinetic Molecular Theory
Element Names and Symbols
Isotope Notation
Element Properties
Glossary Quiz

Exercises Key

Exercise 2.1 - Elements and the Periodic Table: Complete the following
table. (Objs #12 & 15-18)

Name	Symbol	Group number	Metal, nonmetal or metalloid?	Representative element, transition metal, or inner transition metal?	Number for period	Solid, liquid, or gas?
aluminum	Al	13, 3A or IIIA	metal	representative element	3	solid
silicon	Si	14, 4A or IVA	metalloid	representative element	3	solid
nickel	Ni	10, 8B or VIIIB	metal	transition metal	4	solid
sulfur	S	16, 6A or VIA	nonmetal	representative element	3	solid
fluorine	F	17, 7A or VIIA	nonmetal	representative element	2	gas
potassium	K	1, 1A or IA	metal	representative element	4	solid
mercury	Hg	12, 2B or IIB	metal	transition metal	6	liquid
uranium	U	(No group number)	metal	inner transition metal	7	solid
manganese	Mn	7, 7B or VIIB	metal	transition metal	4	solid
calcium	Ca	2, 2A or IIA	metal	representative element	4	solid
bromine	Br	17	nonmetal	representative element	4	liquid
silver	Ag	1B	metal	transition metal	5	solid
carbon	C	14	nonmetal	representative element	2	solid

Exercise 2.2 - Group Names and the Periodic Table: Write the name of
the group on the periodic table to which each of the following elements belongs. (Obj #13)

a. helium   **noble gases**

b. Cl   **halogens**

c. magnesium   **alkaline earth metals**

d. Na   **alkali metals**

Exercise 2.3 - Cations and Anions: Identify each of the following as a cation or an anion, and determine the charge on each. *(Obj #22)*

    a. a magnesium atom with 12 protons and 10 electrons

        $(+12) + (-10) = +2$ This is a **+2 cation.**

    b. a fluorine atom with 9 protons and 10 electrons

        $(+9) + (-10) = -1$ This is a **–1 anion.**

Review Questions Key

1. Define the term matter.

    Matter is anything that occupies space and has mass.

2. Look around you. What do you see that has a length of about a meter? What do you see that has a mass of about a gram?

    The distance between the floor and a typical doorknob is about one meter. A penny weighs about three grams.

Key Ideas Answers

3. Scientific models are like architects' models; they are **simplified but useful** representations of something real.

5. According to the model presented in this chapter, particles of matter are in constant **motion**.

7. Solids, gases, and liquids differ in the freedom of motion of their particles and in how strongly the particles **attract** each other.

9. Particles in a liquid are still close together, but there is generally more **empty space** between them than in a solid. Thus, when a solid substance melts to form a liquid, it usually **expands** to fill a slightly larger volume.

11. When a liquid's temperature is higher, its particles are moving faster and are therefore more likely to **escape** from the liquid.

13. According to our model, each particle in a gas moves freely in a **straight-line path** until it collides with another gas particle or with the particles of a liquid or solid.

15. Elements are substances that cannot be chemically converted into **simpler** ones.

17. The periodic table is arranged in such a way that elements in the same **vertical column** have similar characteristics.

19. At room temperature (20 °C) and normal pressures, most of the elements are **solid**, two of them are **liquid** (Hg and Br), and eleven are **gas** (H, N, O, F, Cl, and the noble gases).

21. A ½-carat diamond contains about 5×10^{21} atoms of carbon. If these atoms, tiny as they are, were arranged in a straight line with each one touching its neighbors, the line would stretch from here to the **sun**.

23. The diameter of a typical nucleus is about 10^{-15} meter.

25. Chemists use a model for electrons in which each electron is visualized as generating a **cloud** of negative charge that surrounds the nucleus.

27. When an atom **gains** one or more electrons, it then has more electrons than protons and more minus charge than plus charge. Thus it becomes an anion, which is an ion with a negative charge.

29. Atoms are assigned to elements on the basis of their **chemical** characteristics.

31. Each noble gas particle consists of a **single atom**.

Problems Key

Section 2.1 Solids, Liquids, and Gases

For each of the questions in this section, illustrate your written answers with simple drawings of the particles that form the structures of the substances mentioned. You do not need to be specific about the nature of the particles. Think of them as simple spheres, and draw them as circles.

33. If you heat white sugar very carefully, it will melt. *(Objs #2, 3, 4, & 6)*

   a. Before you begin to heat the sugar, the sugar granules maintain a constant shape and volume. Why?

   > Strong attractions between the particles keep each particle at the same average distance from other particles and in the same general position with respect to its neighbors.

   b. As you begin to heat the solid sugar, what changes are taking place in its structure?

   > The velocity of the particles increases, causing more violent collisions between them. This causes them to move apart, so the solid expands. See Figure 2.1 on page 61 of the textbook.

   c. What happens to the sugar's structure when sugar melts?

   > The particles break out of their positions in the solid and move more freely throughout the liquid, constantly breaking old attractions and making new ones. Although the particles are still close together in the liquid, they are more disorganized, and there is more empty space between them.

35. Ethylene glycol, an automobile coolant and antifreeze, is commonly mixed with water and added to car radiators. Because it freezes at a lower temperature than water and boils at a higher temperature than water, it helps to keep the liquid in your radiator from freezing or boiling. *(Objs #2, 3, 6, & 8)*

   a. At a constant temperature, liquid ethylene glycol maintains a constant volume but takes on the shape its container. Why?

   > The attractions between liquid particles are not strong enough to keep the particles in position like the solid. The movement of particles allows the liquid to take the shape of its container. The attractions are strong enough to keep the particles at the same average distance, leading to constant volume.

   b. The ethylene glycol-water mixture in your car's radiator heats up as you drive. What is happening to the particles in the liquid?

> *The velocity of the particles increases, so they will move throughout the liquid more rapidly. The particles will collide with more force. This causes them to move apart, so the liquid expands slightly.*

   c. If you spill some engine coolant on your driveway, it evaporates without leaving any residue. Describe the process of evaporation of liquid ethylene glycol, and explain what happens to the ethylene glycol particles that you spilled.

> *Particles that are at the surface of the liquid and that are moving away from the surface fast enough to break the attractions that pull them back will escape to the gaseous form. The gas particles will disperse throughout the neighborhood as they mix with the particles in the air. See Figures 2.3 and 2.4 on pages 63 and 64 of the textbook.*

37. As the summer sun heats up the air at the beach, what is changing for the air particles?

> *The air particles are moving faster.*

39. A gaseous mixture of air and gasoline enters the cylinders of a car engine and is compressed into a smaller volume before being ignited. Explain why gases can be compressed.

> *There is plenty of empty space between particles in a gas.*

Section 2.2 The Chemical Elements and Section 2.3 The Periodic Table of the Elements

40. Write the chemical symbols that represent the following elements. *(Obj #11)*

   a. chlorine **Cl**   c. phosphorus **P**
   b. zinc **Zn**   d. uranium **U**

43. Write the element names that correspond to the following symbols. *(Obj #11)*

   a. C **carbon**   c. Ne **neon**
   b. Cu **copper**   d. K **potassium**

46. Complete the following table. *(Objs #11, 12, 15, 16, & 17)*

Element name	Element symbol	Group number on periodic table	Metal, nonmetal, or metalloid?	Representative element, transition metal, or inner transition metal?	Number of period
sodium	Na	**1 or 1A or IA**	metal	**Representative element**	**3**
tin	**Sn**	**14 or 4A or IVA**	metal	**Representative element**	**5**
helium	He	**18 or 8A or VIIIA**	nonmetal	**Representative element**	**1**
nickel	**Ni**	**10 or 8B or VIIIB**	metal	**Transition metal**	**4**
silver	Ag	**11 or 1B or IB**	metal	**Transition metal**	**5**
aluminum	**Al**	**13 or 3A or IIIA**	metal	**Representative element**	**3**
silicon	Si	**14 or 4A or IVA**	metalloid	**Representative element**	**3**
sulfur	S	16	nonmetal	**Representative element**	3
mercury	**Hg**	2B	metal	**Transition metal**	6

48. Write the name of the group to which each of the following belongs. *(Obj #13)*
    a. bromine **halogens**
    b. neon **noble gases**
    c. potassium **alkali metals**
    d. beryllium **alkaline earth metals**

50. Identify each of the following elements as a solid, a liquid, or a gas at room temperature and pressure. *(Obj #18)*
    a. Kr **gas**
    b. bromine **liquid**
    c. Sb **solid**
    d. fluorine **gas**
    e. Ge **solid**
    f. sulfur **solid**

52. Which two of the following elements would you expect to be most similar: lithium, aluminum, iodine, oxygen, and potassium?

    lithium and potassium

54. Write the name and symbol for the elements that fit the following descriptions.
    a. the halogen in the third period **chlorine, Cl**
    b. the alkali metal in the fourth period **potassium, K**
    c. the metalloid in the third period **silicon, Si**

56. Which element would you expect to be malleable, manganese or phosphorus? Why?

    Because manganese is a metal, we expect it to be malleable.

Section 2.4 The Structure of the Elements

58. Describe the nuclear model of the atom, including the general location of the protons, neutrons, and electrons, the relative size of the nucleus compared to the size of the atom, and the modern description of the electron. *(Obj #20)*

    Protons and neutrons are in a tiny core of the atom called the nucleus, which has a diameter of about 1/100,000 the diameter of the atom. The position and motion of the electrons are uncertain, but they generate a negative charge that is felt in the space that surrounds the nucleus.

60. Identify each of the following as a cation or an anion, and determine the charge on each. *(Obj #22)*
    a. a lithium atom with 3 protons and 2 electrons

       $(+3) + (-2) = +1$ This is a **cation** with a **+1** charge.
    b. a sulfur atom with 16 protons and 18 electrons

       $(+16) + (-18) = -2$ This is an **anion** with a **−2** charge.

63. Write the atomic number for each of the following elements.
    a. Oxygen **8**
    b. Mg **12**
    c. uranium **92**
    d. Li **3**
    e. lead **82**
    f. Mn **25**

66. Write the name and symbol for the elements that fit the following descriptions.
    a. 27 protons in the nucleus of each atom **cobalt, Co**
    b. 50 electrons in each uncharged atom **tin, Sn**
    c. 18 electrons in each +2 cation **calcium, Ca**
    d. 10 electrons in each -1 anion **fluorine, F**

Section 2.5 Common Elements

68. Describe the hydrogen molecule, including a rough sketch of the electron-charge cloud created by its electrons. *(Obj #25)*

    See the image of the hydrogen molecule on page 80 of the textbook. The cloud around the two hydrogen nuclei above represents the negative charge cloud generated by the two electrons in the covalent bond that holds the atoms together in the H_2 molecule.

70. Describe the structure of each of the following substances, including a description of the nature of the particles that form each structure. *(Obj #24)*

    a. neon gas

        Neon is composed of separate neon atoms. Its structure is very similar to the structure of He shown in Figure 2.12 on page 80 of the textbook.

    b. bromine liquid

        Bromine is composed of Br_2 molecules. See Figure 2.16 on page 82 of the textbook.

    c. nitrogen gas

        Nitrogen is composed of N_2 molecules. Its structure is very similar to the structure of H_2 shown in Figure 2.15 on page 81 of the textbook.

72. Describe the "sea-of-electrons" model for metallic solids. *(Obj #27)*

    Each atom in a metallic solid has released one or more electrons, allowing the electrons to move freely throughout the solid. When the atoms lose these electrons, they become cations, which form the organized structure we associate with solids. The released electrons flow between the stationary cations like water flows between islands in the ocean. See Figure 2.18 on page 83 of the textbook.

Chapter 3
Chemical Compounds

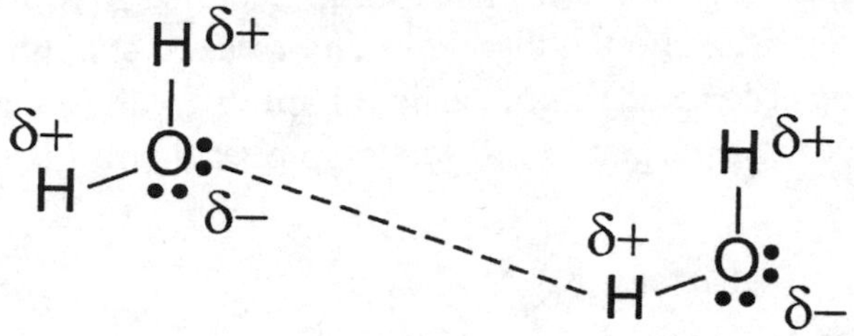

♦ Review Skills

3.1 Classification of Matter

3.2 Compounds and Chemical Bonds
- Equal and Unequal Sharing of Electrons
- Transfer of Electrons
- Summary of Covalent and Ionic Bond Formation
- Predicting Bond Type
- Classifying Compounds

3.3 Molecular Compounds
- Molecular Shape
- Liquid Water

   Special Topic 3.1: Molecular Shapes, Intoxicating Liquids, and the Brain

   Internet: The Structure of Water

3.4 Naming Binary Covalent Compounds
- Memorized Names
- Systematic Names
- Converting Names of Binary Covalent Compounds to Formulas

3.5 Ionic Compounds
- Cations and Anions
- Predicting Ion Charges
- Naming Monatomic Anions and Cations
- Structure of Ionic Compounds
- Polyatomic Ions

   Internet: Oxyanions:
- Converting Formulas to Names
- Converting Names of Ionic Compounds to Formulas

♦ Chapter Glossary

   Internet: Glossary Quiz

♦ Chapter Objectives

Review Questions

Key Ideas

Chapter Problems

Section Goals and Introductions

The Review Skills section for this chapter is very important. Because students are so busy, it's always a battle to keep up, but if you don't keep up in chemistry, you're in real trouble. For this reason, the Review Skills section at the beginning of each chapter tells you what you really need to know (or be able to do) from earlier chapters to understand the chapter you are about to read. They are very important; don't just skip over them; give them the necessary attention.

Section 3.1 Classification of Matter

Goal: To show how forms of matter can be classified as elements, compounds, and mixtures.

This section begins the process of teaching you how to classify matter into the categories of element, compound, and mixture. The distinctions among these categories will become increasingly clear as you study this chapter and Chapter 4. This section also describes how compounds are represented by chemical formulas.

Section 3.2 Compounds and Chemical Bonds

Goals

- *To show how atoms of different elements form links (chemical bonds) between them and to introduce three types of chemical bonding: nonpolar covalent, polar covalent, and ionic.*

- *To show how you can predict whether a pair of atoms will form a covalent bond or an ionic bond.*

- *To show how you can predict whether a chemical formula for a compound represents an ionic compound or a molecular compound.*

We can now take what you have learned about elements in Chapter 2 and expand on it to explain the formation of chemical bonds and chemical compounds. Be sure that you understand the similarities and differences among nonpolar covalent bonds, covalent bonds, and ionic bonds. The most important skills to develop from studying this section are (1) the ability to predict whether atoms of two elements would be expected to form an ionic bond or a covalent bond and (2) the ability to predict whether a chemical formula for a compound represents an ionic compound or a molecular compound. The ability to make these predictions is extremely important for many of the tasks in this chapter and many of the chapters that follow.

Section 3.3 Molecular Compounds

Goals

- *To explain the bonding patterns of the nonmetallic elements.*
- *To introduce the concept of molecules, to show how they form, to show how they are described, and to show how the atoms in some molecules are arranged in space.*
- *To describe the structure of liquid water.*

This section begins to describe the formation of molecules, which are collections of atoms held together by covalent bonds. You will find that much of what is introduced here is explained in much more detail in Chapter 12. From this section, you want to gain some understanding of why atoms of the nonmetallic elements form the covalent bonds that they do. Knowing the most common bonding patterns of these atoms (Table 3.1) will help you to predict how nonmetallic atoms combine to form molecules and will help you to draw Lewis structures that represent these molecules. You will learn how the Lewis structures can be used to predict the geometric arrangement of atoms in molecules.

It is very important that you know the structure of each water molecule and how this affects the nature of liquid water. To understand the process of forming water solutions described in Chapter 4, you need to have a good mental image of the structure of liquid water. The animation on *The Structure of Water* in Chapter 3 of our Web site will help.

www.chemplace.com/college/

Section 3.4 Naming Binary Covalent Compounds

Goal: To show how to convert between names and formulas for binary covalent compounds.

This section begins the process of describing how to convert between names and formulas for compounds. How important this is depends on whether or not you are going to take more chemistry classes. For example, if you are going on to take general college chemistry, it's very important that you master this skill. It's much less important if the course for which you are using this text is the last chemistry course you plan to take. Be sure to ask your instructor how much weight will be given to this topic on your exams.

Section 3.5 Ionic Compounds

Goals

- *To show why some atoms gain or lose electrons to form charged particles called ions.*
- *To show how the charges on ions can be predicted.*
- *To describe the structure of ionic solids.*
- *To describe polyatomic ions, which are charged collections of atoms held together by covalent bonds.*

This section is an important one. It provides more detailed information about ions than that found in Chapter 2, including how to predict their charges, how to convert between their names and symbols, and how they combine to form ionic compounds. All of this information is used extensively in the rest of the book. The section also includes a description of the structure of ionic solids.

Polyatomic ions (charged collections of atoms held together by covalent bonds) are also described. Your instructor may want to expand on the list of polyatomic ions that you are expected to know. The text assumes that you can convert between the names and symbols for those that are found in Table 3.6. The section on *Oxyanions* on our Web site gives you a more complete description of polyatomic ions.

www.chemplace.com/college/

The conversion between names and formulas for ionic compounds is described. Be sure to ask your instructor how much weight will be given to this topic on your exams.

Chapter 3 Map

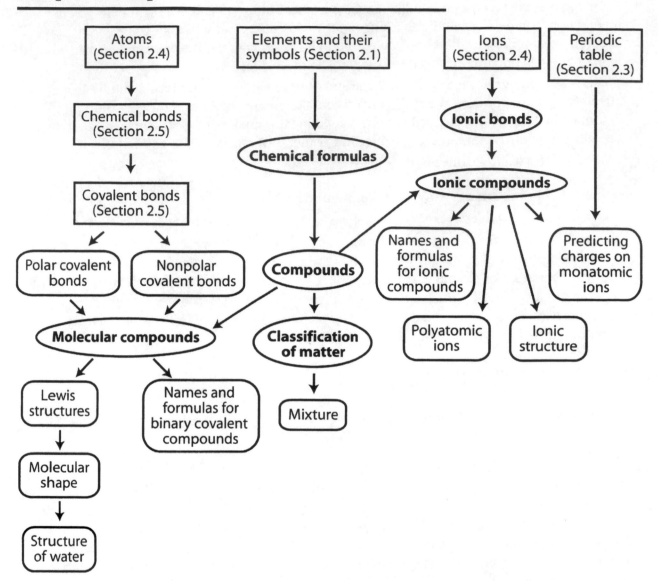

Chapter Checklist

- ☐ Read the Review Skills section. If there is any skill mentioned that you have not yet mastered, review the material on that topic before reading this chapter.
- ☐ Read the chapter quickly before the lecture that describes it.
- ☐ Attend class meetings, take notes, and participate in class discussions.
- ☐ Work the Chapter Exercises, perhaps using the Chapter Examples as guides.
- ☐ Study the Chapter Glossary and test yourself on our Web site:
    www.chemplace.com/college/
- ☐ Study all of the Chapter Objectives. You might want to write a description of how you will meet each objective. (Although it is best to master all of the objectives, the following

objectives are especially important because they pertain to skills that you will need while studying other chapters of this text: 3, 4, 8, 9, 13-16, 21-28, 30, 31, 34, 35, and 37.)

☐ Reread *Sample Study Sheet 3.1: Classification of Matter* and decide whether you will use it or some variation on it to complete the task it describes.

☐ Memorize the following. (Be sure to check with your instructor to determine how much you are expected to know of the following.)

• The usual numbers of covalent bonds and lone pairs for the nonmetallic elements.

Group 4A		Group 5A		Group 6A		Group 7A	
4 valence electrons ·Ẋ·		5 valence electrons ·Ẍ·		6 valence electrons ·Ẍ:		7 valence electrons ·Ẍ:	
4 bonds	No lone pairs	3 bonds	1 lone pair	2 bonds	2 lone pairs	1 bond	3 lone pairs
carbon – C $-\overset{\mid}{\underset{\mid}{C}}-$		nitrogen- N $-\overset{..}{\underset{\mid}{N}}-$		oxygen – O $-\overset{..}{\underset{..}{O}}-$		fluorine – F $-\overset{..}{\underset{..}{F}}:$	
		phosphorus – P $-\overset{..}{\underset{\mid}{P}}-$		sulfur – S $-\overset{..}{\underset{..}{S}}-$		chlorine – Cl $-\overset{..}{\underset{..}{Cl}}:$	
				selenium – Se $-\overset{..}{\underset{..}{Se}}-$		bromine – Br $-\overset{..}{\underset{..}{Br}}:$	
						iodine – I $-\overset{..}{\underset{..}{I}}:$	

• Names and formulas of some binary covalent compounds.

Name	Formula	Name	Formula
water	H_2O	methane	CH_4
ammonia	NH_3	ethane	C_2H_6
		propane	C_3H_8

- Prefixes

Number of atoms	Prefix	Number of atoms	Prefix
1	mon(o)-	6	hex(a)-
2	di-	7	hept(a)-
3	tri-	8	oct(a)-
4	tetr(a)-	9	non(a)-
5	pent(a)-	10	dec(a)-

- Roots of nonmetallic elements

Element	Root	Element	Root	Element	Root	Element	Root
C	carb-	N	nitr-	O	ox-	F	fluor-
		P	phosph-	S	sulf-	Cl	chlor-
		As	arsen-	Se	selen-	Br	brom-
						I	iod-

- Charges on monatomic ions

1 1A	2 2A	3 3B	4 4B	5 5B	6 6B	7 7B	8 8B	9 8B	10 8B	11 1B	12 2B	13 3A	14 4A	15 5A	16 6A	17 7A	18 8A
										H^-							
Li^+	Be^{2+}													N^{3-}	O^{2-}	F^-	
Na^+	Mg^{2+}											Al^{3+}		P^{3-}	S^{2-}	Cl^-	
K^+	Ca^{2+}	Sc^{3+}					Fe^{2+} Fe^{3+}			Cu^+ Cu^{2+}	Zn^{2+}				Se^{2-}	Br^-	
Rb^+	Sr^{2+}	Y^{3+}								Ag^+	Cd^{2+}					I^-	
Cs^+	Ba^{2+}																
Fr^+	Ra^{2+}																

- Names and formulas for polyatomic ions

Ion	Name	Ion	Name	Ion	Name
NH_4^+	ammonium	PO_4^{3-}	phosphate	SO_4^{2-}	sulfate
OH^-	hydroxide	NO_3^-	nitrate	$C_2H_3O_2^-$	acetate
CO_3^{2-}	carbonate				

☐ To get a review of the most important topics in the chapter, fill in the blanks in the Key Ideas section.

☐ Work all of the selected problems at the end of the chapter, and check your answers with the solutions provided in this chapter of the study guide.

☐ Ask for help if you need it.

Web Resources   www.chemplace.com/college/

The Structure of Water
Oxyanions
Glossary Quiz

Exercises Key

✍ **Exercise 3.1 - Classification of Matter:**   The label on a container of double-acting baking powder tells us that it contains cornstarch, bicarbonate of soda (also called sodium hydrogen carbonate, $NaHCO_3$), sodium aluminum sulfate, and acid phosphate of calcium (which chemists call calcium dihydrogen phosphate, $Ca(H_2PO_4)_2$). Classify each of the following as a pure substance or a mixture. If it is a pure substance, is it an element or a compound? *(Objs #3 & 4)*

    a.   calcium   **element**

    b.   calcium dihydrogen phosphate   **compound**

    c.   double-acting baking powder   **mixture**

✍ **Exercise 3.2 - Classifying Compounds:**   Classify each of the following substances as either a molecular compound or an ionic compound. *(Obj #9)*

    a.   formaldehyde, CH_2O (used in embalming fluids)

        all nonmetal atoms - molecular

    b.   magnesium chloride, $MgCl_2$ (used in fireproofing wood and in paper manufacturing)

        metal-nonmetal - ionic

✍ **Exercise 3.3 - Drawing Lewis Structures from Formulas:**   Draw a Lewis structure for each of the following formulas: *(Obj #14)*

    a.   nitrogen triiodide, NI_3 (explodes at the slightest touch)

        Nitrogen atoms usually have 3 covalent bonds and 1 lone pair, and iodine atoms usually have 1 covalent bond and 3 lone pairs.

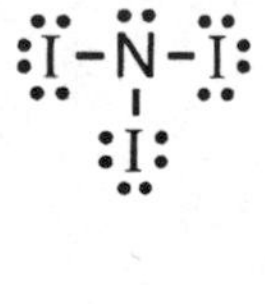

$$:\ddot{I}-\overset{..}{N}-\ddot{I}:$$
$$|$$
$$:\ddot{I}:$$

b. hexachloroethane, C_2Cl_6 (used to make explosives)

Carbon atoms usually have 4 covalent bonds and no lone pairs, and chlorine atoms usually have 1 covalent bond and 3 lone pairs.

c. hydrogen peroxide, H_2O_2 (a common antiseptic)

Hydrogen atoms always have 1 covalent bond and no lone pairs, and oxygen atoms usually have 2 covalent bonds and 2 lone pairs.

d. ethylene (or ethene), C_2H_4 (used to make polyethylene)

Carbon atoms form 4 bonds with no lone pairs, and hydrogen atoms form 1 bond with no lone pairs. To achieve these bonding patterns, there must be a double bond between the carbon atoms.

Exercise 3.4 - Naming Binary Covalent Compounds: Write names that correspond to the following formulas: *(Obj #21)*

a. P_2O_5 **diphosphorus pentoxide**
b. PCl_3 **phosphorus trichloride**
c. CO   **carbon monoxide**
d. H_2S **dihydrogen monosulfide or hydrogen sulfide**
e. NH_3 **ammonia**

Exercise 3.5 - Writing Formulas for Binary Covalent Compounds: Write formulas that correspond to the following names: *(Obj #21)*

a. disulfur decafluoride S_2F_{10}
b. nitrogen trifluoride NF_3
c. propane C_3H_8
d. hydrogen chloride   **HCl**

Exercise 3.6 - Naming Monatomic Ions: Write names that correspond to the following formulas for monatomic ions: *(Obj #24)*

a. Mg^{2+} **magnesium ion**   c. Sn^{2+} **tin(II) ion**
b. F^- **fluoride ion**

✍ Exercise 3.7 - Formulas for Monatomic Ions: Write formulas that correspond to the following names for monatomic ions: *(Obj #24)*

a. bromide ion   **Br^-**

b. aluminum ion   **Al^{3+}**

c. gold(I) ion   **Au^+**

✍ Exercise 3.8 - Naming Ionic Compounds: Write the names that correspond to the following formulas: *(Obj #30)*

a. LiCl   **lithium chloride**

b. $Cr_2(SO_4)_3$ **chromium(III) sulfate**

c. NH_4HCO_3 **ammonium hydrogen carbonate**

✍ Exercise 3.9 - Formulas for Ionic Compounds: Write the formulas that correspond to the following names: *(Obj #30)*

a. aluminum oxide   **Al_2O_3**

b. cobalt(III) fluoride   **CoF_3**

c. iron(II) sulfate   **$FeSO_4$**

d. ammonium hydrogen phosphate   **$(NH_4)_2HPO_4$**

e. potassium bicarbonate   **$KHCO_3$**

Review Questions Key

Write in each blank the word or words that best complete each sentence.

1. An atom or group of atoms that has lost or gained one or more electrons to create a charged particle is called a(n) **ion**.

2. An atom or collection of atoms with an overall positive charge is a(n) **cation**.

3. An atom or collection of atoms with an overall negative charge is a(n) **anion**.

4. A(n) **covalent** bond is a link between atoms that results from the sharing of two electrons.

5. A(n) **molecule** is an uncharged collection of atoms held together with covalent bonds.

6. A molecule like H_2, which is composed of two atoms, is called **diatomic**.

7. Describe the particle nature of solids, liquids, and gases. Your description should include the motion of the particles and the attractions between the particles.

   See Figures 2.1, 2.2, and 2.4 on pages 61, 62, and 64 of the textbook.

8. Describe the nuclear model of the atom.

   Protons and neutrons are in a tiny core of the atom called the nucleus, which has a diameter about 1/100,000 the diameter of the atom. The position and motion of the electrons are uncertain, but they generate a negative charge that is felt in the space that surrounds the nucleus.

9. Describe the hydrogen molecule, H_2. Your description should include the nature of the link between the hydrogen atoms and a sketch that shows the two electrons in the molecule.

> The hydrogen atoms are held together by a covalent bond formed due to the sharing of two electrons. See the image of H_2 on page 68 of the textbook.

10. Complete the following table.

Element name	Element symbol	Group number on periodic table	Metal, nonmetal, or metalloid?
lithium	Li	1 or 1A	**metal**
carbon	**C**	14 or 4A	**nonmetal**
chlorine	Cl	17 or 7A	nonmetal
oxygen	**O**	16 or 6A	**nonmetal**
copper	Cu	11 or 1B	**metal**
calcium	**Ca**	2 or 2A	metal
scandium	Sc	3 or 3B	metal

11. Write the name of the group to which each of the following belongs.

     a. chlorine   **halogens**   c. sodium   **alkali metals**

     b. xenon   **noble gases**   d. magnesium   **alkaline earth metals**

Key Ideas Answers

12. A compound is a substance that contains two or more elements, the atoms of those elements always combining in the same **whole-number ratio**.

14. A chemical formula is a concise written description of the components of a chemical compound. It identifies the elements in the compound by their **symbols** and indicates the relative number of atoms of each element with **subscripts**.

16. Mixtures are samples of matter that contain two or more pure substances and have **variable** composition.

18. Because particles with opposite charges attract each other, there is an attraction between **cations** and **anions**. This attraction is called an ionic bond.

20. The atom in a chemical bond that attracts electrons more strongly acquires a **negative** charge, and the other atom acquires a **positive** charge. If the electron transfer is significant but not enough to form ions, the atoms acquire **partial negative** and **partial positive** charges. The bond in this situation is called a polar covalent bond.

22. When a metallic atom bonds to a nonmetallic atom, the bond is **usually ionic**.

24. The noble gases (group 8A) have an **octet** of electrons (except for helium, which has only two electrons total), and they are so stable that they rarely form chemical bonds with other atoms.

26. The sum of the numbers of covalent bonds and lone pairs for the most common bonding patterns of the atoms of nitrogen, phosphorus, oxygen, sulfur, selenium, and the halogens is **four**.

28. Lewis structures are useful for showing how the atoms in a molecule are connected by covalent bonds, but they do not always give a clear description of how the atoms are **arranged in space**.

30. A space-filling model provides the most accurate representation of the **electron-charge clouds** for the atoms in CH_4.

32. As in other liquids, the attractions between water molecules are strong enough to keep them the **same average distance apart** but weak enough to allow each molecule to be **constantly breaking** the attractions that momentarily connect it to some molecules and **forming new attractions** to other molecules.

34. You can recognize binary covalent compounds from their formulas, which contain symbols for only two **nonmetallic** elements.

36. Nonmetallic atoms form anions to get the same number of electrons as the nearest **noble gas**.

38. When atoms gain electrons and form anions, they get **larger**. When atoms lose electrons and form cations, they get significantly **smaller**.

40. It is common for hydrogen atoms to be transferred from one ion or molecule to another ion or molecule. When this happens, the hydrogen atom is usually transferred without its electron, as H^+.

Problems Key

Section 3.1 Classification of Matter

42. Classify each of the following as a pure substance or a mixture. If it is a pure substance, is it an element or a compound? Explain your answer. *(Objs #3 & 4)*

   a. apple juice

      mixture – *variable composition*

   b. potassium (A serving of one brand of apple juice provides 6% of the recommended daily allowance of potassium.)

      element and pure substance – *The symbol K is on the periodic table of the elements. All elements are pure substances.*

   c. ascorbic acid (vitamin C), $C_6H_8O_6$, in apple juice

      The formula shows the constant composition, so ascorbic acid is a **pure substance.** *The three element symbols in the formula indicate a* **compound.**

44. Write the chemical formula for each of the following compounds. List the symbols for the elements in the order that the elements are mentioned in the description.

   a. a compound with molecules that consist of 2 nitrogen atoms and 3 oxygen atoms.

      N_2O_3

b. a compound with molecules that consist of 1 sulfur atom and 4 fluorine atoms.

   SF$_4$

c. a compound that contains 1 aluminum atom for every 3 chlorine atoms.

   AlCl$_3$

d. a compound that contains 2 lithium atoms and 1 carbon atom for every 3 oxygen atoms.

   Li$_2$CO$_3$

Section 3.2 Chemical Compounds and Chemical Bonds

46. Hydrogen bromide, HBr, is used to make pharmaceuticals that require bromine in their structure. Each hydrogen bromide molecule has one hydrogen atom bonded to one bromine atom by a polar covalent bond. The bromine atom attracts electrons more than does the hydrogen atom. Draw a rough sketch of the electron cloud that represents the electrons involved in the bond. *(Obj #5)*

48. Atoms of potassium and fluorine form ions and ionic bonds in a very similar way to atoms of sodium and chlorine. Each atom of one of these elements loses one electron, and each atom of the other element gains one electron. Describe the process that leads to the formation of the ionic bond between potassium and fluorine atoms in potassium fluoride. Your answer should include mention of the charges that form on the atoms.
   (Obj #6)

   The metallic potassium atoms lose one electron and form +1 cations, and the nonmetallic fluorine atoms gain one electron and form −1 anions.

   K   →   K$^+$ + e^-

   19p/19e$^-$ 19p/18e$^-$

   F   + e^- →   F$^-$

   9p/9e$^-$ 9p/10e$^-$

   The ionic bonds are the attractions between K$^+$ cations and F$^-$ anions.

50. Explain how a nonpolar covalent bond, a polar covalent bond, and an ionic bond differ. Your description should include rough sketches of the electron clouds that represent the electrons involved in the formation of each bond. *(Obj #7)*

   See Figure 3.6 on page 103 of the textbook.

52. Would you expect the bonds between the following atoms to be ionic or covalent bonds?
   (Obj #8)

   a. N–O   **covalent**…nonmetal-nonmetal

   b. Al-Cl   **ionic**…metal-nonmetal

54. Classify each of the following as either a molecular compound or an ionic compound. *(Obj #9)*

 a. acetone, CH_3COCH_3 (a common paint solvent)

   all nonmetallic atoms - **molecular**

 b. sodium sulfide, Na_2S (used in sheep dips) metal-nonmetal - **ionic**

Section 3.3 Molecular Compounds

56. How many valence electrons does each atom of the following elements have? *(Obj #10)*

 a. Cl **7**

 b. C **4**

58. Draw electron-dot symbols for each of the following elements and use them to explain why each element has the bonding pattern listed in Table 3.1 on page 109 of the textbook. *(Obj #11)*

   Each of the following answers is based on the assumption that nonmetallic atoms tend to form covalent bonds in order to get an octet (8) of electrons around each atom, like the very stable noble gases (other than helium). Covalent bonds (represented by lines in Lewis structures) and lone pairs each contribute two electrons to the octet.

 a. oxygen

   If oxygen atoms form 2 covalent bonds, they will have an octet of electrons around them. Water is an example:

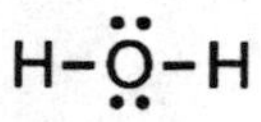

   H–Ö–H

 b. fluorine

   If fluorine atoms form 1 covalent bond, they will have an octet of electrons around them. Hydrogen fluoride, HF, is an example:

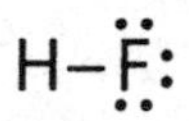

   H–F̈:

 c. carbon

   If carbon atoms form 4 covalent bonds, they will have an octet of electrons around them. Methane, CH_4, is an example:

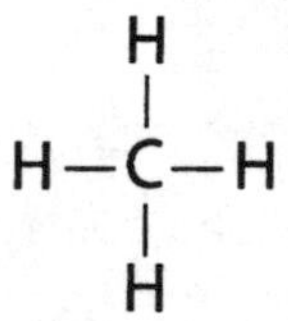

   H–C(–H)(–H)–H

 d. phosphorus

   If phosphorus atoms form 3 covalent bonds, they will have an octet of electrons around them. Phosphorus trichloride, PCl_3, is an example:

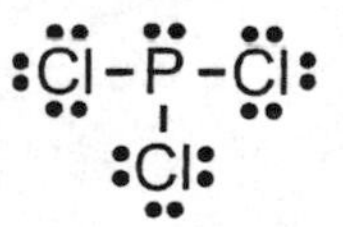

   :Cl–P–Cl:
      |
     :Cl:

60. The following Lewis structure is for CFC-12, which is one of the ozone-depleting chemicals that has been used as an aerosol can propellant and as a refrigerant. Describe the information given in this Lewis structure. *(Obj #12)*

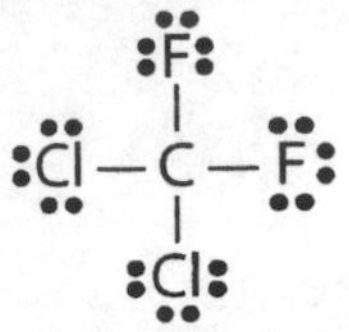

$$:\ddot{F}:$$
$$|$$
$$:\ddot{C}l - C - \ddot{F}:$$
$$|$$
$$:\ddot{C}l:$$

The molecule contains a carbon atom, 2 chlorine atoms, and 2 fluorine atoms. There are 2 covalent C–Cl bonds and 2 covalent C–F bonds. The Cl and F atoms have 3 lone pairs each.

62. Write the most common number of covalent bonds and lone pairs for atoms of each of the following nonmetallic elements. *(Obj #13)*

   a. H – **1 bond, no lone pairs**

   b. iodine – **1 bond, 3 lone pairs**

   c. sulfur – **2 bonds, 2 lone pairs**

   d. N - **3 bonds, 1 lone pair**

64. Draw a Lewis structure for each of the following formulas.   *(Obj #16)*

   a. oxygen difluoride, OF_2 (an unstable, colorless gas)

   Oxygen atoms usually have 2 covalent bonds and 2 lone pairs, and fluorine atoms have 1 covalent bond and 3 lone pairs.

   $$:\ddot{F} - \ddot{O} - \ddot{F}:$$

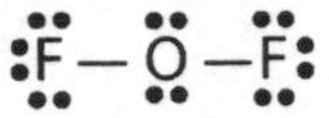

   b. bromoform, $CHBr_3$ (used as a sedative)

   Carbon atoms usually have 4 covalent bonds and no lone pairs, hydrogen atoms always have 1 covalent bond and no lone pairs, and bromine atoms usually have 1 covalent bond and 3 lone pairs. The hydrogen atom can be put in any of the 4 positions.

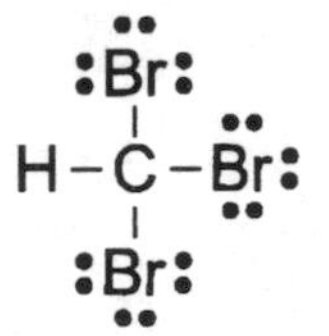

   $$:\ddot{Br}:$$
   $$|$$
   $$H - C - \ddot{Br}:$$
   $$|$$
   $$:\ddot{Br}:$$

   c. phosphorus triiodide, PI_3 (used to make organic compounds)

   Phosphorus atoms usually have 3 covalent bonds and 1 lone pair, and iodine atoms usually have 1 covalent bond and 3 lone pairs. The lone pair can be placed in any one of the 4 positions around the phosphorus atom.

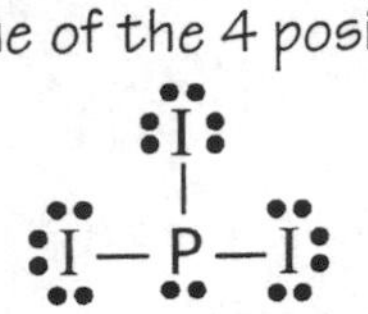

   $$:\ddot{I}:$$
   $$|$$
   $$:\ddot{I} - P - \ddot{I}:$$

66. Draw Lewis structures for the following compounds by adding any necessary lines and dots to the skeletons given. *(Obj #16)*

  a. hydrogen cyanide, HCN (used to manufacture dyes and pesticides)

   H−C≡N:

  b. dichloroethene, C_2Cl_4 (used to make perfumes)

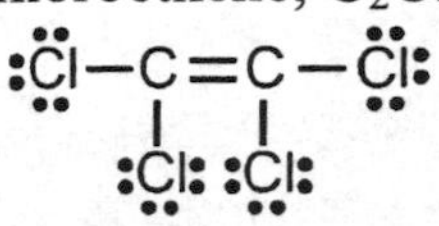

   :Cl−C=C−Cl:
       | |
      :Cl: :Cl:

68. Write two different names for each of the following alcohols.   *(Objs #14 & 15)*

           H
           | ..
      H−C−O−H
           | ..
  a. 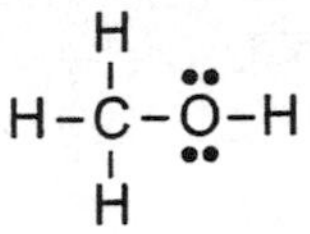   H   **methanol and methyl alcohol**

       H H
       | | ..
      H−C−C−O−H
       | | ..
  b. 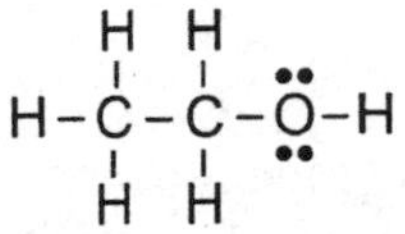   H H   **ethanol and ethyl alcohol**

          :O−H
       H | H
       | | |
      H−C−C−C−H
       | | |
  c.   H H H   **2-propanol and isopropyl alcohol**

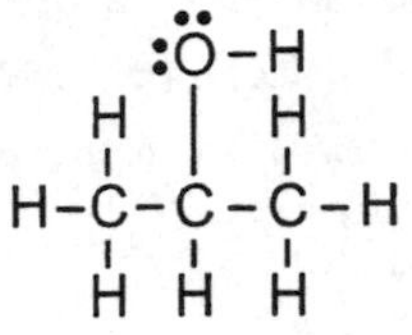

70. Compare and contrast the information given in the Lewis structure, the space-filling model, the ball-and-stick model, and the geometric sketch of a methane molecule, CH_4. *(Objs #19 & 20)*

   See Figure 3.10 on page 113 of the textbook. The Lewis structure shows the 4 covalent bonds between the carbon atoms and the hydrogen atoms. The space-filling model provides the most accurate representation of the electron-charge clouds for the atoms in CH_4. The ball-and-stick model emphasizes the molecule's correct molecular shape and shows the covalent bonds more clearly. Each ball represents an atom, and each stick represents a covalent bond between two atoms. The geometric sketch shows the three-dimensional tetrahedral structures with a two-dimensional drawing. Picture the hydrogen atoms connected to the central carbon atom with solid lines as being in the same plane as the carbon atom. The hydrogen atom connected to the central carbon with a solid wedge comes out of the plane toward you. The hydrogen atom connected to the carbon atom by a dashed wedge is located back behind the plane of the page.

72. Compare and contrast the information given in the Lewis structure, the space-filling model, the ball-and-stick model, and the geometric sketch of a water molecule, H_2O. *(Objs #19 & 20)*

   See Figure 3.12 on page 114 of the textbook. The Lewis structure shows the 2 O−H covalent bonds and the 2 lone pairs on the oxygen atom. The space-filling model provides the most accurate representation of the electron-charge clouds for the atoms and the bonding electrons. The ball-and-stick model emphasizes the molecule's correct molecular shape and shows the covalent bonds more clearly. The geometric sketch shows the structure with a two-dimensional drawing.

74. Describe the structure of liquid water. *(Obj #22)*

Water is composed of H_2O molecules. Pairs of these molecules are attracted to each other by the attraction between the partially positive hydrogen atom of one molecule and the partially negative oxygen atom of the other molecule. See Figure 3.13 on page 115 of the textbook. Each water molecule is moving constantly, breaking the attractions to some molecules, and making new attractions to other molecules. The image below shows the model we will use to visualize liquid water. See Figure 3.14 on page 115 of the textbook.

Section 3.4 Naming Binary Covalent Compounds

76. The compound represented by the ball-and-stick model that follows is used in the processing of nuclear fuels. Although bromine atoms most commonly form one covalent bond, they can form five bonds, as in the molecule shown here, in which the central sphere represents a bromine atom. The other atoms are fluorine atoms. Write this compound's chemical formula and name. List the bromine atom first in the chemical formula. *(Obj #28)*

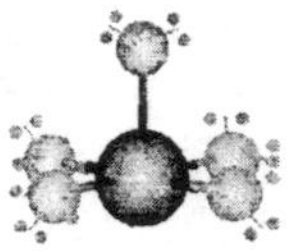

BrF_5 – bromine pentafluoride

78. The compound represented by the space-filling model that follows is used to vulcanize rubber and harden softwoods. Write its chemical formula and name. The central ball represents a sulfur atom, and the other atoms are chlorine atoms. List the sulfur atom first in the chemical formula. *(Obj #28)*

SCl_2 – sulfur dichloride

80. Write the name for each of the following chemical formulas. *(Obj #28)*
    a. I_2O_5 (an oxidizing agent)   **diiodine pentoxide**
    b. BrF_3 (adds fluorine atoms to other compounds)   **bromine trifluoride**
    c. IBr (used in organic synthesis)   **iodine monobromide**
    d. CH_4 (a primary component of natural gas)   **methane**
    e. HBr (used to make pharmaceuticals)   **hydrogen bromide or hydrogen monobromide**

82. Write the chemical formula for each of the following names. *(Obj #28)*
    a. propane (a fuel in heating torches)   **C_3H_8**
    b. chlorine monofluoride (a fluorinating agent)   **ClF**
    c. tetraphosphorus heptasulfide (a dangerous fire risk)   **P_4S_7**
    d. carbon tetrabromide (used to make organic compounds)   **CBr_4**
    e. hydrogen fluoride (an additive to liquid rocket propellants)   **HF**

Section 3.5 Ionic Compounds

84. Explain why metals usually combine with nonmetals to form ionic bonds. *(Obj #29)*

   Because metallic atoms hold some of their electrons relatively loosely, they tend to lose electrons and form cations. Because nonmetallic atoms attract electrons more strongly than metallic atoms, they tend to gain electrons and form anions. Thus, when a metallic atom and a nonmetallic atom combine, the nonmetallic atom often pulls one or more electrons far enough away from the metallic atom to form ions and an ionic bond.

85. How may protons and electrons do each of the following ions have?
   a. Be^{2+} **4 protons and 2 electrons**
   b. S^{2-} **16 protons and 18 electrons**

87. Write the name for each of these monatomic ions. *(Obj #31)*
   a. Ca^{2+} **calcium ion**   e. Ag^+ **silver ion or silver(I) ion**
   b. Li^+ **lithium ion**   f. Sc^{3+} **scandium ion**
   c. Cr^{2+} **chromium(II) ion**   g. P^{3-} **phosphide**
   d. F^- **fluoride**   h. Pb^{2+} **lead(II) ion**

89. Write the formula for each of these monatomic ions. *(Objs #30 & 31)*
   a. magnesium ion Mg^{2+} e. scandium ion Sc^{3+}
   b. sodium ion Na^+ f. nitride ion N^{3-}
   c. sulfide ion S^{2-} g. manganese(III) ion Mn^{3+}
   d. iron(III) ion Fe^{3+} h. zinc ion Zn^{2+}

91. Silver bromide, AgBr, is the compound on black and white film that causes the color change when the film is exposed to light. It has a structure similar structure to that of sodium chloride. What are the particles that form the basic structure of silver bromide? What type of attraction holds these particles together? Draw a rough sketch of the structure of solid silver bromide. *(Obj #32)*

   The metallic silver atoms form cations, and the nonmetallic bromine atoms form anions. The anions and cations alternate in the ionic solid with each cation surrounded by six anions and each anion surrounded by six cations. See Figure 3.18 on page 127 of the textbook, picturing Ag^+ ions in the place of the Na^+ ions and Br^- in the place of the Cl^- ions.

93. Write the name for each of these polyatomic ions. *(Objs #34 & 35)*
   a. NH_4^+ **ammonium**   c. HSO_4^- **hydrogen sulfate**
   b. $C_2H_3O_2^-$ **acetate**

95. Write the formula for each of these polyatomic ions. *(Objs #34-36)*
   a. ammonium NH_4^+ c. hydrogen sulfate ion HSO_4^-
   b. bicarbonate ion HCO_3^-

96. Write the name for each of these chemical formulas. *(Obj #37)*
   a. Na_2O (a dehydrating agent)   **sodium oxide**
   b. Ni_2O_3 (in storage batteries)   **nickel(III) oxide**
   c. $Pb(NO_3)_2$ (in matches and explosives)   **lead(II) nitrate**

    d. $Ba(OH)_2$ (an analytical reagent)   **barium hydroxide**

    e. $KHCO_3$ (in baking powder and fire-extinguishing agents) **potassium hydrogen carbonate**

98. Write the chemical formula for each of the following names. *(Obj #37)*

    a. potassium sulfide   (a depilatory) K_2S

    b. zinc phosphide   (a rodenticide) Zn_3P_2

    c. nickel(II) chloride   (used in nickel electroplating) $NiCl_2$

    d. magnesium dihydrogen phosphate   (used in fireproofing wood) $Mg(H_2PO_4)_2$

    e. lithium bicarbonate   (in mineral waters) $LiHCO_3$

100. The ionic compounds CuF_2, NH_4Cl, CdO, and $HgSO_4$ are all used to make batteries. Write the name for each of these compounds. *(Obj #37)*

    copper(II) fluoride (CuF_2), ammonium chloride (NH_4Cl), cadmium oxide (CdO), and mercury(II) sulfate ($HgSO_4$)

102. The ionic compounds copper(II) chloride, lithium nitrate, and cadmium sulfide are all used to make fireworks. Write the chemical formulas for these compounds. *(Obj #37)*

    copper(II) chloride ($CuCl_2$), lithium nitrate ($LiNO_3$), and cadmium sulfide (CdS)

Chapter 4
An Introduction to Chemical Reactions

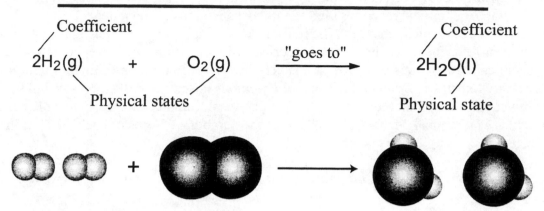

$$2H_2(g) \quad + \quad O_2(g) \quad \xrightarrow{\text{"goes to"}} \quad 2H_2O(l)$$

Coefficient — Coefficient
Physical states — Physical state

- ◆ Review Skills
- 4.1 Chemical Reactions and Chemical Equations
  - Interpreting a Chemical Equation
  - Balancing Chemical Equations
- 4.2 Solubility of Ionic Compounds and Precipitation Reactions
  - Water Solutions of Ionic Compounds
    Internet: Dissolving NaCl
  - Precipitation Reactions
    Internet: Precipitation Reaction
  - Predicting Water Solubility
    Internet: Writing Net Ionic Equations for Precipitation Reactions
    Special Topic 4.1: Hard Water and Your Hot Water Pipes

Having Trouble?
- ◆ Chapter Glossary
  - *Internet: Glossary Quiz*
- ◆ Chapter Objectives

Review Questions
Key Ideas
Chapter Problems

Section Goals and Introductions

Now that you know about atoms, elements, chemical bonds, and chemical compounds, you are ready to be introduced to chemical changes and the ways that we describe them.

Section 4.1 Chemical Reactions and Chemical Equations
Goals
- *To describe the nature of chemical reactions.*

- *To show how chemical reactions can be described with chemical equations.*

- *To show how to balance chemical equations.*

This section starts with a brief description of chemical reactions. You will have a much better understanding of chemical reactions after reading Chapters 4-6, which describe several different types of chemical changes. The introduction to chemical reactions is followed by a description of how chemical changes are described with chemical equations. You will see many chemical equations in this text, so it is very important that you be able to interpret them. *Sample Study Sheet 4.1: Balancing Chemical Equations* describes the important skill of balancing equations so that they reflect the fact that the number of atoms of each element for the products of chemical changes always equals the number of atoms of each element in the initial reactants.

Section 4.2 Solubility of Ionic Compounds and Precipitation Reactions
Goals

- *To describe the process by which ionic compounds dissolve in water.*

- *To describe the changes that take place on the molecular level during precipitation reactions.*

- *To provide guidelines for predicting water solubility of ionic compounds.*

- *To describe the process for predicting precipitation reactions and writing chemical equations for them.*

This section begins by describing solutions in general, describing solutions of ionic compounds in particular, and most important, describing the process by which ionic compounds dissolve in water. If you develop the ability to *see* this process in your mind's eye, it will help you understand the process of chemical changes between substances in solution. The animation on our Web site on *Dissolving NaCl* will help.

www.chemplace.com/college/

The definition of the nature of solutions leads to the description of a simple chemical change that takes place when two solutions of ionic compounds are mixed, which leads to the formation of an ionic compound that is insoluble in water and therefore comes out of solution as a solid. You will learn how to visualize the changes that take place in this type of reaction (called a precipitation reaction), how to predict whether mixtures of two solutions of ionic compounds will lead to a precipitation reaction, and how to write chemical equations for those reactions that do take place. The animation on our Web site on *Precipitation Reactions* will help.

www.chemplace.com/college/

You might want to look closely at the *Having Trouble* section at the end of Chapter 4. It describes all of the skills from Chapters 2-4 that are necessary for writing chemical equations for precipitation reactions. Students often have trouble with writing these equations, not because of the new components to the process found in Chapter 4, but because they are still having trouble with skills from Chapters 2 and 3.

Chapter 4 Map

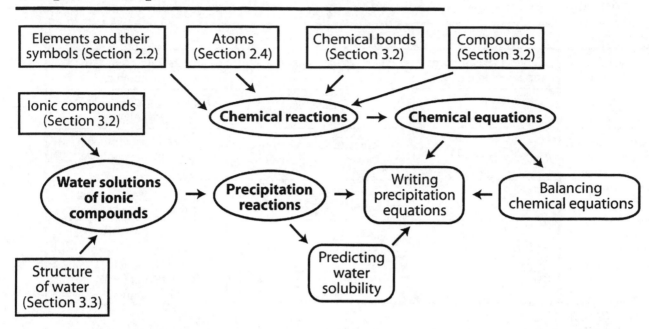

Chapter Checklist

- ☐ Read the Review Skills section. If there is any skill mentioned that you have not yet mastered, review the material on that topic before reading this chapter.
- ☐ Read the chapter quickly before the lecture that describes it.
- ☐ Attend class meetings, take notes, and participate in class discussions.
- ☐ Work the Chapter Exercises, perhaps using the Chapter Examples as guides.
- ☐ Study the Chapter Glossary and test yourself at our Web site:

  www.chemplace.com/college/

- ☐ Study all of the Chapter Objectives. You might want to write a description of how you will meet each objective. (Although it is best to master all of the objectives, the following objectives are especially important because they pertain to skills that you will need while studying other chapters of this text: 3, 4, and 10.)
- ☐ Reread the Study Sheets in this chapter and decide whether you will use them or some variation on them to complete the tasks they describe.

  Sample Study Sheet 4.1: Balancing Chemical Equations

  Sample Study Sheet 4.2: Predicting Precipitation Reactions and Writing Precipitation Equations

☐ Memorize the following solubility guidelines. Be sure to check with your instructor to determine how much you are expected to know of the following guidelines.

Category	Ions	Except with these ions	Examples
soluble cations	Group 1 metallic ions and ammonium, NH_4^+	No exceptions	Na_2CO_3, $LiOH$, and $(NH_4)_2S$ are soluble.
soluble anions	NO_3^- and $C_2H_3O_2^-$	No exceptions	$Bi(NO_3)_3$, and $Co(C_2H_3O_2)_2$ are soluble.
usually soluble anions	Cl^-, Br^-, and I^-	Soluble with some exceptions, including with Ag^+ and Pb^{2+}	$CuCl_2$ is water soluble, but $AgCl$ is insoluble.
	SO_4^{2-}	Soluble with some exceptions, including with Ba^{2+} and Pb^{2+}	$FeSO_4$ is water soluble, but $BaSO_4$ is insoluble.
usually insoluble	CO_3^{2-}, PO_4^{3-}, and OH^-	Insoluble with some exceptions, including with group 1 elements and NH_4^+	$CaCO_3$, $Ca_3(PO_4)_2$, and $Mn(OH)_2$ are insoluble in water, but $(NH_4)_2CO_3$, Li_3PO_4, and $CsOH$ are soluble.

☐ To get a review of the most important topics in the chapter, fill in the blanks in the Key Ideas section.

☐ Work all of the selected problems at the end of the chapter, and check your answers with the solutions provided in this chapter of the study guide.

☐ Ask for help if you need it.

Web Resources   www.chemplace.com/college/

Dissolving NaCl
Precipitation Reaction
Writing Complete and Net-Ionic Equations for Precipitation Reactions
Glossary Quiz

Exercises Key

✍ **Exercise 4.1 - Balancing Equations:**   Balance the following chemical equations. *(Obj #4)*

    a. $P_4(s) + 6Cl_2(g) \rightarrow 4PCl_3(l)$

    b. $3PbO(s) + 2NH_3(g) \rightarrow 3Pb(s) + N_2(g) + 3H_2O(l)$

    c. $P_4O_{10}(s) + 6H_2O(l) \rightarrow 4H_3PO_4(aq)$

    d. $3Mn(s) + 2CrCl_3(aq) \rightarrow 3MnCl_2(aq) + 2Cr(s)$

    e. $C_2H_2(g) + 5/2 O_2(g) \rightarrow 2CO_2(g) + H_2O(l)$

        or $2C_2H_2(g) + 5O_2(g) \rightarrow 4CO_2(g) + 2H_2O(l)$

    f. $3Co(NO_3)_2(aq) + 2Na_3PO_4(aq) \rightarrow Co_3(PO_4)_2(s) + 6NaNO_3(aq)$

    g. $2CH_3NH_2(g) + 9/2 O_2(g) \rightarrow 2CO_2(g) + 5H_2O(l) + N_2(g)$

        or $4CH_3NH_2(g) + 9O_2(g) \rightarrow 4CO_2(g) + 10H_2O(l) + 2N_2(g)$

    h. $2FeS(s) + 9/2 O_2(g) + 2H_2O(l) \rightarrow Fe_2O_3(s) + 2H_2SO_4(aq)$

        or $4FeS(s) + 9O_2(g) + 4H_2O(l) \rightarrow 2Fe_2O_3(s) + 4H_2SO_4(aq)$

✍ **Exercise 4.2 - Predicting Water Solubility:**   Predict whether each of the following is soluble or insoluble in water. *(Obj #9)*

    a. $Hg(NO_3)_2$ (used to manufacture felt)   **soluble**

    b. $BaCO_3$ (used to make radiation-resistant glass for color TV tubes) **insoluble**

    c. K_3PO_4 (used to make liquid soaps)   **soluble**

    d. $PbCl_2$ (used to make other lead salts)   **insoluble**

    e. $Cd(OH)_2$ (storage battery electrodes)   **insoluble**

✍ **Exercise 4.3 - Precipitation Reactions:**   Predict whether a precipitate will form when each of the following pairs of water solutions is mixed. If there is a precipitation reaction, write the complete equation that describes the reaction. *(Obj #10)*

    a. $3CaCl_2(aq) + 2Na_3PO_4(aq) \rightarrow \textbf{Ca}_3\textbf{(PO}_4\textbf{)}_2\textbf{(s)} + \textbf{6NaCl(aq)}$

    b. $3KOH(aq) + Fe(NO_3)_3(aq) \rightarrow \textbf{3KNO}_3\textbf{(aq)} + \textbf{Fe(OH)}_3\textbf{(s)}$

    c. $NaC_2H_3O_2(aq) + CaSO_4(aq)$ **No Reaction**

    d. $K_2SO_4(aq) + Pb(NO_3)_2(aq) \rightarrow \textbf{2KNO}_3\textbf{(aq)} + \textbf{PbSO}_4\textbf{(s)}$

Review Questions Key

1. Write the formulas for all of the diatomic elements. $\textbf{H}_2, \textbf{N}_2, \textbf{O}_2, \textbf{F}_2, \textbf{Cl}_2, \textbf{Br}_2, \textbf{I}_2$

2. Predict whether atoms of each of the following pairs of elements would be expected to form ionic or covalent bonds.

    a. Mg and F   **ionic**   c. Fe and O   **ionic**

    b. O and H   **covalent**   d. N and Cl   **covalent**

3. Describe the structure of liquid water, including a description of water molecules and the attractions between them.

> Water is composed of H_2O molecules that are attracted to each other due to the attraction between partially positive hydrogen atoms and the partially negative oxygen atoms of other molecules. See Figure 3.13 on page 115 of the textbook.

> Each water molecule is moving constantly, breaking the attractions to some molecules, and making new attractions to other molecules. See Figure 3.14 on page 115 of the textbook.

4. Write formulas that correspond to the following names.
    a. ammonia   **NH$_3$**
    b. methane   **CH$_4$**
    c. propane   **C$_3$H$_8$**
    d. water   **H$_2$O**

5. Write formulas that correspond to the following names.
    a. nitrogen dioxide   **NO$_2$**
    b. carbon tetrabromide   **CBr$_4$**
    c. dibromine monoxide   **Br$_2$O**
    d. nitrogen monoxide   **NO**

6. Write formulas that correspond to the following names.
    a. lithium fluoride   **LiF**
    b. lead(II) hydroxide   **Pb(OH)$_2$**
    c. potassium oxide   **K$_2$O**
    d. sodium carbonate   **Na$_2$CO$_3$**
    e. chromium(III) chloride   **CrCl$_3$**
    f. sodium hydrogen phosphate   **Na$_2$HPO$_4$**

Key Ideas Answers

7. A chemical change or chemical reaction is a process in which one or more pure substances are **converted into** one or more different pure substances.

9. A chemical equation is a **shorthand description** of a chemical reaction.

11. To indicate that a chemical reaction requires the **continuous** addition of heat in order to proceed, we place an upper-case Greek **delta, Δ,** above the arrow in the reaction's chemical equation.

13. When balancing chemical equations, we do not change the **subscripts** in the formulas.

15. Every part of a water solution of an ionic compound has the **same proportions** of water molecules and ions as every other part.

17. In solutions of solids dissolved in liquids, we call the solid the **solute** and the liquid the **solvent**.

19. In solutions of two liquids, we call the **minor** component the solute and the **major** component the solvent.

21. Crystals are solid particles whose component atoms, ions, or molecules are arranged in an **organized, repeating** pattern.

23. Because spectator ions are not involved in the reaction, they are often **left out** of the chemical equation.

Problems Key

Section 4.1 Chemical Reactions and Chemical Equations

25. Describe the information given in the following chemical equation. *(Objs #2 & 3)*

$$2CuHCO_3(s) \xrightarrow{\Delta} Cu_2CO_3(s) + H_2O(l) + CO_2(g)$$

For each 2 particles of solid copper(I) hydrogen carbonate that are heated strongly, 1 particle of solid copper(I) carbonate, 1 molecule of liquid water, and 1 molecule of gaseous carbon dioxide are formed.

27. Balance the following equations. *(Obj #4)*

   a. $N_2(g) + 3H_2(g) \rightarrow 2NH_3(g)$

   b. $4Cl_2(g) + 2CH_4(g) + O_2(g) \rightarrow 8HCl(g) + 2CO(g)$

   c. $B_2O_3(s) + 6NaOH(aq) \rightarrow 2Na_3BO_3(aq) + 3H_2O(l)$

   d. $2Al(s) + 2H_3PO_4(aq) \rightarrow 2AlPO_4(s) + 3H_2(g)$

   e. $CO(g) + \frac{1}{2}O_2(g) \rightarrow CO_2(g)$ or $2CO(g) + O_2(g) \rightarrow 2CO_2(g)$

   f. $C_6H_{14}(l) + 19/2O_2(g) \rightarrow 6CO_2(g) + 7H_2O(l)$

      or $2C_6H_{14}(l) + 19O_2(g) \rightarrow 12CO_2(g) + 14H_2O(l)$

   g. $Sb_2S_3(s) + 9/2O_2(g) \rightarrow Sb_2O_3(s) + 3SO_2(g)$

      or $2Sb_2S_3(s) + 9O_2(g) \rightarrow 2Sb_2O_3(s) + 6SO_2(g)$

   h. $2Al(s) + 3CuSO_4(aq) \rightarrow Al_2(SO_4)_3(aq) + 3Cu(s)$

   i. $3P_2H_4(l) \rightarrow 4PH_3(g) + \frac{1}{2}P_4(s)$ or $6P_2H_4(l) \rightarrow 8PH_3(g) + P_4(s)$

29. Because of its toxicity, carbon tetrachloride is prohibited in products intended for home use, but it is used industrially for a variety of purposes, including the production of chlorofluorocarbons (CFCs). It is made in three steps. Balance their equations:

   $CS_2 + 3Cl_2 \rightarrow S_2Cl_2 + CCl_4$

   $4CS_2 + 8S_2Cl_2 \rightarrow 3S_8 + 4CCl_4$

   $S_8 + 4C \rightarrow 4CS_2$

31. Hydrochlorofluorocarbons (HCFCs), which contain hydrogen as well as carbon, fluorine, and chlorine, are less damaging to the ozone layer than the chlorofluorocarbons (CFCs) described in problem 30. HCFCs are therefore used instead of CFCs for many purposes. Balance the following equation that shows how the HCFC chlorodifluoromethane, $CHClF_2$, is made.

   $2HF + CHCl_3 \rightarrow CHClF_2 + 2HCl$

Section 4.2 Solubility of Ionic Compounds and Precipitation Reactions

33. Describe the process for dissolving the ionic compound lithium iodide, LiI, in water, including the nature of the particles in solution and the attractions between the particles in the solution. *(Obj #5)*

> When solid lithium iodide is added to water, all of the ions at the surface of the solid can be viewed as vibrating back and forth between moving out into the water and returning to the solid surface. Sometimes when an ion vibrates out into the water, a water molecule collides with it, helping to break the ionic bond, and pushing it out into the solution. Water molecules move into the gap between the ion in solution and the solid and shield the ion from the attraction to the solid.
>
> The ions are kept stable and held in solution by attractions between them and the polar water molecules. The negatively charged oxygen ends of water molecules surround the lithium ions, and the positively charged hydrogen ends of water molecules surround the iodide ions. (See Figures 4.4 and 4.5 on pages 161 and 162 in the textbook with Li^+ in the place of Na^+ and I^- in the place of Cl^-.)

35. Describe the process for dissolving the ionic compound sodium sulfate, Na_2SO_4, in water. Include the nature of the particles in solution and the attractions between the particles in the solution. *(Obj #5)*

> Same answer as problem 33 but with sodium ions, Na^+, instead of lithium ions and sulfate ions, $SO_4{}^{2-}$, in the place of iodide ions. The final solution will have two times as many sodium ions as sulfate ions.

37. Solid camphor and liquid ethanol mix to form a solution. Which of these substances is the solute and which is the solvent? *(Obj #6)*

> In a solution of a solid in a liquid, the solid is generally considered the solute, and the liquid is the solvent. Therefore, camphor is the solute in this solution, and ethanol is the solvent.

40. Black-and-white photographic film has a thin layer of silver bromide deposited on it. Wherever light strikes the film, silver ions are converted to uncharged silver atoms, creating a dark image on the film. Describe the precipitation reaction that takes place between water solutions of silver nitrate, $AgNO_3(aq)$, and sodium bromide, $NaBr(aq)$, to form solid silver bromide, $AgBr(s)$, and aqueous sodium nitrate, $NaNO_3(aq)$. Include the nature of the particles in the system before and after the reaction, a description of the cause of the reaction, and a description of the attractions between the particles before and after the reaction. *(Obj #8)*

> At the instant that the solution of silver nitrate is added to the aqueous sodium bromide, there are 4 different ions in solution surrounded by water molecules, Ag^+, $NO_3{}^-$, Na^+, and Br^-. The oxygen ends of the water molecules surround the silver and sodium cations, and the hydrogen ends of water molecules surround the nitrate and bromide anions.

When silver ions and bromide ions collide, they stay together long enough for other silver ions and bromide ions to collide with them, forming clusters of ions that precipitate from the solution.

The sodium and nitrate ions are unchanged in the reaction. They were separate and surrounded by water molecules at the beginning of the reaction, and they are still separate and surrounded by water molecules at the end of the reaction. See Figures 4-7 to 4-9 on pages 161 and 162 in the textbook with silver ions in the place of calcium ions and bromide ions in the place of carbonate ions.

42. Predict whether each of the following substances is soluble or insoluble in water. *(Obj #9)*
   a. Na_2SO_3 (water treatment)   **soluble**
   b. iron(III) acetate   (wood preservative)   **soluble**
   c. $CoCO_3$ (red pigment)   **insoluble**
   d. lead(II) chloride   (preparation of lead salts)   **insoluble**

44. Predict whether each of the following substances is soluble or insoluble in water. *(Obj #9)*
   a. zinc phosphate   (dental cements)   **insoluble**
   b. $Mn(C_2H_3O_2)_2$ (cloth dyeing)   **soluble**
   c. nickel(II) sulfate   (nickel plating)   **soluble**
   d. AgCl   (silver plating)   **insoluble**

46. For each of the following pairs of formulas, predict whether the substances they represent would react in a precipitation reaction. The products formed in the reactions that take place are used in ceramics, cloud seeding, photography, electroplating, and paper coatings. If there is no reaction, write, "No Reaction". If there is a reaction, write the complete equation for the reaction. *(Obj #10)*

   a. $Co(NO_3)_2(aq) + Na_2CO_3(aq)$ → $CoCO_3(s) + 2NaNO_3(aq)$
   b. $2KI(aq) + Pb(C_2H_3O_2)_2(aq)$ → $2KC_2H_3O_2(aq) + PbI_2(s)$
   c. $CuSO_4(aq) + LiNO_3(aq)$ **no reaction**
   d. $3Ni(NO_3)_2(aq) + 2Na_3PO_4(aq)$ → $Ni_3(PO_4)_2(s) + 6NaNO_3(aq)$
   e. $K_2SO_4(aq) + Ba(NO_3)_2(aq)$ → $2KNO_3(aq) + BaSO_4(s)$

48. Phosphate ions find their way into our water system from the fertilizers dissolved in the runoff from agricultural fields and from detergents that we send down our drains. Some of these phosphate ions can be removed by adding aluminum sulfate to the water and precipitating the phosphate ions as aluminum phosphate. Write the net ionic equation for the reaction that forms the aluminum phosphate.

   $Al^{3+}(aq) + PO_4^{3-}(aq)$ → $AlPO_4(s)$

50. Cadmium hydroxide is used in storage batteries. It is made from the precipitation reaction of cadmium acetate and sodium hydroxide. Write the complete equation for this reaction.

   $Cd(C_2H_3O_2)_2(aq) + 2NaOH(aq)$ → $Cd(OH)_2(s) + 2NaC_2H_3O_2(aq)$

Additional Problems

52. Balance the following chemical equations.

    a. $SiCl_4 + 2H_2O \rightarrow SiO_2 + 4HCl$

    b. $2H_3BO_3 \rightarrow B_2O_3 + 3H_2O$

    c. $I_2 + 3Cl_2 \rightarrow 2ICl_3$

    d. $2Al_2O_3 + 3C \rightarrow 4Al + 3CO_2$

54. Balance the following chemical equations.

    a. $4NH_3 + Cl_2 \rightarrow N_2H_4 + 2NH_4Cl$

    b. $Cu + 2AgNO_3 \rightarrow Cu(NO_3)_2 + 2Ag$

    c. $Sb_2S_3 + 6HNO_3 \rightarrow 2Sb(NO_3)_3 + 3H_2S$

    d. $Al_2O_3 + 3Cl_2 + 3C \rightarrow 2AlCl_3 + 3CO$

56. Phosphoric acid, H_3PO_4, is an important chemical used to make fertilizers, detergents, pharmaceuticals, and many other substances. High purity phosphoric acid is made in a two-step process called the furnace process. Balance its two equations:

$$2Ca_3(PO_4)_2 + 6SiO_2 + 10C \rightarrow P_4 + 10CO + 6CaSiO_3$$

$$P_4 + 5O_2 + 6H_2O \rightarrow 4H_3PO_4$$

58. Predict whether each of the following substances is soluble or insoluble in water.

    a. manganese(II) chloride (used as a dietary supplement)   **soluble**

    b. $CdSO_4$ (used in pigments) **soluble**

    c. copper(II) carbonate (used in fireworks)   **insoluble**

    d. $Co(OH)_3$ (used as a catalyst)   **insoluble**

60. For each of the following pairs of formulas, predict whether the substances they represent would react to yield a precipitate. (The products formed in the reactions that take place are used to coat steel, as a fire-proofing filler for plastics, in cosmetics, and as a topical antiseptic.) If there is no reaction, write, "No Reaction." If there is a reaction, write the complete equation for the reaction. *(Obj #10)*

    a. $NaCl(aq) + Al(NO_3)_3(aq)$ **no reaction**

    b. $Ni(NO_3)_2(aq) + 2NaOH(aq) \rightarrow Ni(OH)_2(s) + 2NaNO_3(aq)$

    c. $3MnCl_2(aq) + 2Na_3PO_4(aq) \rightarrow Mn_3(PO_4)_2(s) + 6NaCl(aq)$

    d. $Zn(C_2H_3O_2)_2(aq) + Na_2CO_3(aq) \rightarrow ZnCO_3(s) + 2NaC_2H_3O_2(aq)$

Before working problems 62 through 78, you might want to review the procedures for writing chemical formulas that are described in Chapter 3. Remember that some elements are described with formulas containing subscripts (as in O_2).

62. Hydrochloric acid is used in the cleaning of metals (called pickling). Hydrogen chloride, used to make hydrochloric acid, is made industrially by combining hydrogen and chlorine. Write a balanced equation, without including states, for this reaction.

$$H_2 + Cl_2 \rightarrow 2HCl$$

64. Aluminum sulfate, commonly called alum, is used to coat paper made from wood pulp. (It fills in tiny holes in the paper and thus keeps the ink from running.) Alum is made in the reaction of aluminum oxide with sulfuric acid, H_2SO_4, which produces aluminum sulfate and water. Write a balanced equation, without including states, for this reaction.

$$Al_2O_3 + 3H_2SO_4 \rightarrow Al_2(SO_4)_3 + 3H_2O$$

66. Hydrogen fluoride is used to make chlorofluorocarbons (CFCs) and in uranium processing. Calcium fluoride reacts with sulfuric acid, H_2SO_4, to form hydrogen fluoride and calcium sulfate. Write a balanced equation, without including states, for this reaction.

$$CaF_2 + H_2SO_4 \rightarrow 2HF + CaSO_4$$

68. Sodium hydroxide, which is often called caustic soda, is used to make paper, soaps, and detergents. For many years, it was made from the reaction of sodium carbonate with calcium hydroxide (also called slaked lime). The products are sodium hydroxide and calcium carbonate. Write a balanced equation, without including states, for this reaction.

$$Na_2CO_3 + Ca(OH)_2 \rightarrow 2NaOH + CaCO_3$$

71. All of the equations for the Solvay process described in problem 70 can be summarized by a single equation, called a net equation, that describes the overall change for the process. This equation shows calcium carbonate reacting with sodium chloride to form sodium carbonate and calcium chloride. Write a balanced equation, without including states, for this net reaction.

$$CaCO_3 + 2NaCl \rightarrow Na_2CO_3 + CaCl_2$$

73. All of the equations for the production of nitric acid described in problem 72 can be summarized in a single equation, called a net equation, that describes the overall change for the complete process. This equation shows ammonia combining with oxygen to yield nitric acid and water. Write a balanced equation, without including states, for this net reaction.

$$NH_3 + 2O_2 \rightarrow HNO_3 + H_2O$$

75. Hydrogen gas has many practical uses, including the conversion of vegetable oils into margarine. One way the gas is produced by the chemical industry is by reacting propane gas with gaseous water to form carbon dioxide gas and hydrogen gas. Write a balanced equation for this reaction, showing the states of reactants and products.

$$C_3H_8(g) + 6H_2O(g) \rightarrow 3CO_2(g) + 10H_2(g)$$

77. Pig iron is iron with about 4.3% carbon in it. The carbon lowers the metal's melting point and makes it easier to shape. To produce pig iron, iron(III) oxide is combined with carbon and oxygen at high temperature. Three changes then take place to form molten iron with carbon dispersed in it. Write a balanced equation, without including states, for each of these changes:

  a. Carbon combines with oxygen to form carbon monoxide.

$$C + \tfrac{1}{2}O_2 \rightarrow CO$$
$$\text{or} \quad 2C + O_2 \rightarrow 2CO$$

  b. Iron(III) oxide combines with the carbon monoxide to form iron and carbon dioxide.

$$Fe_2O_3 + 3CO \rightarrow 2Fe + 3CO_2$$

  c. Carbon monoxide changes into carbon (in the molten iron) and carbon dioxide.

$$2CO \rightarrow C + CO_2$$

79. Assume you are given a water solution that contains either sodium ions or aluminum ions. Describe how you could determine which of these is in solution.

> Add a water solution of an ionic compound that forms an insoluble substance with aluminum ions and a soluble substance with sodium ions. For example, a potassium carbonate solution would precipitate the aluminum ions as $Al_2(CO_3)_3$.

81. Write a complete, balanced chemical equation for the reaction between water solutions of iron(III) chloride and silver nitrate.

$$FeCl_3(aq) + 3AgNO_3(aq) \rightarrow Fe(NO_3)_3(aq) + 3AgCl(s)$$

83. When the solid amino acid methionine, $C_5H_{11}NSO_2$, reacts with oxygen gas, the products are carbon dioxide gas, liquid water, sulfur dioxide gas, and nitrogen gas. Write a complete, balanced equation for this reaction.

$$2C_5H_{11}NSO_2(s) + 31/2O_2(g)$$
$$\rightarrow 10CO_2(g) + 11H_2O(l) + 2SO_2(g) + N_2(g)$$
$$or\ 4C_5H_{11}NSO_2(s) + 31O_2(g)$$
$$\rightarrow 20CO_2(g) + 22H_2O(l) + 4SO_2(g) + 2N_2(g)$$

Chapter 5
Acids, Bases, and Acid-Base Reactions

This proton, H⁺, is transferred to a hydroxide ion.

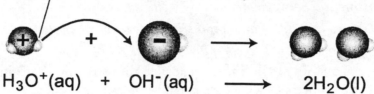

$$H_3O^+(aq) \; + \; OH^-(aq) \; \longrightarrow \; 2H_2O(l)$$

♦ Review Skills

5.1 Acids
- Arrhenius Acids
- Types of Arrhenius Acids
- Strong and Weak Acids

Special Topic 5.1: Acid Rain

Internet: Acid Animation

5.2 Acid Nomenclature
- Names and Formulas of Binary Acids
- Names and Formulas of Oxyacids

Internet: Acid Nomenclature

5.3 Summary of Chemical Nomenclature

5.4 Strong and Weak Arrhenius Bases

Special Topic 5.2: Chemicals and Your Sense of Taste

Internet: Identification of Strong and Weak Acids and Bases

5.5 pH and Acidic and Basic Solutions

5.6 Arrhenius Acid-Base Reactions
- Reactions of Aqueous Strong Arrhenius Acids and Aqueous Strong Arrhenius Bases
- Writing Equations for Reactions Between Acids and Bases

Special Topic 5.3: Precipitation, Acid-Base Reactions, and Tooth Decay

- Reactions of Arrhenius Acids and Ionic Compounds That Contain Carbonate or Hydrogen Carbonate

Special Topic 5.4: Saving Valuable Books

Special Topic 5.5: Be Careful with Bleach

Internet: Acid-Base Reaction Animation

5.7 Brønsted-Lowry Acids and Bases

♦ Chapter Glossary

Internet: Glossary Quiz

♦ Chapter Objectives

Review Questions

Key Ideas

Chapter Problems

Section Goals and Introductions

Section 5.1 Acids

Goals

- *To describe acids*
- *To make the distinction between strong and weak acids.*
- *To show the changes that take place on the particle level when acids dissolve in water.*
- *To show how you can recognize strong and weak acids.*

This section introduces one way to define acids, called the Arrhenius definition. The most important skills to develop in this section are (1) to be able to recognize acids from names or formulas and (2) to be able to describe the changes that take place at the particle level when strong and weak acids dissolve in water. Be sure to give special attention to Figures 5.1, 5.5, and 5.6. Visit our Web site to view an animation that shows acid solutions.

www.chemplace.com/college/

Section 5.2 Acid Nomenclature

Goal: To describe how to convert between names and chemical formulas for acids.

This section adds acids to the list of compounds for which you should be able to convert between names and formulas. Visit our Web site for more information about *Acid Nomenclature.*

www.chemplace.com/college/

Section 5.3 Summary of Chemical Nomenclature

Goal: To review the process for converting between names and formulas for binary covalent compounds, binary ionic compounds, ionic compounds with polyatomic ions, binary acids, and oxyacids.

Although chemical nomenclature may not be your favorite topic, it is an important one. The ability to convert among names and formulas for chemical compounds is crucial to communication between chemists and chemistry students. This section collects the nomenclature guidelines from Chapters 3 and 5 and gives you a chance to review them. Table 5.6 provides your most concise summary of these guidelines.

Section 5.4 Strong and Weak Arrhenius Bases

Goals

- *To describe bases and to make the distinction between strong and weak bases.*
- *To show how you can recognize strong and weak bases.*
- *To show the changes that take place on the particle level when bases dissolve in water.*

This section does for bases what Section 5.1 does for acids: (1) it states the Arrhenius definition of base, (2) it provides you with the information necessary to identify strong and weak bases, and (3) it describes the changes that take place when one weak base (ammonia) dissolves in water (Figure 5.8). *Sample Study Sheet 5.1* summarizes the steps for identification of strong and weak acids and bases. Visit our Web site for more information about *Identification of Strong and Weak Acids and Bases.*

www.chemplace.com/college/

Section 5.5 pH and Acidic and Basic Solutions

Goal: To explain the pH scale used to describe acidic and basic solutions.

This section provides an introduction to the pH scale used to describe acidic and basic solutions. Figure 5.10 contains the most important information.

Section 5.6 Arrhenius Acid-Base Reactions

Goals

- *To describe acid-base reactions, with an emphasis on developing the ability to visualize the changes that take place on the particle level.*
- *To show how you can predict whether two reactants will react in an acid-base reaction.*
- *To show how to write equations for acid-base reactions.*

This section does for acid-base reactions what Section 4.2 does for precipitation reactions. It might help to consider the similarities and differences between these two types of chemical changes. Be sure that you can visualize the changes that take place at the particle level for both types of chemical reactions. Pay special attention to Figures 5.12, 5.13, and 5.15. Visit our Web site to see an animation showing an *Acid-Base Reaction*.

www.chemplace.com/college/

Section 5.7 Brønsted-Lowry Acids and Bases

Goal: To describe a second set of definitions for acid, base, and acid-base reactions, called the Brønsted-Lowry definitions.

Although the Arrhenius definitions of acid, base, and acid-base reactions provided in Sections 5.1, 5.4, and 5.6 are very important, especially to the beginning chemistry student, chemists have found it useful to extend these definitions to include new substances as acids and bases that would not be classified as such according to the Arrhenius definitions. The new definitions, called the Brønsted-Lowry definitions, are described in this section.

Chapter 5 Map

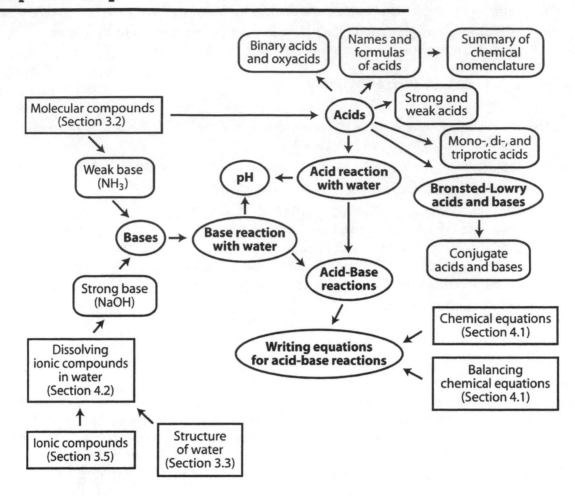

Chapter Checklist

- ☐ Read the Review Skills section. If there is any skill mentioned that you have not yet mastered, review the material on that topic before reading this chapter.
- ☐ Read the chapter quickly before the lecture that describes it.
- ☐ Attend class meetings, take notes, and participate in class discussions.
- ☐ Work the Chapter Exercises, perhaps using the Chapter Examples as guides.
- ☐ Study the Chapter Glossary and test yourself on our Web site:

   www.chemplace.com/college/

- ☐ Study all of the Chapter Objectives. You might want to write a description of how you will meet each objective. (Although it is best to master all of the objectives, the following objectives are especially important because they pertain to skills that you will need while studying other chapters of this text: 13, 14, 18-21, and 24.)
- ☐ Reread *Sample Study Sheet 5.1: Identification of Strong and Weak Acids and Bases* and decide whether you will use it or some variation on it to complete the task it describes.

☐ Memorize the following. Be sure to check with your instructor to determine how much you are expected to know of the following.

- Guidelines for writing names and formulas for compounds

Type of compound	General formula	Examples	General name	Examples
Binary covalent Section 3.4	A_aB_b	N_2O_5 or CO_2	(prefix unless mono)(name of first element in formula) (prefix)(root of second element)ide	dinitrogen pentoxide or carbon dioxide
Binary ionic Section 3.5	M_aA_b	NaCl or $FeCl_3$	(name of metal) (root of nonmetal)ide or (name of metal)(Roman #) (root of nonmetal)ide	sodium chloride or iron(III) chloride
Ionic with polyatomic ion(s) Section 3.5	M_aX_b or $(NH_4)_aX_b$ X = recognized formula of polyatomic ion	Li_2HPO_4 or $CuSO_4$ or NH_4Cl or $(NH_4)_2SO_4$	(name of metal) (name of polyatomic ion) or (name of metal)(Roman #) (name of polyatomic ion) or ammonium (root of nonmetal)ide or ammonium (name of polyatomic ion)	lithium hydrogen phosphate or copper(II) sulfate or ammonium chloride or ammonium sulfate
Binary acid Section 5.2	HX(*aq*)	HCl(*aq*)	hydro(root)ic acid	hydrochloric acid
Oxyacid Section 5.2	$H_aX_bO_c$	HNO_3 or H_2SO_4 or H_3PO_4	(root)ic acid	nitric acid or sulfuric acid or phosphoric acid

Notes: M = symbol of metal
A and B = symbols of nonmetals
X = some element other than H or O
a, b, & c indicate subscripts

- The significance of the numbers in the pH scale

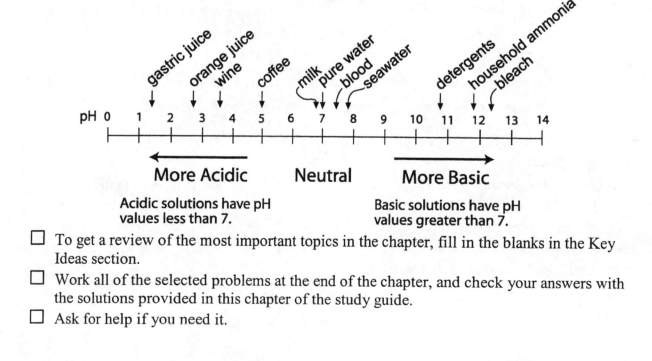

☐ To get a review of the most important topics in the chapter, fill in the blanks in the Key Ideas section.

☐ Work all of the selected problems at the end of the chapter, and check your answers with the solutions provided in this chapter of the study guide.

☐ Ask for help if you need it.

Web Resources   www.chemplace.com/college/

Acids
Acid Nomenclature
Identification of Strong and Weak Acids and Bases
Acid-Base Reaction
Glossary Quiz

Exercises Key

Exercise 5.1 - Formulas for Acids: Write the chemical formulas that correspond to the names   *(Obj #12)*

(a) hydrofluoric acid

This name has the form of a binary acid, hydro(root)ic acid, so its formula is **HF(aq)**.

(b) phosphoric acid

This name has the form of an oxyacid, (root)ic acid, so it contains hydrogen, phosphorus, and oxygen. Phosphate is PO_4^{3-}, so phosphoric acid is **H_3PO_4**.

Exercise 5.2 - Naming Acids: Write the names that correspond to the chemical formulas   *(Obj #12)*

(a) HI(*aq*)

This is a binary acid, so its name has the form hydro(root)ic acid. HI(*aq*) is **hydriodic acid**. (The "o" in hydro- is usually left off.)

(b) $HC_2H_3O_2$.

The name of the oxyanion $C_2H_3O_2^-$ is acetate, so $HC_2H_3O_2$ is **acetic acid**. CH_3CO_2H and CH_3COOH are also commonly used as formulas for acetic acid.

Exercise 5.3 - Formulas to Names: Write the names that correspond to the following chemical formulas.   *(Obj #14)*

a. AlF_3 **aluminum fluoride**
b. PF_3 **phosphorus trifluoride**
c. H_3PO_4 **phosphoric acid**
d. $CaSO_4$ **calcium sulfate**
e. $Ca(HSO_4)_2$ **calcium hydrogen sulfate**

f. $CuCl_2$ **copper(II) chloride**
g. NH_4F **ammonium fluoride**
h. HCl(*aq*)   **hydrochloric acid**
i. $(NH_4)_3PO_4$ **ammonium phosphate**

Exercise 5.4 - Names to Formulas: Write the chemical formulas that correspond to the following names. *(Obj #14)*

a. ammonium nitrate NH_4NO_3
b. acetic acid $HC_2H_3O_2$
c. sodium hydrogen sulfate $NaHSO_4$
d. potassium bromide KBr
e. magnesium hydrogen phosphate $MgHPO_4$

f. hydrofluoric acid $HF(aq)$
g. diphosphorus tetroxide P_2O_4
h. aluminum carbonate $Al_2(CO_3)_3$
i. sulfuric acid H_2SO_4

Exercise 5.5 - Identification of Acids and Bases: Identify each of the following as an Arrhenius strong acid, an Arrhenius weak acid, an Arrhenius strong base, or an Arrhenius weak base. *(Obj #18)*

a. HNO_3 **strong acid**
b. lithium hydroxide **strong base**

c. K_2CO_3 **weak base**
d. hydrofluoric acid **weak acid**

Exercise 5.6 - Neutralization Reactions: Write the complete equation for the neutralization reactions that take place when the following water solutions are mixed. (If an acid has more than one acidic hydrogen, assume that there is enough base to remove all of them. Assume that there is enough acid to neutralize all of the basic hydroxide ions.) *(Obj #24)*

a. $HCl(aq) + NaOH(aq) \rightarrow H_2O(l) + NaCl(aq)$
b. $HF(aq) + LiOH(aq) \rightarrow H_2O(l) + LiF(aq)$
c. $H_3PO_4(aq) + 3LiOH(aq) \rightarrow 3H_2O(l) + Li_3PO_4(aq)$
d. $Fe(OH)_3(s) + 3HNO_3(aq) \rightarrow Fe(NO_3)_3(aq) + 3H_2O(l)$

Exercise 5.7 - Neutralization Reactions with Compounds Containing Carbonate: Write the complete equation for the neutralization reaction that takes place when water solutions of sodium carbonate, Na_2CO_3, and hydrobromic acid, HBr, are mixed. *(Obj #24)*

$Na_2CO_3(aq) + 2HBr(aq) \rightarrow 2NaBr(aq) + H_2O(l) + CO_2(g)$

Exercise 5.8 - Conjugate Acids: Write the formula for the conjugate acid of (a) NO_2^-, (b) HCO_3^-, (c) H_2O, and (d) PO_4^{3-}. *(Obj #27)*

a. HNO_2 b. H_2CO_3 c. H_3O^+ d. HPO_4^{2-}

Exercise 5.9 - Conjugate Bases: Write the formula for the conjugate base of (a) $H_2C_2O_4$, (b) $HBrO_4$, (c) NH_3, and (d) $H_2PO_4^-$. *(Obj #28)*

a. $HC_2O_4^-$ b. BrO_4^- c. NH_2^- d. HPO_4^{2-}

Exercise 5.10 - Brønsted-Lowry Acids and Bases: Identify the Brønsted-Lowry acid and base in each of the following equations. *(Obj #31)*

a. $HNO_2(aq)$ + $NaBrO(aq)$ → $HBrO(aq)$ + $NaNO_2(aq)$
　　　B/L acid　　　　**B/L base**

b. $H_2AsO_4^-(aq)$ + $HNO_2(aq)$ ⇌ $H_3AsO_4(aq)$ + $NO_2^-(aq)$
　　　B/L base　　　　**B/L acid**

c. $H_2AsO_4^-(aq)$ + $2OH^-(aq)$ → $AsO_4^{3-}(aq)$ + $2H_2O(l)$
　　　B/L acid　　　　**B/L base**

Review Questions Key

1. Define the following terms.
    a. aqueous

        Water solutions are called aqueous solutions.

    b. spectator ion

        Ions that are important for delivering other ions into solution to react, but do not actively participate in the reaction themselves are called spectator ions.

    c. double-displacement reaction

        A chemical reaction that has the following form is called a double-displacement reaction.

        AB + CD → AD + CB

    d. net ionic equation

        A net ionic equation is a chemical equation for which the spectator ions have been eliminated leaving only the substances actively involved in the reaction.

2. Write the name of the polyatomic ions represented by the formulas CO_3^{2-} and HCO_3^-.
    a. CO_3^{2-} **carbonate**   b. HCO_3^- **hydrogen carbonate**

3. Write the formulas for the polyatomic ions dihydrogen phosphate ion and acetate ion.
    a. dihydrogen phosphate ion $H_2PO_4^-$ b. acetate ion $C_2H_3O_2^-$

4. Which of the following formulas represents an ionic compound?
    a. $MgCl_2$ **ionic**   d. Na_2SO_4 **ionic**
    b. PCl_3 **not ionic**   e. H_2SO_4 **not ionic**
    c. $KHSO_4$ **ionic**

5. Write the names that correspond to the formulas KBr, $Cu(NO_3)_2$, and $(NH_4)_2HPO_4$.
    a. KBr **potassium bromide**   c. $(NH_4)_2HPO_4$ **ammonium hydrogen**
    b. $Cu(NO_3)_2$ **copper(II) nitrate**   **phosphate**

6. Write the formulas that correspond to the names nickel(II) hydroxide, ammonium chloride, and calcium hydrogen carbonate.
    a. nickel(II) hydroxide $Ni(OH)_2$ c. calcium hydrogen carbonate
    b. ammonium chloride NH_4Cl $Ca(HCO_3)_2$

7. Predict whether each of the following is soluble or insoluble in water.
    a. iron(III) hydroxide **insoluble**   c. aluminum nitrate **soluble**
    b. barium sulfate **insoluble**   d. copper(II) chloride **soluble**
8. Describe the process by which the ionic compound sodium hydroxide dissolves in water.

> When solid sodium hydroxide, NaOH, is added to water, all of the sodium ions, Na⁺, and hydroxide ions, OH⁻, at the surface of the solid can be viewed as shifting back and forth between moving out into the water and returning to the solid surface. Sometimes when an ion moves out into the water, a water molecule collides with it, helping to break the ionic bond, and pushing it out into the solution. Water molecules move into the gap between the ion in solution and the solid and shield the ion from the attraction to the solid.

> The ions are kept stable and held in solution by attractions between them and the polar water molecules. The negatively charged oxygen ends of water molecules surround the sodium ions, and the positively charged hydrogen ends of water molecules surround the hydroxide ions. (See Figures 4.4 and 4.5 on pages 161 and 162 of the textbook with OH⁻ in the place of Cl⁻.)

9. Write the complete equation for the precipitation reaction that takes place when water solutions of zinc chloride and sodium phosphate are mixed.

$$3ZnCl_2(aq) \; + \; 2Na_3PO_4(aq) \; \rightarrow \; Zn_3(PO_4)_2(s) \; + \; 6NaCl(aq)$$

Key Ideas Answers

10. Any substance that has a **sour** taste is an acid.
12. On the basis of the Arrhenius definitions, an **acidic** solution is a solution with a significant concentration of H_3O^+.
14. Oxyacids (often called oxoacids) are molecular substances that have the general formula $\mathbf{H_aX_bO_c}$.
16. Strong acids form **nearly one** H_3O^+ ion in solution for each acid molecule dissolved in water, whereas weak acids yield **significantly less than one** H_3O^+ ion in solution for each acid molecule dissolved in water.
18. A **weak** acid is a substance that is incompletely ionized in water because of the reversibility of its reaction with water that forms hydronium ion, H_3O^+.
20. Binary acids are named by writing **hydro** followed by the root of the name of the halogen, then **–ic**, and finally **acid**.
22. According to the modern version of the Arrhenius theory of acids and bases, a base is a substance that produces **hydroxide ions, OH⁻,** when it is added to water.
24. Compounds that contain hydroxide ions are often called **hydroxides**.
26. A weak base is a base that produces **fewer** hydroxide ions in water solution than there are particles of base dissolved.

28. Basic solutions have pH values **greater than 7**, and the more basic the solution is, the **higher** its pH.

30. When an Arrhenius acid is combined with an Arrhenius base, we say that they **neutralize** each other.

32. Most Arrhenius neutralization reactions, like the reaction between nitric acid and sodium hydroxide, are **double-displacement** reactions.

34. A Brønsted-Lowry acid is a proton (H^+) **donor**, a Brønsted-Lowry base is a proton **acceptor**, and a Brønsted-Lowry acid-base reaction is a proton **transfer**.

36. The conjugate base of a molecule or ion is the molecule or ion that forms when one H^+ ion is **removed**.

38. The **Brønsted-Lowry** system is often used to describe specific acid-base reactions, but the **Arrhenius** system is used to describe whether isolated substances are acids, bases, or neither.

Problems Key

Section 5.1 Acids

39. Describe how the strong monoprotic acid nitric acid, $HNO_3(aq)$ (used in the reprocessing of spent nuclear fuels) acts when it is added to water, including a description of the nature of the particles in solution before and after the reaction with water. If there is a reversible reaction with water, describe the forward and the reverse reactions. *(Obj #3)*

> When HNO_3 molecules dissolve in water, each HNO_3 molecule donates a proton, H^+, to water forming hydronium ion, H_3O^+, and nitrate ion, NO_3^-. This reaction goes to completion, and the solution of the HNO_3 contains essentially no uncharged acid molecules. Once the nitrate ion and the hydronium ion are formed, the negatively charged oxygen atoms of the water molecules surround the hydronium ion and the positively charged hydrogen atoms of the water molecules surround the nitrate ion. Figure 5.12 on page 209 of the textbook shows you how you can picture this solution.

41. Describe how the strong diprotic acid sulfuric acid, H_2SO_4 (used to make industrial explosives) acts when it is added to water, including a description of the nature of the particles in solution before and after the reaction with water. If there is a reversible reaction with water, describe the forward and the reverse reactions. *(Obj #10)*

> Each sulfuric acid molecule loses its first hydrogen ion completely.
>
> $$H_2SO_4(aq) + H_2O(l) \rightarrow H_3O^+(aq) + HSO_4^-(aq)$$
>
> The second hydrogen ion is not lost completely.
>
> $$HSO_4^-(aq) + H_2O(l) \rightleftharpoons H_3O^+(aq) + SO_4^{2-}(aq)$$
>
> In a typical solution of sulfuric acid, for each 100 sulfuric acid molecules added to water, the solution contains about 101 hydronium ions, H_3O^+, 99 hydrogen sulfate ions, HSO_4^-, and 1 sulfate ion, SO_4^{2-}.

43. Explain why weak acids produce fewer H_3O^+ ions in water than strong acids, even when the same number of acid molecules are added to equal volumes of water. *(Obj #7)*

> A weak acid is a substance that is incompletely ionized in water due to a reversible reaction with water that forms hydronium ion, H_3O^+. A strong acid is a substance that is completely ionized in water due to a completion reaction with water that forms hydronium ions, H_3O^+.

45. Identify each of the following as a strong or a weak acid.   *(Obj #11)*
    a. sulfurous acid (for bleaching straw) **weak**
    b. H_2SO_4 (used to make plastics) **strong**
    c. oxalic acid (in car radiator cleaners) **weak**

47. Identify each of the following as a strong or a weak acid.   *(Obj #11)*
    a. H_3PO_4 (added to animal feeds) **weak**
    b. hypophosphorous acid (in electroplating baths) **weak**
    c. HF(*aq*) (used to process uranium) **weak**

49. For each of the following, write the chemical equation for its reaction with water.
    a. The monoprotic weak acid nitrous acid, HNO_2

$$HNO_2(aq) + H_2O(l) \rightleftharpoons H_3O^+(aq) + NO_2^-(aq)$$

    b. The monoprotic strong acid hydrobromic acid, HBr

$$HBr(aq) + H_2O(l) \rightarrow H_3O^+(aq) + Br^-(aq)$$

Section 5.2 and 5.3 Acid Nomenclature and Summary of Chemical Nomenclature

51. Write the formulas and names of the acids that are derived from adding enough H^+ ions to the following ions to neutralize their charge.
    a. NO_3^- **HNO_3 nitric acid**
    b. CO_3^{2-} **H_2CO_3 carbonic acid**
    c. PO_4^{3-} **H_3PO_4 phosphoric acid**

53. Classify each of the following compounds as (1) a binary ionic compound, (2) an ionic compound with polyatomic ion(s), (3) a binary covalent compound, (4) a binary acid, or (5) an oxyacid. Write the chemical formula that corresponds to each name. *(Objs #12-14)*
    a. phosphoric acid   **oxyacid H_3PO_4**
    b. ammonium bromide   **ionic compound with polyatomic ion NH_4Br**
    c. diphosphorus tetriodide   **binary covalent compound P_2I_4**
    d. lithium hydrogen sulfate   **ionic compound with polyatomic ion $LiHSO_4$**
    e. hydrochloric acid   **binary acid   HCl(aq)**
    f. magnesium nitride   **binary ionic compound Mg_3N_2**
    g. acetic acid   **oxyacid $HC_2H_3O_2$**
    h. lead(II) hydrogen phosphate   **ionic compound with polyatomic ion $PbHPO_4$**

55. Classify each of the following formulas as (1) a binary ionic compound, (2) an ionic compound with polyatomic ion(s), (3) a binary covalent compound, (4) a binary acid, or (5) an oxyacid. Write the name that corresponds to each formula. *(Objs #12-14)*

    a. HBr(*aq*)   **binary acid   hydrobromic acid**
    b. ClF$_3$ **binary covalent compound   chlorine trifluoride**
    c. CaBr$_2$ **binary ionic compound   calcium bromide**
    d. Fe$_2$(SO$_4$)$_3$ **ionic compound with polyatomic ion   iron(III) sulfate**
    e. H$_2$CO$_3$ **oxyacid   carbonic acid**
    f. (NH$_4$)$_2$SO$_4$ **ionic compound with polyatomic ion   ammonium sulfate**
    g. KHSO$_4$ **ionic compound with polyatomic ion   potassium hydrogen sulfate**

Section 5.4 Strong and Weak Arrhenius Bases

58. Classify each of the following substances as a weak acid, strong acid, weak base, or strong base in the Arrhenius acid-base sense. *(Obj #18)*

    a. H$_2$CO$_3$ **weak acid**
    b. cesium hydroxide   **strong base**
    c. HF(*aq*)   **weak acid**
    d. sodium carbonate   **weak base**
    e. NH$_3$ **weak base**
    f. chlorous acid   **weak acid**
    g. HCl(*aq*)   **strong acid**
    h. benzoic acid   **weak acid**

Section 5.5 pH and Acidic and Basic Solutions

60. Classify each of the following solutions as acidic, basic, or neutral. *(Obj #19)*

    a. tomato juice with a pH of 4.53   pH < 7, so **acidic**

    b. milk of magnesia with a pH of 10.4   pH > 7, so **basic**

    c. urine with a pH of 6.8   pH about 7, so **essentially neutral (or more specifically, very slightly acidic)**

62. Which is more acidic, carbonated water with a pH of 3.95 or milk with a pH of 6.3? *(Obj #20)*

    The lower the pH is, the more acidic the solution. **Carbonated water** is more acidic than milk.

64. Identify each of the following characteristics as associated with acids or bases. *(Objs #2 & 22)*

    a. tastes sour   **acid**
    b. turns litmus red   **acid**
    c. reacts with HNO$_3$ **base**

Section 5.6 Arrhenius Acid-Base Reactions

Describe the process that takes place between the participants in each of the following neutralization reactions, mentioning the nature of the particles in the solution before and after the reaction. (Obj #23)

66. The strong acid hydrochloric acid, $HCl(aq)$, and the strong base sodium hydroxide, $NaOH(aq)$, form water and sodium chloride, $NaCl(aq)$.

Because hydrochloric acid, $HCl(aq)$, is an acid, it reacts with water to form hydronium ions, H_3O^+, and chloride ions, Cl^-. Because it is a strong acid, the reaction is a completion reaction, leaving only H_3O^+ and Cl^- in solution with no HCl remaining.

$$HCl(aq) + H_2O(l) \rightarrow H_3O^+(aq) + Cl^-(aq)$$

$$or \quad HCl(aq) \rightarrow H^+(aq) + Cl^-(aq)$$

Because $NaOH$ is a water-soluble ionic compound, it separates into sodium ions, Na^+, and hydroxide ions, OH^-, when it dissolves in water. Thus, at the instant that the two solutions are mixed, the solution contains water molecules, hydronium ions, H_3O^+, chloride ions, Cl^-, sodium ions, Na^+, and hydroxide ions, OH^-.

When the hydronium ions collide with the hydroxide ions, they react to form water. If an equivalent amount of acid and base are added together, the H_3O^+ and the OH^- will be completely reacted.

$$H_3O^+(aq) + OH^-(aq) \rightarrow 2H_2O(l)$$

$$or \quad H^+(aq) + OH^-(aq) \rightarrow H_2O(l)$$

The sodium ions and chloride ions remain in solution with the water molecules.

68. The strong acid nitric acid, $HNO_3(aq)$, and water-insoluble nickel(II) hydroxide, $Ni(OH)_2(s)$, form nickel(II) nitrate, $Ni(NO_3)_2(aq)$, and water.

A solution with an insoluble ionic compound, like $Ni(OH)_2$, at the bottom has a constant escape of ions from the solid into the solution balanced by the constant return of ions to the solid due to collisions of ions with the surface of the solid. Thus, even though $Ni(OH)_2$ has very low solubility in water, there are always a few Ni^{2+} and OH^- ions in solution.

If a nitric acid solution is added to water with solid $Ni(OH)_2$ at the bottom, a neutralization reaction takes place. Because the nitric acid is a strong acid, it is ionized in solution, so the nitric acid solution contains hydronium ions, H_3O^+, and nitrate ions, NO_3^-. The hydronium ions will react with the basic hydroxide ions in solution to form water molecules.

Because the hydronium ions remove the hydroxide anions from solution, the return of ions to the solid is stopped. The nickel(II) cations cannot return to the solid unless they are accompanied by anions to balance their charge. The escape of ions from the surface of the solid continues. When hydroxide ions escape, they react with the hydronium ions and do not return to the solid. Thus, there is a steady movement of ions into solution, and the solid that contains the basic anion dissolves. The complete equation for this reaction is below.

$$Ni(OH)_2(s) + 2HNO_3(aq) \rightarrow Ni(NO_3)_2(aq) + 2H_2O(l)$$

70. The strong acid hydrochloric acid, HCl(*aq*), and the weak base potassium carbonate, $K_2CO_3(aq)$, form water, carbon dioxide, $CO_2(g)$, and potassium chloride, KCl(*aq*).

Because hydrochloric acid, HCl(*aq*), is an acid, it reacts with water to form hydronium ions, H_3O^+, and chloride ions, Cl^-. Because it is a strong acid, the reaction is a completion reaction, leaving only H_3O^+ and Cl^- in an HCl(*aq*) solution with no HCl remaining.

$$HCl(aq) + H_2O(l) \rightarrow H_3O^+(aq) + Cl^-(aq)$$

or $HCl(aq) \rightarrow H^+(aq) + Cl^-(aq)$

Because K_2CO_3 is a water-soluble ionic compound, it separates into potassium ions, K^+, and carbonate ions, CO_3^{2-}, when it dissolves in water. The carbonate ions are weakly basic, so they react with water in a reversible reaction to form hydrogen carbonate, HCO_3^- and hydroxide, OH^-.

$$CO_3^{2-}(aq) + H_2O(l) \rightleftharpoons HCO_3^-(aq) + OH^-(aq)$$

Thus, at the instant that the two solutions are mixed, the solution contains water molecules, hydronium ions, H_3O^+, chloride ions, Cl^-, potassium ions, K^+, carbonate ions, CO_3^{2-}, hydrogen carbonate ions, HCO_3^-, and hydroxide ions, OH^-.

The hydronium ions react with hydroxide ions, carbonate ions, and hydrogen carbonate ions. When the hydronium ions collide with the hydroxide ions, they react to form water. When the hydronium ions collide with the carbonate ions or hydrogen carbonate ions, they react to form carbonic acid, H_2CO_3. The carbonate with its minus two charge requires two H^+ ions to yield a neutral compound, and the hydrogen carbonate requires one H^+ to neutralize its minus one charge.

$$2H_3O^+(aq) + CO_3^{2-}(aq) \rightarrow H_2CO_3(aq) + 2H_2O(l)$$

or $2H^+(aq) + CO_3^{2-}(aq) \rightarrow H_2CO_3(aq)$

$$H_3O^+(aq) + HCO_3^-(aq) \rightarrow H_2CO_3(aq) + H_2O(l)$$

or $H^+(aq) + HCO_3^-(aq) \rightarrow H_2CO_3(aq)$

The carbonic acid is unstable in water and decomposes to form carbon dioxide gas and water.

$$H_2CO_3(aq) \rightarrow CO_2(g) + H_2O(l)$$

If an equivalent amount of acid and base are added together, the H_3O^+, OH^-, CO_3^{2-}, and HCO_3^-, will be completely reacted.

The potassium ions and chloride ions remain in solution with the water molecules.

72. Write the complete equation for the neutralization reactions that take place when the following water solutions are mixed. (If an acid has more than one acidic hydrogen, assume that there is enough base to remove all of them. Assume that there is enough acid to neutralize all of the basic hydroxide ions.) *(Obj #24)*

    a. $HCl(aq) + LiOH(aq) \rightarrow H_2O(l) + LiCl(aq)$

    b. $H_2SO_4(aq) + 2NaOH(aq) \rightarrow 2H_2O(l) + Na_2SO_4(aq)$

    c. $KOH(aq) + HF(aq) \rightarrow KF(aq) + H_2O(l)$

    d. $Cd(OH)_2(s) + 2HCl(aq) \rightarrow CdCl_2(aq) + 2H_2O(l)$

74. Write the complete equation for the reaction between $HI(aq)$ and water-insoluble solid $CaCO_3$. *(Objs #24 & 25)*

$$2HI(aq) + CaCO_3(s) \rightarrow H_2O(l) + CO_2(g) + CaI_2(aq)$$

76. Iron(III) sulfate is made in industry by the neutralization reaction between solid iron(III) hydroxide and aqueous sulfuric acid. The iron(III) sulfate is then added with sodium hydroxide to municipal water in water treatment plants. These compounds react to form a precipitate that settles to the bottom of the holding tank, taking impurities with it. Write the complete equations for both the neutralization reaction that forms iron(III) sulfate and the precipitation reaction between water solutions of iron(III) sulfate and sodium hydroxide. *(Obj #24)*

$$2Fe(OH)_3(s) + 3H_2SO_4(aq) \rightarrow Fe_2(SO_4)_3(aq) + 6H_2O(l)$$

$$Fe_2(SO_4)_3(aq) + 6NaOH(aq) \rightarrow 2Fe(OH)_3(s) + 3Na_2SO_4(aq)$$

78. Complete the following equations by writing the formulas for the acid and base that could form the given products.

    a. $\mathbf{HCl(aq) + NaOH(aq)} \rightarrow H_2O(l) + NaCl(aq)$

    b. $\mathbf{H_2SO_4(aq) + 2LiOH(aq)} \rightarrow 2H_2O(l) + Li_2SO_4(aq)$

    c. $\mathbf{2HCl(aq) + K_2CO_3(aq)} \rightarrow H_2O(l) + CO_2(g) + 2KCl(aq)$

Section 5.7 Brønsted-Lowry Acids and Bases

81. Write the formula for the conjugate acid of each of the following. *(Obj #27)*

    a. IO_3^- $\mathbf{HIO_3}$ b. HSO_3^- $\mathbf{H_2SO_3}$ c. PO_4^{3-} $\mathbf{HPO_3^{2-}}$ d. H^- $\mathbf{H_2}$

83. Write the formula for the conjugate base of each of the following. *(Obj #28)*

    a. $HClO_4$ $\mathbf{ClO_4^-}$ b. HSO_3^- $\mathbf{SO_3^{2-}}$ c. H_3O^+ $\mathbf{H_2O}$ d. H_3PO_2 $\mathbf{H_2PO_2^-}$

85. Explain why a substance can be a Brønsted-Lowry acid in one reaction and a Brønsted-Lowry base in a different reaction. Give an example to illustrate your explanation. *(Obj #29)*

The same substance can donate an H^+ in one reaction (and act as a Brønsted-Lowry acid) and accept an H^+ in another reaction (and act as a Brønsted-Lowry base). For example, consider the following net ionic equations for the reaction of dihydrogen phosphate ion.

$$H_2PO_4^-(aq) \;+\; HCl(aq) \;\rightarrow\; H_3PO_4(aq) \;+\; Cl^-(aq)$$
    B/L base   **B/L acid**

$$H_2PO_4^-(aq) \;+\; 2\,OH^-(aq) \;\rightarrow\; PO_4^{3-}(aq) \;+\; 2H_2O(l)$$
    B/L acid   **B/L base**

87. For each of the following equations, identify the Brønsted-Lowry acid and base for the forward reaction. *(Obj #31)*

    a. $NaCN(aq) \;+\; HC_2H_3O_2(aq) \;\rightarrow\; NaC_2H_3O_2(aq) \;+\; HCN(aq)$
        B/L base   **B/L acid**

    b. $H_2PO_3^-(aq) \;+\; HF(aq) \;\rightleftharpoons\; H_3PO_3(aq) \;+\; F^-(aq)$
        B/L base   **B/L acid**

    c. $H_2PO_3^-(aq) \;+\; 2OH^-(aq) \;\rightarrow\; PO_3^{3-}(aq) \;+\; 2H_2O(l)$
        B/L acid   **B/L base**

    d. $3NaOH(aq) \;+\; H_3PO_3(aq) \;\rightarrow\; 3H_2O(g) \;+\; Na_3PO_3(aq)$
        B/L base   **B/L acid**

89. Butanoic acid, $CH_3CH_2CH_2CO_2H$, is a monoprotic weak acid that is responsible for the smell of rancid butter. Write the formula for the conjugate base of this acid. Write the equation for the reaction between this acid and water, and indicate the Brønsted-Lowry acid and base for the forward reaction. (The acidic hydrogen atom is on the right side of the formula.)

    Conjugate base - $CH_3CH_2CH_2CO_2^-$

$$CH_3CH_2CH_2CO_2H(aq) \;+\; H_2O(l) \;\rightleftharpoons\; H_3O^+(aq) \;+\; CH_3CH_2CH_2CO_2^-(aq)$$
    B/L acid   **B/L base**

91. Identify the amphoteric substance in each of the following equations.

$$HCl(aq) + HS^-(aq) \;\rightarrow\; Cl^-(aq) + H_2S(aq)$$

$$HS^-(aq) + OH^-(aq) \;\rightarrow\; S^{2-}(aq) + H_2O(l)$$

    $HS^-(aq)$

Additional Problems

93. For each of the following pairs of compounds, write the complete equation for the neutralization reaction that takes place when the substances are mixed. (You can assume that there is enough base to remove all of the acidic hydrogen atoms, that there is enough acid to neutralize all of the basic hydroxide ions, and that each reaction goes to completion.)

    a. $HBr(aq) + NaOH(aq) \rightarrow \textbf{H}_2\textbf{O(l)} + \textbf{NaBr(aq)}$

    b. $H_2SO_3(aq) + 2LiOH(aq) \rightarrow \textbf{2H}_2\textbf{O(l)} + \textbf{Li}_2\textbf{SO}_3\textbf{(aq)}$

    c. $KHCO_3(aq) + HF(aq) \rightarrow \textbf{KF(aq)} + \textbf{H}_2\textbf{O(l)} + \textbf{CO}_2\textbf{(g)}$

    d. $Al(OH)_3(s) + 3HNO_3(aq) \rightarrow \textbf{Al(NO}_3\textbf{)}_3\textbf{(aq)} + \textbf{3H}_2\textbf{O(l)}$

95. Classify each of the following substances as acidic, basic, or neutral.

    a.   An apple with a pH of 2.9   **acidic**

    b.   Milk of Magnesia with a pH of 10.4   **basic**

    c.   Fresh egg white with a pH of 7.6   **very slightly basic (essentially neutral)**

97. The pH of processed cheese is kept at about 5.7 to prevent it from spoiling. Is this acidic, basic, or neutral?

    acidic

99. The walls of limestone caverns are composed of solid calcium carbonate. The ground water that makes its way down from the surface into these caverns is often acidic. The calcium carbonate and the H^+ ions from the acidic water react to dissolve the limestone. If this happens to the ceiling of the cavern, the ceiling can collapse, leading to what is called a sinkhole. Write the net ionic equation for the reaction between the solid calcium carbonate and the aqueous H^+ ions.

     $\textbf{CaCO}_3\textbf{(s)} + \textbf{2H}^+\textbf{(aq)} \rightarrow \textbf{Ca}^{2+}\textbf{(aq)} + \textbf{CO}_2\textbf{(g)} + \textbf{H}_2\textbf{O(l)}$

100. Magnesium sulfate, a substance used for fireproofing and paper sizing, is made in industry from the reaction of aqueous sulfuric acid and solid magnesium hydroxide. Write the complete equation for this reaction.

     $\textbf{H}_2\textbf{SO}_4\textbf{(aq)} + \textbf{Mg(OH)}_2\textbf{(s)} \rightarrow \textbf{2H}_2\textbf{O(l)} + \textbf{MgSO}_4\textbf{(aq)}$

102. The smell of Swiss cheese is, in part, due to the monoprotic weak acid propanoic acid, $CH_3CH_2CO_2H$. Write the equation for the complete reaction between this acid and sodium hydroxide. (The acidic hydrogen atom is on the right.)

     $\textbf{CH}_3\textbf{CH}_2\textbf{CO}_2\textbf{H(aq)} + \textbf{NaOH(aq)} \rightarrow \textbf{NaCH}_3\textbf{CH}_2\textbf{CO}_2\textbf{(aq)} + \textbf{H}_2\textbf{O(l)}$

104. Malic acid, $HO_2CCH_2CH(OH)CO_2H$, is a diprotic weak acid found in apples and watermelon. Write the equation for the complete reaction between this acid and sodium hydroxide. (The acidic hydrogen atoms are on each end of the formula.)

     $\textbf{HO}_2\textbf{CCH}_2\textbf{CH(OH)CO}_2\textbf{H(aq)} + \textbf{2NaOH(aq)}$

                                   $\rightarrow \textbf{Na}_2\textbf{O}_2\textbf{CCH}_2\textbf{CH(OH)CO}_2\textbf{(aq)} + \textbf{2H}_2\textbf{O(l)}$

106. For the following equation, identify the Brønsted-Lowry acid and base for the forward reaction, and write the formulas for the conjugate acid-base pairs.

     $NaHS(aq) + NaHSO_4(aq) \rightarrow H_2S(g) + Na_2SO_4(aq)$

    B/L base   **B/L acid**

    Conjugate acid-base pairs: $\textbf{HS}^-\textbf{/H}_2\textbf{S}$ **and** $\textbf{HSO}_4^-\textbf{/SO}_4^{2-}$

                or $\textbf{NaHS/H}_2\textbf{S}$ **and** $\textbf{NaHSO}_4^-\textbf{/Na}_2\textbf{SO}_4$

Chapter 6
Oxidation-Reduction Reactions

sulfur atom

The methane thiol added to natural gas warns us when there is a leak.

◆ Review Skills

6.1 An Introduction to Oxidation-Reduction Reactions
- Oxidation, Reduction, and the Formation of Binary Ionic Compounds
- Oxidation-Reduction and Molecular Compounds

    Special Topic 6.1: Oxidizing Agents and Aging

6.2 Oxidation Numbers

    Internet: Balancing Redox Reactions

6.3 Types of Chemical Reactions
- Combination Reactions
- Decomposition Reactions
- Combustion Reactions

    Special Topic 6.2: Air Pollution and Catalytic Converters
- Single-Displacement Reactions

    Internet: Single-Displacement Reaction

6.4 Voltaic Cells
- Dry Cells
- Electrolysis
- Nickel-Cadmium Batteries

    Special Topic 6.3: Zinc-Air Batteries

◆ Chapter Glossary

    Internet: Glossary Quiz

◆ Chapter Objectives

Review Questions

Key Ideas

Chapter Problems

Section Goals and Introductions

Section 6.1 An Introduction to Oxidation-Reduction Reactions

Goals

- *To describe what oxidation means to the chemist.*
- *To describe what reduction means to the chemist.*
- *To describe chemical reactions for which electrons are transferred (oxidation-reduction reactions).*
- *To describe oxidizing agents and reducing agents.*

In many chemical reactions, electrons are completely or partially transferred from one atom to another. These reactions are called oxidation-reduction reactions (or redox reactions). This section provides examples of these reactions and introduces the terms oxidation, reduction, oxidizing agent, and reducing agent, which are summarized in Figure 6.2.

Section 6.2 Oxidation Numbers

Goal: To describe what oxidation numbers are, how they can be determined, and how they can be used to determine (1) whether the reaction is an oxidation-reduction reaction, (2) what is oxidized in an oxidation-reduction reaction, (3) what is reduced in an oxidation-reduction reaction, (4) what the oxidizing agent is in an oxidation-reduction reaction, and (5) what the reducing agent is in an oxidation-reduction reaction.

Oxidation numbers, which are described in this section, provide a tool that allows you to determine the things mentioned in the goal above. *Sample Study Sheet 6.1: Assignment of Oxidation Numbers* and Tables 6.1 and 6.2, which support the study sheet, summarize the process. The section on our Web site called *Balancing Equations for Redox Reaction* describes how you can use oxidation numbers.

www.chemplace.com/college/

Section 6.3 Types of Chemical Reactions

Goal: To describe different types of chemical reactions.

Chemical reactions can be classified into types of similar reactions. This section describes some of these types: combination, decomposition, combustion, and single-displacement reactions. You will learn how to identify each type of reaction from chemical equations. The section also describes how you can write chemical equations for combustion reactions (*Study Sheet 6.2*). The attempt to help you visualize chemical changes on the particle level continues with a description of the single-displacement reaction between solid zinc and aqueous copper(II) sulfate (Figure 6.4). The animation on our Web site called *Single-Displacement Reactions* will help you visualize single-displacement reactions.

www.chemplace.com/college/

Section 6.4 Voltaic Cells

Goal: To show how oxidation-reduction reactions can be used to create voltaic cells—that is, batteries.

This is one of the sections that are spread throughout the text that show you how what you are learning relates to the real world. In this case, you see how electrons transferred in oxidation-reduction reactions can be made to pass through a wire that separates the reactants, thus creating a voltaic cell, which is often called a battery. This section describes the fundamental components of voltaic cells and describes several different types.

Chapter 6 Map

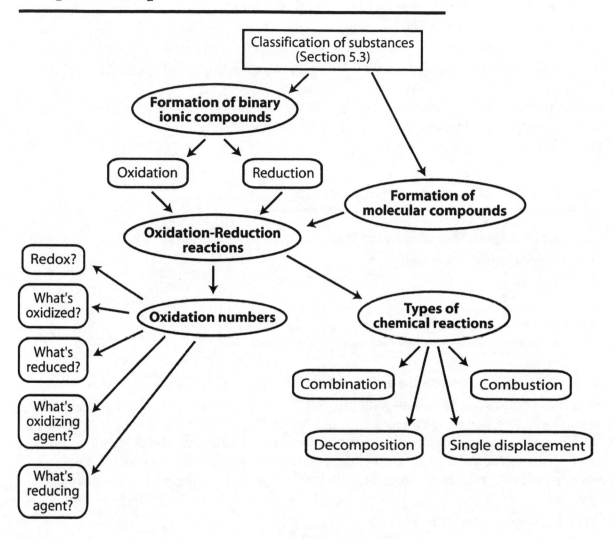

Chapter Checklist

- ☐ Read the Review Skills section. If there is any skill mentioned that you have not yet mastered, review the material on that topic before reading this chapter.
- ☐ Read the chapter quickly before the lecture that describes it.
- ☐ Attend class meetings, take notes, and participate in class discussions.
- ☐ Work the Chapter Exercises, perhaps using the Chapter Examples as guides.
- ☐ Study the Chapter Glossary and test yourself on our Web site:
     www.chemplace.com/college/
- ☐ Study all of the Chapter Objectives. You might want to write a description of how you will meet each objective. (Although it is best to master all of the objectives, the following

objectives are especially important because they pertain to skills that you will need while studying other chapters of this text: 7 and 8.)

☐ Reread the Study Sheets in this chapter and decide whether you will use them or some variation on them to complete the tasks they describe.

Sample Study Sheet 6.1: Assignment of Oxidation Numbers

Sample Study Sheet 6.2: Writing Equations for Combustion Reactions

☐ Memorize the guidelines in Table 6.2 for assigning oxidation numbers.

☐ To get a review of the most important topics in the chapter, fill in the blanks in the Key Ideas section.

☐ Work all of the selected problems at the end of the chapter, and check your answers with the solutions provided in this chapter of the study guide.

☐ Ask for help if you need it.

Web Resources   www.chemplace.com/college/

Balancing Equations for Redox Reactions
Single-Displacement Reaction
Glossary Quiz

Exercises Key

Exercise 6.1 - Oxidation Numbers: In one part of the steel manufacturing process, carbon is combined with iron to form pig iron. Pig iron is easier to work with than pure iron because it has a lower melting point (about 1130 °C, compared to 1539 °C for pure iron) and is more pliable. The following equations describe its formation. Determine the oxidation number for each atom in the formulas. Decide whether each reaction is a redox reaction, and if it is, identify what is oxidized, what is reduced, what the oxidizing agent is, and what the reducing agent is. *(Obj #3)*

$$2C(s) + O_2(g) \rightarrow 2CO(g)$$
$$Fe_2O_3(s) + 3CO(g) \rightarrow 2Fe(l) + 3CO_2(g)$$
$$2CO(g) \rightarrow C(\text{in iron}) + CO_2(g)$$

Solution:

Atoms in pure elements have an oxidation number of zero, so the C atoms in $C(s)$ and C(in iron), the O atoms in O_2, and the Fe atoms in $Fe(s)$ have an oxidation number of zero.

Oxygen atoms combined with other elements in compounds have an oxidation number of -2 (except in peroxides), so each O atom in CO, Fe_2O_3, and CO_2 has an oxidation number of -2.

Each Fe ion in Fe_2O_3 has a $+3$ charge, so each has an oxidation number of $+3$.

Because the sum of the oxidation number of the atoms in an uncharged molecule is zero, each C atom in CO must have an oxidation number of +2, and each C atom in CO_2 must have an oxidation number of +4.

The oxidation numbers for each atom are above their symbols in the equations below.

$$2C(s) + O_2(g) \rightarrow 2CO(g)$$

$$Fe_2O_3(s) + CO(g) \rightarrow 2Fe(l) + 3CO_2(g)$$

$$2CO(g) \rightarrow C(\text{in iron}) + CO_2(g)$$

They are all redox reactions.

In the first reaction, each carbon atom increases its oxidation number from 0 to +2, so **C(s) is oxidized and is the reducing agent.** Each oxygen atom decreases its oxidation number from 0 to –2, so **O_2 is reduced and is the oxidizing agent.**

In the second reaction, each carbon atom increases its oxidation number from +2 to +4, so **each carbon atom in CO(g) is oxidized and CO(g) is the reducing agent.** Each iron atom in Fe_2O_3 decreases its oxidation number from +3 to 0, so **each Fe atom in Fe_2O_3 is reduced and Fe_2O_3 is the oxidizing agent.**

Because there is only one reactant in the third reaction, it is different from the other two. Some of the carbon atoms in CO(g) are oxidized from +2 to +4, and some of them are reduced from +2 to 0. Thus, **carbon atoms in CO are both oxidized and reduced, and CO is both the oxidizing agent and the reducing agent.**

✍ **Exercise 6.2 - Combustion Reactions:**   Write balanced equations for the complete combustion of (a) $C_4H_{10}(g)$, (b) $C_3H_7OH(l)$, and (c) $C_4H_9SH(l)$. *(Obj #8)*

Solution:

a. $2C_4H_{10}(g) + 13O_2(g) \rightarrow 8CO_2(g) + 10H_2O(l)$

b. $2C_3H_7OH(l) + 9O_2(g) \rightarrow 6CO_2(g) + 8H_2O(l)$

c. $2C_4H_9SH(l) + 15O_2(g) \rightarrow 8CO_2(g) + 10H_2O(l) + 2SO_2(g)$

✍ **Exercise 6.3 - Classification of Chemical Reactions:**   Classify each of these reactions as a combination reaction, a decomposition reaction, a combustion reaction, or a single-displacement reaction. *(Obj #7)*

a. $2HgO(s) \overset{\Delta}{\rightarrow} 2Hg(l) + O_2(g)$ **decomposition**

b. $C_{12}H_{22}O_{11}(s) + 12O_2(g) \rightarrow 12CO_2(g) + 11H_2O(l)$ **combustion**

c. $B_2O_3(s) + 3Mg(s) \overset{\Delta}{\rightarrow} 2B(s) + 3MgO(s)$ **single-displacement**

d. $C_2H_4(g) + H_2(g) \rightarrow C_2H_6(g)$ **combination**

Review Questions Key

1. For each of the following ionic formulas, write the formula for the cation and the formula for the anion.

   a. $FeBr_3$ **Fe^{3+} and Br^-**

   c. AgCl **Ag^+ and Cl^-**

   b. $Co_3(PO_4)_2$ **Co^{2+} and PO_4^{3-}**

   d. $(NH_4)_2SO_4$ **NH_4^+ and SO_4^{2-}**

2. Classify each of the following formulas as representing a binary ionic compound, an ionic compound with polyatomic ions, or a molecular compound.

   a. CF_4 **molecular**

   e. H_2S **molecular**

   b. $Pb(C_2H_3O_2)_2$ **ionic with polyatomic ion**

   f. ClF **molecular**

   g. $Cr(OH)_3$ **ionic with polyatomic ion**

   c. $CoCl_2$ **binary ionic**

   h. H_3PO_4 **molecular**

   d. C_2H_5OH **molecular**

3. Balance the following equations. (C_8H_{18} is a component of gasoline, and P_2S_5 is used to make the insecticides parathion and malathion.)

   a. $C_8H_{18}(l) + 25/2O_2(g) \rightarrow 8CO_2(g) + 9H_2O(l)$

   or $2C_8H_{18}(l) + 25O_2(g) \rightarrow 16CO_2(g) + 18H_2O(l)$

   b. $4P_4(s) + 5S_8(s) \rightarrow 8P_2S_5(s)$

Key Ideas Answers

4. According to the modern convention, any chemical change in which an element **loses** electrons is called an oxidation.

6. Electrons are **rarely** found unattached to atoms. Thus, for one element or compound to lose electrons and be **oxidized**, another element or compound must be there to gain the electrons and be **reduced**. In other words, **oxidation** (loss of electrons) must be accompanied by **reduction** (gain of electrons).

8. The separate oxidation and reduction equations are called **half-reactions**.

10. A(n) **oxidizing agent** is a substance that gains electrons, making it possible for another substance to lose electrons and be oxidized.

12. Just think of oxidation numbers as tools for keeping track of the **flow of electrons** in redox reactions.

14. In combination reactions, **two or more** elements or compounds combine to form one compound.

16. In a combustion reaction, oxidation is very rapid and is accompanied by **heat** and usually **light**.

18. When a substance that contains hydrogen is burned completely, the hydrogen forms **water**.

20. In single-displacement reactions, atoms of one element in a compound are displaced (or replaced) by atoms from a(n) **pure element**.

21. Strictly speaking, a battery is a series of **voltaic cells** joined in such a way that they work together. A battery can also be described as a device that converts **electrical energy** into **chemical energy** using redox reactions.

23. Metal strips in voltaic cells are called electrodes, which is the general name for **electrical conductors** placed in half-cells of voltaic cells.

25. The cathode is the electrode in a voltaic cell at which **reduction** occurs. By convention, the cathode is designated the **positive electrode**. Because electrons flow along the wire to the cathode, and because substances gain those electrons to become more negative (or less positive), the cathode surroundings tend to become more negative. Thus, cations are attracted to the cathode.

27. Voltage, a measure of the strength of an electric current, represents the **force** that moves electrons from the anode to the cathode in a voltaic cell. When a greater voltage is applied in the opposite direction, electrons can be pushed from what would normally be the cathode toward the voltaic cell's anode. This process is called **electrolysis**.

Problems Key

Section 6.1 An Introduction to Oxidation-Reduction Reactions

30. Are the electrons in the following redox reactions transferred completely from the atoms of one element to the atoms of another, or are they only partially transferred?
    a. $4Al(s) + 3O_2(g) \rightarrow 2Al_2O_3(s)$ **complete – ionic bonds formed**
    b. $C(s) + O_2(g) \rightarrow CO_2(g)$ **incomplete – polar covalent bonds formed**

32. Are the electrons in the following redox reactions transferred completely from the atoms of one element to the atoms of another, or are they only partially transferred?
    a. $S_8(s) + 8O_2(g) \rightarrow 8SO_2(g)$ **incomplete – polar covalent bonds formed**
    b. $P_4(s) + 6Cl_2(g) \rightarrow 4PCl_3(l)$ **incomplete – polar covalent bonds formed**

34. Aluminum bromide, $AlBr_3$, which is used to add bromine atoms to organic compounds, can be made by passing gaseous bromine over hot aluminum. Which of the following half-reactions for this oxidation-reduction reaction describes the oxidation, and which one describes the reduction?

    $2Al \rightarrow 2Al^{3+} + 6e^-$ **oxidation** – loss of electrons

    $3Br_2 + 6e^- \rightarrow 6Br^-$ **reduction** – gain of electrons

Section 6.2 Oxidation Numbers

36. Determine the oxidation number for the atoms of each element in the following formulas. *(Obj #3)*

    a. S_8 Because this is a pure, uncharged element, each **S** is **zero**.

    b. S^{2-} Because this is a monatomic ion, the **S** is **−2**.

    c. Na_2S This is a binary ionic compound. The oxidation numbers are equal to the charges. Each **Na** is **+1**, and the **S** is **−2**.

    d. FeS This is a binary ionic compound. The oxidation numbers are equal to the charges. The **Fe** is **+2**, and the **S** is **−2**.

38. Determine the oxidation number for the atoms of each element in the following formulas. *(Obj #3)*

   a. Sc_2O_3 This is a binary ionic compound. The oxidation numbers are equal to the charges. Each **Sc** is **+3**, and each **O** is **–2**.

   b. RbH   This is a binary ionic compound. The oxidation numbers are equal to the charges. The **Rb** is **+1**, and the **H** is **–1**.

   c. N_2 Because this is a pure, uncharged element, the oxidation number for each **N** is **zero**.

   d. NH_3 This is a molecular compound. **H** is **+1** in molecular compounds. **N** must be **–3** for the sum to be zero.

40. Determine the oxidation number for the atoms of each element in the following formulas. *(Obj #3)*

   e. CHF_3 This is a molecular compound. **H** is **+1** in molecular compounds. **F** is **–1** when combined with other elements. **C** must be **+2** for the sum to be zero.

   f. H_2O_2 This is a molecular compound. **H** is **+1** in molecular compounds. **O** is **–1** in peroxides.

   g. H_2SO_4 **H** is **+1** in molecular compounds. **O** is **–2** in molecular compounds other than peroxides. **S** must be **+6** for the sum to be 0.

42. Determine the oxidation number for the atoms of each element in the following formulas. *(Obj #3)*

   a. $HPO_4{}^{2-}$ This is a polyatomic ion. **H** is **+1** in polyatomic ions. **O** is **–2** in polyatomic ions other than peroxide. **P** must be **+5** for the sum to be –2.

   b. $NiSO_4$ This is an ionic compound with a polyatomic ion. The **Ni** is in the form of the monatomic ion Ni^{2+}, so it has an oxidation number of **+2**. **O** is **–2** when combined with other elements, except in peroxides. **S** must be **+6** for the sum to be zero.

   c. $N_2O_4{}^{2-}$ This is a polyatomic ion. **O** is **–2** when combined with other elements, except in peroxides. **N** must be **+3** for the sum to be –2.

   d. $Mn_3(PO_2)_2$ This is an ionic compound with a polyatomic ion. The **Mn** is in the form of the monatomic ion Mn^{2+}, so it has an oxidation number of **+2**. **O** is **–2** when combined with other elements, except in peroxides. Each **P** must be **+1** for the sum to be zero.

45. About 47% of the hydrochloric acid produced in the United States is used for cleaning metallic surfaces. Hydrogen chloride, HCl, which is dissolved in water to make the acid, is formed in the reaction of chlorine gas and hydrogen gas, displayed below. Determine the oxidation number for each atom in the equation, and decide whether the reaction is a redox reaction or not. If it is redox, identify which substance is oxidized, which substance is reduced, the oxidizing agent, and the reducing agent. *(Objs #3–6)*

$$\overset{0}{Cl_2}(g) + \overset{0}{H_2}(g) \rightarrow 2\overset{+1\ -1}{HCl}(g)$$

Yes, it's redox.

H atoms in H_2 are oxidized, Cl atoms in Cl_2 are reduced, Cl_2 is the oxidizing agent, and H_2 is the reducing agent.

47. Water and carbon dioxide fire extinguishers should not be used on magnesium fires because both substances react with magnesium and generate enough heat to intensify the fire. Determine the oxidation number for each atom in the equations that describe these reactions (displayed below), and decide whether each reaction is a redox reaction or not. If it is redox, identify which substance is oxidized, which substance is reduced, the oxidizing agent, and the reducing agent. *(Objs #3-6)*

$$\overset{0}{Mg}(s) + 2\overset{+1\ -2}{H_2O}(l) \rightarrow \overset{+2\ -2\ +1}{Mg(OH)_2}(aq) + \overset{0}{H_2}(g) + heat$$

Yes, it's redox.

Mg atoms in Mg(s) are oxidized, H atoms in H_2O are reduced, H_2O is the oxidizing agent, and Mg is the reducing agent.

$$2\overset{0}{Mg}(s) + \overset{+4\ -2}{CO_2}(g) \rightarrow 2\overset{+2\ -2}{MgO}(s) + \overset{0}{C}(s) + heat$$

Yes, it's redox.

Mg atoms in Mg(s) are oxidized, C atoms in CO_2 are reduced, CO_2 is the oxidizing agent, and Mg is the reducing agent.

49. Formaldehyde, CH_2O, which is used in embalming fluids, is made from methanol in the reaction described below. Determine the oxidation number for each atom in this equation, and decide whether the reaction is a redox reaction or not. If it is redox, identify which substance is oxidized, which substance is reduced, the oxidizing agent, and the reducing agent. *(Objs #3-6)*

$$2\overset{-2\ +1\ -2\ +1}{CH_3OH} + \overset{0}{O_2} \rightarrow 2\overset{0\ +1\ -2}{CH_2O} + 2\overset{+1\ -2}{H_2O}$$

Yes, it's redox.

C atoms in CH_3OH are oxidized, O atoms in O_2 are reduced, O_2 is the oxidizing agent, and CH_3OH is the reducing agent.

51. For each of the following equations, determine the oxidation number for each atom in the equation, and indicate whether the reaction is a redox reaction. If the reaction is redox, identify what is oxidized, what is reduced, the oxidizing agent, and the reducing agent. *(Objs #3-6)*

    a. $\overset{0}{Co}(s) + 2\overset{+1\ +5\ -2}{AgNO_3}(aq) \rightarrow \overset{+2\ +5\ -2}{Co(NO_3)_2}(aq) + 2\overset{0}{Ag}(s)$

    Co in Co(s) is oxidized, and Co(s) is the reducing agent. Ag in $AgNO_3$ is reduced, and $AgNO_3$ is the oxidizing agent.

    b. $\overset{+5\ -2}{V_2O_5}(s) + 5\overset{0}{Ca}(l) \overset{\Delta}{\rightarrow} 2\overset{0}{V}(l) + 5\overset{+2\ -2}{CaO}(s)$

    V in V_2O_5 is reduced, and V_2O_5 is the oxidizing agent. Ca in Ca(l) is oxidized, and Ca(l) is the reducing agent.

    c. $\overset{+2\ +4\ -2}{CaCO_3}(aq) + \overset{+4\ -2}{SiO_2}(s) \rightarrow \overset{+2\ +4\ -2}{CaSiO_3}(s) + \overset{+4\ -2}{CO_2}(g)$

    Not redox

    d. $2\overset{+1\ -1}{NaH}(s) \overset{\Delta}{\rightarrow} 2\overset{0}{Na}(s) + \overset{0}{H_2}(g)$

    Na in NaH is reduced, and NaH (or Na^+ in NaH) is the oxidizing agent.

    H in NaH is oxidized, and NaH (or H^- in NaH) is also the reducing agent.

    e. $5\overset{+3\ -2}{As_4O_6}(s) + 8\overset{+1\ +7\ -2}{KMnO_4}(aq) + 18\overset{+1\ -2}{H_2O}(l) + 52\overset{+1\ -1}{KCl}(aq)$

    $\rightarrow 20\overset{+1\ +5\ -2}{K_3AsO_4}(aq) + 8\overset{+2\ -1}{MnCl_2}(aq) + 36\overset{+1\ -1}{HCl}(aq)$

    As in As_4O_6 is oxidized, and As_4O_6 is the reducing agent. Mn in $KMnO_4$ is reduced, and $KMnO_4$ is the oxidizing agent.

53. The following equations summarize the steps in the process used to make about 99% of the sulfuric acid produced in the United States. Determine the oxidation number for each atom in each of the following equations, and indicate whether each reaction is a redox reaction. For the redox reactions, identify what is oxidized, what is reduced, the oxidizing agent, and the reducing agent. *(Objs #3-6)*

    $1/8\overset{0}{S_8}(s) + \overset{0}{O_2}(g) \rightarrow \overset{+4\ -2}{SO_2}(g)$

    Yes, it's redox.

    S atoms in S_8 are oxidized, O atoms in O_2 are reduced, O_2 is the oxidizing agent, and S_8 is the reducing agent.

    $\overset{+4\ -2}{SO_2} + \tfrac{1}{2}\overset{0}{O_2} \rightarrow \overset{+6\ -2}{SO_3}$

    Yes, it's redox.

    S atoms in SO_2 are oxidized, O atoms in O_2 are reduced, O_2 is the oxidizing agent, and SO_2 is the reducing agent.

$$\overset{+6\ -2}{SO_3} + \overset{+1\ -2}{H_2O} \rightarrow \overset{+1\ +6\ -2}{H_2SO_4}$$

Because none of the atoms change their oxidation number, this is not redox.

Section 6.3 Types of Chemical Reactions

55. Classify each of these reactions as a combination reaction, a decomposition reaction, a combustion reaction, or a single-displacement reaction. *(Obj #7)*

    a. $2NaH(s) \rightarrow 2Na(s) + H_2(g)$ **decomposition**

    b. $2KI(aq) + Cl_2(g) \rightarrow 2KCl(aq) + I_2(s)$ **single-displacement**

    c. $2C_2H_5SH(l) + 9O_2(g) \rightarrow 4CO_2(g) + 6H_2O(l) + 2SO_2(g)$ **combustion**

    d. $H_2(g) + CuO(s) \overset{\Delta}{\rightarrow} Cu(s) + H_2O(l)$ **single-displacement**

    e. $P_4(s) + 5O_2(g) \rightarrow P_4O_{10}(s)$ **combination and combustion**

57. Classify each of these reactions as a combination reaction, a decomposition reaction, a combustion reaction, or a single-displacement reaction. *(Obj #7)*

    a. $4B(s) + 3O_2(g) \rightarrow 2B_2O_3(s)$ **combination and combustion**

    b. $(C_2H_5)_2O(l) + 6O_2(g) \rightarrow 4CO_2(g) + 5H_2O(l)$ **combustion**

    c. $2Cr_2O_3(s) + 3Si(s) \overset{\Delta}{\rightarrow} 4Cr(s) + 3SiO_2(s)$ **single-displacement**

    d. $C_6H_{11}SH(l) + 10O_2(g) \rightarrow 6CO_2(g) + 6H_2O(l) + SO_2(g)$ **combustion**

    e. $2NaHCO_3(s) \overset{\Delta}{\rightarrow} Na_2CO_3(s) + H_2O(l) + CO_2(g)$ **decomposition**

59. Write balanced equations for the complete combustion of each of the following substances. *(Obj #8)*

    a. $C_3H_8(g)$

            $C_3H_8(g) + 5O_2(g) \rightarrow 3CO_2(g) + 4H_2O(l)$

    b. $C_4H_9OH(l)$

            $C_4H_9OH(l) + 6O_2(g) \rightarrow 4CO_2(g) + 5H_2O(l)$

    c. $CH_3COSH(l)$

            $CH_3COSH(l) + 7/2O_2(g) \rightarrow 2CO_2(g) + 2H_2O(l) + SO_2(g)$

            or

            $2CH_3COSH(l) + 7O_2(g) \rightarrow 4CO_2(g) + 4H_2O(l) + 2SO_2(g)$

61. The following pairs react in single-displacement reactions that are similar to the reaction between uncharged zinc metal and the copper(II) ions in a copper(II) sulfate solution. Describe the changes in these reactions, including the nature of the particles in the system before the reaction takes place, the nature of the reaction itself, and the nature of the particles in the system after the reaction. Your description should also include the equations for the half-reactions and the net ionic equation for the overall reaction. *(Obj #9)*

a. magnesium metal and nickel(II) nitrate, $Ni(NO_3)_2(aq)$

Because nickel(II) nitrate is a water soluble ionic compound, the $Ni(NO_3)_2$ solution contains Ni^{2+} ions surrounded by the negatively charged oxygen ends of water molecules and separate NO_3^- ions surrounded by the positively charged hydrogen ends of water molecules. These ions move throughout the solution colliding with each other, with water molecules, and with the walls of their container. When solid magnesium is added to the solution, nickel ions begin to collide with the surface of the magnesium. When each Ni^{2+} ion collides with an uncharged magnesium atom, two electrons are transferred from the magnesium atom to the nickel(II) ion. Magnesium ions go into solution, and uncharged nickel solid forms on the surface of the magnesium. See Figure 6.4 on page 252 of the textbook. Picture Ni^{2+} in the place of Cu^{2+} and magnesium metal in the place of zinc metal.

Because the magnesium atoms lose electrons and change their oxidation number from 0 to +2, they are oxidized and act as the reducing agent. The Ni^{2+} ions gain electrons and decrease their oxidation number from +2 to 0, so they are reduced and act as the oxidizing agent. The half reaction equations and the net ionic equation for this reaction are below.

oxidation: $Mg(s) \rightarrow Mg^{2+}(aq) + 2e^-$

reduction: $Ni^{2+}(aq) + 2e^- \rightarrow Ni(s)$

Net ionic equation: $Mg(s) + Ni^{2+}(aq) \rightarrow Mg^{2+}(aq) + Ni(s)$

Section 6.4 Voltaic Cells

62. We know that the following reaction can be used to generate an electric current in a voltaic cell.

$$Zn(s) + CuSO_4(aq) \rightarrow Cu(s) + ZnSO_4(aq)$$

Sketch similar voltaic cells made from each of the reactions presented below, showing the key components of the two half-cells and indicating the cathode electrode and the anode electrode, the negative and positive electrodes, the direction of movement of the electrons in the wire between the electrodes, and the direction of movement of the ions in the system. Show a salt bridge in each sketch, and show the movement of ions out of the salt bridge. *(Objs #10 & 11)*

a. $Mn(s) + Pb(NO_3)_2(aq) \rightarrow Pb(s) + Mn(NO_3)_2\ (aq)$

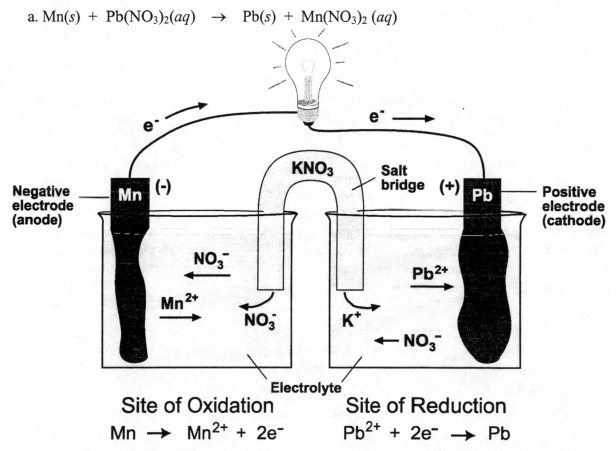

Site of Oxidation   Site of Reduction

$$Mn \rightarrow Mn^{2+} + 2e^-$$ $$Pb^{2+} + 2e^- \rightarrow Pb$$

The voltaic cell that utilizes the redox reaction between manganese metal and lead(II) ions is composed of 2 half-cells. The first half-cell consists of a strip of manganese metal in a solution of manganese(II) nitrate. The second half-cell consists of a strip of lead metal in a solution of lead(II) nitrate. In the Mn/Mn^{2+} half-cell, manganese atoms lose 2 electrons and are converted to manganese ions. The electrons pass through the wire to the Pb/Pb^{2+} half-cell where Pb^{2+} ions gain the 2 electrons to form uncharged lead atoms. Mn is oxidized to Mn^{2+} at the manganese electrode, so this electrode is the anode. Pb^{2+} ions are reduced to uncharged lead atoms at the lead strip, so metallic lead is the cathode.

63. The following equation summarizes the chemical changes that take place in a typical dry cell.

$$Zn(s) + 2MnO_2(s) + 2NH_4^+(aq) \rightarrow Zn^{2+}(aq) + Mn_2O_3(s) + 2NH_3(aq) + H_2O(l)$$

Determine the oxidation number for each atom in this equation, and identify what is oxidized, what is reduced, the oxidizing agent, and the reducing agent. *(Obj. #3-6)*

$$\overset{0}{Zn}(s) + 2\overset{+4\ -2}{MnO_2}(s) + 2\overset{-3\ +1}{NH_4^+}(aq) \rightarrow \overset{+2}{Zn^{2+}}(aq) + \overset{+3\ -2}{Mn_2O_3}(s) + 2\overset{-3\ +1}{NH_3}(aq) + \overset{+1\ -2}{H_2O}(l)$$

Zn is oxidized and is the reducing agent. Mn in MnO_2 is reduced, so MnO_2 is the oxidizing agent.

65. The following equation summarizes the chemical changes that take place in a lead-acid battery. Determine the oxidation number for each atom in the equation, and identify what is oxidized, what is reduced, the oxidizing agent, and the reducing agent. *(Obj. #3-6)*

$$\overset{0}{Pb}(s) + \overset{+4\ -2}{PbO_2}(s) + 2\overset{+1\ +6\ -2}{HSO_4^-}(aq) + 2\overset{+1\ -2}{H_3O^+}(aq) \rightarrow 2\overset{+2\ +6\ -2}{PbSO_4}(s) + 4\overset{+1\ -2}{H_2O}(l)$$

Pb is oxidized and is the reducing agent. Pb in PbO_2 is reduced, so PbO_2 is the oxidizing agent.

Additional Problems

67. Sodium hydrogen carbonate, $NaHCO_3$, best known as the active ingredient in baking soda, is used in several ways in food preparation. It is also added to animal feeds and used to make soaps and detergents. Baking soda can be used to put out small fires on your stovetop. The heat of the flames causes the $NaHCO_3$ to decompose to form carbon dioxide, which displaces the air above the flames depriving the fire of the oxygen necessary for combustion. The equation for this reaction is *(Obj. #3-5)*

$$2\overset{+1\ +1\ +4\ -2}{NaHCO_3}(s) \overset{\Delta}{\rightarrow} \overset{+4\ -2}{CO_2}(g) + \overset{+1\ +4\ -2}{Na_2CO_3}(s) + \overset{+1\ -2}{H_2O}(g)$$

Determine the oxidation number for each atom in this equation, and indicate whether the reaction is a redox reaction. If the reaction is redox, identify what is oxidized and what is reduced. *(Obj. #3-5)*

Because none of the atoms change their oxidation number, this is **not redox**.

69. Not too long ago mercury batteries were commonly used to power electronic watches and small appliances. The overall reaction for this type of battery is

$$\overset{+2\ -2}{HgO}(s) + \overset{0}{Zn}(s) \rightarrow \overset{+2\ -2}{ZnO}(s) + \overset{0}{Hg}(l)$$

Determine the oxidation number for each atom in the equation and decide whether the reaction is a redox reaction or not. If it is redox, identify which substance is oxidized, which substance is reduced, the oxidizing agent, and the reducing agent.

Yes, it's redox.

Zn atoms in Zn(s) are oxidized, Hg atoms in HgO are reduced, HgO is the oxidizing agent, and Zn is the reducing agent.

71. $Mn_3(PO_4)_2$, which is used to make corrosion resistant coatings on steel, aluminum, and other metals, is made from the reaction of $Mn(OH)_2$ with H_3PO_4.

$$\overset{+2 \quad -2 \ +1}{Mn(OH)_2(s)} + \overset{+1 \ +5 \ -2}{H_3PO_4(aq)} \rightarrow \overset{+2 \quad +5 \ -2}{Mn_3(PO_4)_2(s)} + \overset{+1 \ -2}{3H_2O(l)}$$

Determine the oxidation number for each atom in the equation and identify whether the reaction is a redox reaction or not. If it is redox, identify what is oxidized, what is reduced, the oxidizing agent, and the reducing agent.

No, none of the atoms change their oxidation number, so it's not redox.

73. The *noble* gases in group 18 on the periodic table used to be called the *inert* gases because they were thought to be incapable of forming compounds. Their name has been changed to noble gases because although they resist combining with the more common elements to their left on the periodic table, they do mingle with them on rare occasions. The following equations describe reactions that form xenon compounds. Determine the oxidation number for each atom in the reactions, and identify each reaction as redox or not. If it is redox, identify which substance is oxidized, which substance is reduced, the oxidizing agent, and the reducing agent.

$$\overset{0}{Xe} + \overset{0}{F_2} \rightarrow \overset{+6 \ -1}{XeF_6}$$

Yes, it's redox.

Xe atoms are oxidized, F atoms in F_2 are reduced, F_2 is the oxidizing agent, and Xe is the reducing agent.

$$\overset{+6 \ -1}{XeF_6} + \overset{+1 \ -2}{H_2O} \rightarrow \overset{+6 \ -2 \ -1}{XeOF_4} + \overset{+1 \ -1}{2HF}$$

None of the atoms change their oxidation number, so it's not redox.

$$\overset{+6 \ -1}{XeF_6} + \overset{-2 \ +5 \ -1}{OPF_3} \rightarrow \overset{+6 \ -2 \ -1}{XeOF_4} + \overset{+5 \ -1}{PF_5}$$

None of the atoms change their oxidation number, so it's not redox.

75. Sodium perbromate is an oxidizing agent that can be made in the two ways represented by the equations below. The first equation shows the way it was made in the past, and the second equation represents the technique used today. Determine the oxidation number for each atom in each of these equations, and decide whether each reaction is a redox reaction or not. If a reaction is a redox reaction, identify which substance is oxidized, which substance is reduced, the oxidizing agent, and the reducing agent.

$$\overset{+1 \ +5 \ -2}{NaBrO_3} + \overset{+2 \ -1}{XeF_2} + \overset{+1 \ -2}{H_2O} \rightarrow \overset{+1 \ +7 \ -2}{NaBrO_4} + \overset{+1 \ -1}{2HF} + \overset{0}{Xe}$$

Yes, it's redox.

Br atoms in $NaBrO_3$ are oxidized, Xe atoms in XeF_2 are reduced, XeF_2 is the oxidizing agent, and $NaBrO_3$ is the reducing agent.

$$\overset{+1 \ +5 \ -2}{NaBrO_3} + \overset{0}{F_2} + \overset{+1 \ -2 \ +1}{2NaOH} \rightarrow \overset{+1 \ +7 \ -2}{NaBrO_4} + \overset{+1 \ -1}{2NaF} + \overset{+1 \ -2}{H_2O}$$

Yes, it's redox.

Br atoms in $NaBrO_3$ are oxidized, F atoms in F_2 are reduced, F_2 is the oxidizing agent, and $NaBrO_3$ is the reducing agent.

77. The following equations represent reactions that involve only halogen atoms. Iodine pentafluoride, IF_5, is used to add fluorine atoms to other compounds, bromine pentafluoride, BrF_5 is an oxidizing agent in liquid rocket propellants, and chlorine trifluoride, ClF_3, is used to reprocess nuclear reactor fuels. Determine the oxidation number for each atom in these equations, and decide whether each reaction is a redox reaction or not. If a reaction is a redox reaction, identify which substance is oxidized, which substance is reduced, the oxidizing agent, and the reducing agent.

$$\overset{+1\ -1}{IF}(g) + \overset{0}{2F_2}(g) \rightarrow \overset{+5\ -1}{IF_5}(l)$$

Yes, it's redox.

I atoms in IF are oxidized, F atoms in F_2 are reduced, F_2 is the oxidizing agent, and IF is the reducing agent.

$$\overset{+1\ -1}{BrF}(g) + \overset{0}{2F_2}(g) \rightarrow \overset{+5\ -1}{BrF_5}(g)$$

Yes, it's redox.

Br atoms in BrF are oxidized, F atoms in F_2 are reduced, F_2 is the oxidizing agent, and BrF is the reducing agent.

$$\overset{0}{Cl_2}(g) + \overset{0}{3F_2}(g) \rightarrow \overset{+3\ -1}{2ClF_3}(g)$$

Yes, it's redox.

Cl atoms in Cl_2 are oxidized, F atoms in F_2 are reduced, F_2 is the oxidizing agent, and Cl_2 is the reducing agent.

79. Sodium sulfate, which is used to make detergents and glass, is formed in the following reaction.

$$\overset{+1\ -1}{4NaCl} + \overset{+4\ -2}{2SO_2} + \overset{+1\ -2}{2H_2O} + \overset{0}{O_2} \rightarrow \overset{+1\ +6\ -2}{2Na_2SO_4} + \overset{+1\ -1}{4HCl}$$

Determine the oxidation number for each atom in the equation and decide whether the reaction is a redox reaction or not. If a reaction is a redox reaction, identify which substance is oxidized, which substance is reduced, the oxidizing agent, and the reducing agent.

Yes, it's redox.

S atoms in SO_2 are oxidized, O atoms in O_2 are reduced, O_2 is the oxidizing agent, and SO_2 is the reducing agent.

81. Elemental sulfur is produced by the chemical industry from naturally occurring hydrogen sulfide in the following steps. Determine the oxidation number for each atom in these equations and decide whether each reaction is a redox reaction or not. If a reaction is a redox reaction, identify which substance is oxidized, which substance is reduced, the oxidizing agent, and the reducing agent.

$$\overset{+1\ -2}{3H_2S} + \overset{0}{2O_2} \rightarrow \overset{+4\ -2}{3SO_2} + \overset{+1\ -2}{3H_2O}$$

Yes, it's redox.

S atoms in H_2S are oxidized, O atoms in O_2 are reduced, O_2 is the oxidizing agent, and H_2S is the reducing agent.

$$\overset{+4\ -2}{SO_2} + \overset{+1\ -2}{2H_2S} \rightarrow \overset{0}{3S} + \overset{+1\ -2}{3H_2O}$$

Yes, it's redox.

S atoms in H_2S are oxidized, S atoms in SO_2 are reduced, SO_2 is the oxidizing agent, and H_2S is the reducing agent.

83. Leaded gasoline, originally developed to decrease pollution, is now banned because the lead(II) bromide, $PbBr_2$, emitted when it burns decomposes in the atmosphere into two serious pollutants, lead and bromine. The equation for this reaction follows. Determine the oxidation number for each atom in the equation, and indicate whether the reaction is a redox reaction. If the reaction is redox, identify what is oxidized and what is reduced. *(Objs #3-5)*

$$\overset{+2\ -1}{PbBr_2} \overset{sunlight}{\rightarrow} \overset{0}{Pb} + \overset{0}{Br_2}$$

Yes, it's redox. Br atoms in $PbBr_2$ are oxidized, and Pb atoms in $PbBr_2$ are reduced.

85. When the calcium carbonate, $CaCO_3$, in limestone is heated to a high temperature, it decomposes into calcium oxide (called lime or quicklime) and carbon dioxide. Lime was used by the early Romans, Greeks, and Egyptians to make cement, and it is used today to make over 150 different chemicals. In another reaction, calcium oxide and water form calcium hydroxide, $Ca(OH)_2$ (called slaked lime), which is used to remove the sulfur dioxide from smokestacks above power plants that burn high-sulfur coal. The equations for all these reactions follow. Determine the oxidation number for each atom in the equations, and indicate whether the reactions are redox reactions. For each redox reaction, identify what is oxidized and what is reduced. *(Objs #3-5)*

$$\overset{+2\ +4\ -2}{CaCO_3} \rightarrow \overset{+2\ -2}{CaO} + \overset{+4\ -2}{CO_2}$$

$$\overset{+2\ -2}{CaO} + \overset{+1\ -2}{H_2O} \rightarrow \overset{+2\ -2\ +1}{Ca(OH)_2}$$

$$\overset{+4\ -2}{SO_2} + \overset{+1\ -2}{H_2O} \rightarrow \overset{+1\ +4\ -2}{H_2SO_3}$$

$$\overset{+2\ -2\ +1}{Ca(OH)_2} + \overset{+1\ +4\ -2}{H_2SO_3} \rightarrow \overset{+2\ +4\ -2}{CaSO_3} + \overset{+1\ -2}{2H_2O}$$

Because none of the atoms change their oxidation number, **none of these reactions are redox reactions**.

87. The space shuttle's solid rocket boosters get their thrust from the reaction of aluminum metal with ammonium perchlorate, NH_4ClO_4, which generates a lot of gas and heat. The billowy white smoke is due to the formation of very finely divided aluminum oxide solid. One of the reactions that takes place is

$$\overset{0}{10Al(s)} + \overset{-3\ +1\ +7\ -2}{6NH_4ClO_4(s)} \rightarrow \overset{+3\ -2}{5Al_2O_3(s)} + \overset{+1\ -1}{6HCl(g)} + \overset{0}{3N_2(g)} + \overset{+1\ -1}{9H_2O(g)}$$

Is this a redox reaction? What is oxidized and what is reduced? *(Obj #3-5)*

> Yes, it's redox. Al atoms in Al and N atoms in NH_4ClO_4 are oxidized, and Cl atoms in NH_4ClO_4 are reduced.

89. Determine the oxidation number for each atom in the following equations and decide whether each reaction is a redox reaction or not. If a reaction is redox, identify which substance is oxidized, which is reduced, the oxidizing agent, and the reducing agent.

a. $$\overset{+1\ +6\ \ -2}{K_2Cr_2O_7(aq)} + \overset{+1\ -1}{14HCl(aq)} \rightarrow \overset{+1\ -1}{2KCl(aq)} + \overset{+3\ -1}{2CrCl_3(aq)} + \overset{+1\ -2}{7H_2O(l)} + \overset{0}{3Cl_2(g)}$$

> Yes, it is redox. Cl in HCl is oxidized, Cr in $K_2Cr_2O_7$ is reduced, HCl is the reducing agent, and $K_2Cr_2O_7$ is the oxidizing agent.

b. $$\overset{0}{Ca(s)} + \overset{+1\ -2}{2H_2O(l)} \rightarrow \overset{+2\ -2\ +1}{Ca(OH)_2(s)} + \overset{0}{H_2(g)}$$

> Yes, it is redox. Ca is oxidized, H in H_2O is reduced, Ca is the reducing agent, and H_2O is the oxidizing agent.

91. For each of the following equations, determine the oxidation number for each atom in the equation, and indicate whether the reaction is a redox reaction. If the reaction is redox, identify what is oxidized, what is reduced, the oxidizing agent, and the reducing agent. *(Objs #3-6)*

a. $$\overset{0}{Ca(s)} + \overset{0}{F_2(g)} \rightarrow \overset{+2\ -1}{CaF_2(s)}$$

> Yes, it's redox. Ca is oxidized and is the reducing agent. F in F_2 is reduced, and F_2 is the oxidizing agent.

b. $$\overset{0}{2Al(s)} + \overset{+1\ -2}{3H_2O(g)} \xrightarrow{\Delta} \overset{+3\ -2}{Al_2O_3(s)} + \overset{0}{3H_2(g)}$$

> Yes, it's redox. Al is oxidized and is the reducing agent. H in H_2O is reduced, so H_2O is the oxidizing agent.

93. The following equations represent reactions used by the U.S. chemical industry. Classify each as a combination reaction, a decomposition reaction, a combustion reaction, or a single-displacement reaction. *(Obj #7)*

a. $P_4 + 5O_2 + 6H_2O \rightarrow 4H_3PO_4$　　**combination**

b. $TiCl_4 + O_2 \rightarrow TiO_2 + 2Cl_2$　　**single-displacement**

c. $CH_3CH_3 \xrightarrow{\Delta} CH_2CH_2 + H_2$　　**decomposition**

95. Write a balanced equation for the redox reaction of carbon dioxide gas and hydrogen gas to form carbon solid and water vapor.

> $CO_2(g) + 2H_2(g) \rightarrow C(s) + 2H_2O(g)$

97. Titanium metal is used to make metal alloys for aircraft, missiles, and artificial hip joints. It is formed in the reaction of titanium(IV) chloride with magnesium metal. The other product is magnesium chloride. Write a balanced equation, without including states, for this redox reaction.

$$TiCl_4 + 2Mg \rightarrow Ti + 2MgCl_2$$

99. Write a balanced equation for the redox reaction of solid potassium with liquid water to form aqueous potassium hydroxide and hydrogen gas.

$$2K(s) + 2H_2O(l) \rightarrow 2KOH(aq) + H_2(g)$$

101. Write a balanced equation for the redox reaction at room temperature of calcium metal and bromine liquid to form solid calcium bromide.

$$Ca(s) + Br_2(l) \rightarrow CaBr_2(s)$$

103. Magnesium chloride is used to make disinfectants, fire extinguishers, paper, and floor sweeping compounds. It is made from the redox reaction of hydrochloric acid with solid magnesium hydroxide. Write a balanced equation for this reaction, which yields aqueous magnesium chloride and liquid water.

$$2HCl(aq) + Mg(OH)_2(s) \rightarrow MgCl_2(aq) + 2H_2O(l)$$

Chapter 7
Energy and Chemical Reactions

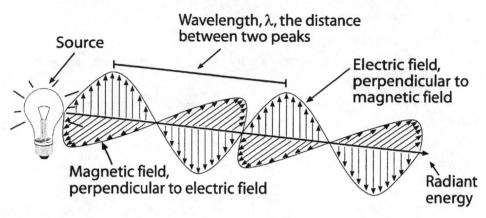

Source

Wavelength, λ, the distance between two peaks

Electric field, perpendicular to magnetic field

Magnetic field, perpendicular to electric field

Radiant energy

♦ Review Skills

7.1 Energy
- Kinetic Energy
- Potential Energy
- Units of Energy
- Thermal Energy and Heat
- Radiant Energy

7.2 Chemical Changes and Energy

7.3 Ozone: Pollutant and Protector
- Removal of UV Radiation by Oxygen and Ozone Molecules
- The Natural Destruction of Ozone

7.4 Chlorofluorocarbons: A Chemical Success Story Gone Wrong
- CFCs and the Ozone Layer

  Special Topic 7.1: Green Chemistry - Substitutes for Chlorofluorocarbons

  Special Topic 7.2: Other Ozone-Depleting Chemicals

  Internet: Ozone and CFCs

♦ Chapter Glossary

  Internet: Glossary Quiz

♦ Chapter Objectives

Review Questions

Key Ideas

Chapter Problems

Section Goals and Introductions

Section 7.1 Energy
Goals

- *To introduce the terms energy, kinetic energy, and potential energy.*
- *To introduce the Law of Conservation of Energy.*
- *To describe the relationships between stability, capacity to do work, and potential energy.*
- *To explain why breaking chemical bonds requires energy and why the formation of chemical bonds releases energy.*
- *To show how energy can be classified as kinetic energy or potential energy.*
- *To describe the units used to describe energy: joules, calories, and dietary calories.*
- *To describe thermal energy, temperature, and heat.*
- *To describe radiant energy.*

Chemical changes are accompanied by energy changes. This section begins to develop your understanding of this relationship by introducing some important terms that relate to energy. An understanding of what potential energy is and how it is related to stability is probably the most important (and perhaps the most difficult) part of this section. Understanding potential energy will help you to understand why energy is absorbed in the breaking of chemical bonds and why it is released in the making of chemical bonds. Be sure that you understand the distinctions between kinetic energy and potential energy and between thermal energy and heat energy. The section ends with a description of radiant energy, including descriptions of what it is, how it is described, and what its different forms are.

Section 7.2 Chemical Changes and Energy
Goals

- *To describe the relationship between energy and chemical reactions.*
- *To explain why some chemical changes release energy as they proceed and why others need to absorb energy to take place.*

This section uses what you have learned in Section 7.1 to explain why some chemical reactions absorb heat energy from their surroundings and why others release heat energy to their surroundings.

Section 7.3 Ozone: Pollutant and Protector
Goals

- *To describe why ozone is a pollutant in the lower atmosphere and why it helps to protect us from potentially harmful radiant energy when it is in the stratosphere.*
- *To describe how ozone is created in the lower atmosphere and to explain why it is created in greater quantities in large industrial cities with lots of cars and lots of sun.*

Because the issues relating to ozone are so important, this section is dedicated to explaining why ozone is either a pollutant or a protector, depending on where it is found. You will discover why the levels of the pollutant ozone are much higher in Los Angeles than in other places in the United States and why we are concerned by decreased levels of ozone in the stratosphere.

Section 7.4 Chlorofluorocarbons: A Chemical Success Story Gone Wrong

Goal: To describe what chlorofluorocarbons (CFCs) are and explain how they are thought to destroy ozone in the stratosphere.

This section continues the ozone story by describing how chlorofluorocarbons (CFCs) (used as aerosol propellants, solvents, expansion gases in foams, heat-exchanging fluids in air conditioners, and temperature-reducing fluids in refrigerators) can destroy the ozone molecules in the stratosphere that help to protect us from high-energy ultraviolet radiation. More information on this topic can be found on our Web site in a section called *Ozone and CFCs*.

www.chemplace.com/college/

Chapter 7 Map

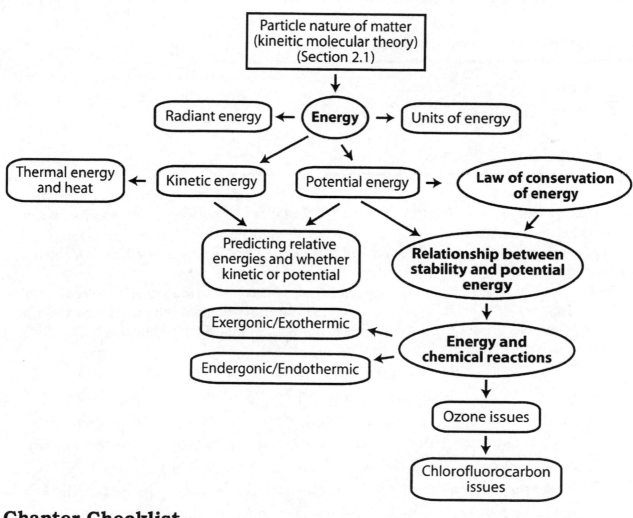

Chapter Checklist

☐ Read the Review Skills section. If there is any skill mentioned that you have not yet mastered, review the material on that topic before reading this chapter.

☐ Read the chapter quickly before the lecture that describes it.

☐ Attend class meetings, take notes, and participate in class discussions.

☐ Work the Chapter Exercises, perhaps using the Chapter Examples as guides.

☐ Study the Chapter Glossary and test yourself at the following Web address:
   www.chemplace.com/college/

☐ Study all of the Chapter Objectives. You might want to write a description of how you will meet each objective. (Although it is best to master all of the objectives, the following objectives are especially important because they pertain to skills that you will need while studying other chapters of this text: 5, 7, 8, and 22.)

☐ To get a review of the most important topics in the chapter, fill in the blanks in the Key Ideas section.

☐ Work all of the selected problems at the end of the chapter, and check your answers with the solutions provided in this chapter of the study guide.

☐ Ask for help if you need it.

Web Resources   www.chemplace.com/college/

Ozone and CFCs
Glossary Quiz

Exercises Key

Exercise 7.1 - Energy: For each of the following situations, you are asked which of two objects or substances has greater energy. Explain your answer with reference to the capacity of each to do work, and indicate whether the energy that distinguishes them is kinetic energy or potential energy.

a. Nitric acid molecules, HNO_3, in the upper atmosphere decompose to form HO molecules and NO_2 molecules by the breaking of a bond between the nitrogen atom and one of the oxygen atoms. Which has greater energy, a nitric acid molecule or the HO molecule and NO_2 molecule that come from its decomposition?

$$HNO_3(g) \rightarrow HO(g) + NO_2(g)$$

> HO and NO_2 have higher potential energy than HNO_3. Separated atoms are less stable and have higher potential energy than atoms in a chemical bond, so energy is required to break a chemical bond. Thus, energy is required to separate the nitrogen and oxygen atoms being held together by mutual attraction in a chemical bond. The energy supplied goes to an increased potential energy of the separated HO and NO_2 molecules compared to HNO_3. If the bond is reformed, the potential energy is converted into a form of energy that could be used to do work.

b. Nitrogen oxides, NO(g) and $NO_2(g)$, are released into the atmosphere in the exhaust of our cars. Which has greater energy, an NO_2 molecule moving at 439 m/s or the same NO_2 molecule moving at 399 m/s? (These are the average velocities of NO_2 molecules at 80 °C and 20 °C, respectively.)

A nitrogen dioxide molecule with a velocity of 439 m/s has greater kinetic energy than the same molecule with a velocity of 399 m/s. Any object in motion can collide with another object and move it, so any object in motion has the capacity to do work. This capacity to do work resulting from the motion of an object is called kinetic energy, KE. The particle with the higher velocity will move another object (like another molecule) farther, so it can do more work. It must therefore have more energy.

c. Which has greater energy, a nitrogen monoxide molecule, NO, emitted from your car's tail pipe at 450 m/s or a nitrogen dioxide molecule, NO_2, moving at the same velocity?

The more massive nitrogen dioxide molecule has greater kinetic energy than the less massive nitrogen monoxide molecule with the same velocity. The moving particle with the higher mass can move another object (like another molecule) farther, so it can do more work. It must therefore have more energy.

d. Liquid nitrogen is used for a number of purposes, including the removal (by freezing) of warts. Assume that the temperature remains constant. Which has greater energy, liquid nitrogen or gaseous nitrogen?

Gaseous nitrogen has higher potential energy than liquid nitrogen. When nitrogen goes from liquid to gas, the attractions that link the N_2 molecules together are broken. The energy that the nitrogen liquid must absorb to break these attractions goes to an increased potential energy of the nitrogen gas. If the nitrogen returns to the liquid form, attractions are re-formed, and potential energy is converted into a form of energy that could be used to do work.

e. Halons, such as halon-1301 (CF_3Br) and halon-1211 (CF_2ClBr), which have been used as fire-extinguishing agents, are a potential threat to Earth's protective ozone layer, partly because they lead to the production of $BrONO_2$ in the upper atmosphere. Which has greater energy, separate BrO and NO_2 molecules or the $BrONO_2$ that they form?

Separate BrO and NO_2 molecules have a higher potential energy than the $BrONO_2$ molecule that they form. Atoms in a chemical bond are more stable and have lower potential energy than separated atoms, so energy is released when chemical bonds form. When BrO and NO_2 are converted into $BrONO_2$, a new bond is formed, and some of the potential energy of the BrO and NO_2 is released. The energy could be used to do some work. For example, if some of the potential energy is converted into increased kinetic energy of a molecule like N_2, the faster moving molecule could bump into something and move it and therefore do work.

$$BrO(g) + NO_2(g) \rightarrow BrONO_2(g)$$

f. The so-called alpha particles released by large radioactive elements such as uranium are helium nuclei consisting of 2 protons and 2 neutrons. Which has greater energy, an uncharged helium atom or an alpha particle and two separate electrons?

An alpha particle and 2 separate electrons have higher potential energy than an uncharged helium atom. The attraction between the alpha particle and the electrons will pull them together, and as they move together, they could bump into something, move it, and do work.

Review Questions Key

For questions 1 and 2, illustrate your answers with simple drawings of the particles that form the structures of the substances mentioned. You do not need to be specific about the nature of the particles: Think of them as simple spheres, and draw them as circles. Provide a general description of the arrangement of the particles, the degree of interaction between them, and their freedom of movement.

1. A pressurized can of a commercial product used to blow the dust off computer components contains tetrafluoroethane, $C_2H_2F_4$. At room temperature, this substance is a liquid at pressures slightly above normal pressure and a gas at normal pressures. Although most of the tetrafluoroethane in the can is in the liquid form, $C_2H_2F_4$ evaporates rapidly, resulting in a significant amount of vapor above the liquid. When the valve on the top of the can is pushed, the tetrafluoroethane gas rushes out, blowing dust off the computer. When the valve closes, more of the liquid $C_2H_2F_4$ evaporates to replace the vapor released. If the can is heated, the liquid evaporates more quickly, and the increase in gas causes the pressure to build up to possibly dangerous levels.

    a. Describe the general structure of liquid tetrafluoroethane.

    The particles in the liquid are close together, but there is generally more empty space between them than in the solid. The particles of a liquid fill about 70% of the total volume. Because the particles in a liquid are moving faster than in a solid, and because there is more empty space between them, they break the attractions to the particles around them and constantly move into new positions to form new attractions. This leads to a less organized arrangement of particles compared to that of the solid. The following image shows the structure of a typical liquid. See Figure 2.2 on page 62 of the textbook.

    b. Describe the general structure of gaseous tetrafluoroethane.

    The particles of a gas are much farther apart than in the solid or liquid. For a typical gas, the average distance between particles is about 10 times the diameter of each particle. This leads to the gas particles themselves taking up only about 0.1% of the total volume. The other 99.9% of the total volume is empty space. According to our model, each particle in a gas moves freely in a straight-line path until it collides with another gas particle or with a liquid or solid. The particles are usually moving fast enough to break any attraction that might form between them, so after two particles collide, they bounce off each other and continue on their way alone. The following image shows the structure of a typical gas. See Figure 2.4 on page 64 of the textbook.

c. Describe the process by which particles move from the liquid form to the gaseous form.

> Particles that are at the surface of the liquid and that are moving away from the surface fast enough to break the attractions that pull them back will escape to the gaseous form. See Figure 2.3 on page 63 of the textbook.

d. Describe the changes that take place in the liquid when it is heated, and explain why these changes lead to a greater rate of evaporation of the liquid.

> Increased temperature leads to an increase in the average velocity of the particles in the liquid. This makes it easier for the particles to break the attractions between them and move from one position to another, including away from the surface into the gaseous form.

2. Sodium metal can be made by running an electric current through molten sodium chloride.

a. Describe the general structure of solid sodium chloride.

> According to our model, the particles of a solid can be pictured as spheres that are packed as closely together as possible. The spheres for NaCl are alternating Na$^+$ cations and Cl$^-$ anions. Strong attractions hold these particles in the same general position, but the particles are still constantly moving (See Figure 2.1 on page 61). Each particle is constantly changing its direction and speeding up and slowing down. Despite the constant changes in direction and velocity, at a constant temperature, the strong attractions between particles keep them the same average distance apart and in the same general orientation to each other.

b. Describe the changes that take place when the temperature of NaCl solid increases.

> When a solid is heated, the average velocity of the particles increases. The more violent collisions between the faster moving particles usually cause each particle to push its neighbors farther away. Therefore, increased temperature usually leads to an expansion of solids (See Figure 2.1 on page 61).

c. Describe the changes that take place when sodium chloride melts.

> The sodium ions and chloride ions break out of their positions in the solid and move more freely throughout the liquid, constantly breaking old attractions and making new ones. Although the particles are still close together in the liquid, they are more disorganized, and there is more empty space between them.

c. Describe the changes that take place when sodium chloride melts.

> As the heating of a solid continues, the movement of each particle eventually becomes powerful enough to enable it to push the other particles around it completely out of position. Because there is very little empty space between the particles, each one that moves out of position has to push its neighbors out of their positions too. Therefore, for one particle to move out of its general position,

all of the particles must be able to move. The organized structure collapses, and the solid becomes a liquid.

Key Ideas Answers

3. The simplest definition of **energy** is that it is the capacity to do work. Work, in this context, may be defined as what is done to move an object against some sort of **resistance**.

5. If two objects are moving at the same velocity, the one with the greater **mass** will have a greater capacity to do work and thus a greater kinetic energy.

7. The Law of Conservation of Energy states that energy can be neither **created** nor **destroyed**, but it can be **transferred** from one system to another and **changed** from one form to another.

9. A system's stability is a measure of its tendency to **change**.

11. Any time a system shifts from a more stable state to a less stable state, the potential energy of the system **increases**.

13. Because less stable separate atoms have higher potential energy than the more stable atoms that participate in a bond, the change from separate atoms to atoms in a bond corresponds to a(n) **decrease** in potential energy.

15. The U.S. National Institute of Standards and Technology defines the calorie as **4.184** joules.

17. The energy associated with internal motion of particles that compose an object can be called either internal kinetic energy or **thermal** energy.

19. Heat is the thermal energy that is transferred from a region of higher temperature to a region of lower temperature as a consequence of the **collisions** of particles.

21. Radiant energy can be viewed as a stream of tiny, **massless** packets of energy called photons.

23. One distinguishing characteristic of the waves of radiant energy is wavelength, λ, the distance between two **peaks** on the wave of electromagnetic radiation. A more specific definition of wavelength is the distance in space over which a wave completes one **cycle** of its repeated form.

25. If in a chemical reaction, more energy is released in the formation of new bonds than was necessary to break old bonds, energy is **released** overall, and the reaction is exergonic.

27. If less energy is released in the formation of the new bonds than is necessary to break the old bonds, energy must be **absorbed** from the surroundings for the reaction to proceed.

29. Two forms of the element oxygen are found in nature: the life-sustaining diatomic **oxygen, O_2,** and **ozone, O_3,** which is a pale blue gas with a strong odor.

31. The highest concentrations of O_3 in the air we breathe are found in large industrial cities with lots of **cars** and lots of **sun**.

33. Radiant energy of wavelengths **shorter** than 400 nm has enough energy to break N–O bonds in NO_2 molecules, but radiant energy with wavelengths **longer** than 400 nm does not supply enough energy to separate the atoms.

35. The stratosphere extends from about **10** km to about **50** km above sea level.

37. **UV-B** radiation has wavelengths from about 290 to 320 nm. Some of it is removed by the gases in the stratosphere, but some of it reaches the surface of the earth. Radiation in this portion of the spectrum has energy great enough that excessive exposure can cause **sunburn**, **premature aging**, and **skin cancer**.

39. Oxygen molecules, O_2, and ozone molecules, O_3, work together to absorb high-energy UV radiation. Oxygen molecules absorb UV radiation with wavelengths less than **242 nm**, and ozone molecules absorb radiant energy with wavelengths from **240 nm to 320 nm**.

41. One of the reasons why CFCs were so successful is that they are extremely **stable** compounds; very few substances react with them.

Problems Key

Section 7.1 Energy

43. For each of the following situations, you are asked which of two objects or substances has greater energy. Explain your answer with reference to the capacity of each to do work, and indicate whether the energy that distinguishes them is kinetic energy or potential energy. *(Objs #2, 3, & 5)*

   a. An ozone molecule, O_3, with a velocity of 393 m/s or the same molecule moving with a velocity of 410 m/s. (These are the average velocities of ozone molecules at 25 °C and 50 °C.)

   > An ozone molecule, O_3, with a velocity of 410 m/s has greater kinetic energy than the same molecule with a velocity of 393 m/s. Any object in motion can collide with another object and move it, so any object in motion has the capacity to do work. This capacity to do work resulting from the motion of an object is called kinetic energy, KE. The particle with the higher velocity will move another object (like another molecule) farther, so it can do more work. It must therefore have more energy.

   b. An ozone molecule, O_3, moving at 300 m/s or an oxygen molecule, O_2, moving at the same velocity.

   > An ozone molecule, O_3, has greater kinetic energy than an O_2 molecule with the same velocity. The moving particle with the higher mass can move another object (like another molecule) farther, so it can do more work. It must therefore have more energy.

   c. A proton and an electron close together or a proton and an electron farther apart.

   > The attraction between the separated electron and a proton will pull them together, and as they move together, they could bump into something, move it, and do work. Therefore, a proton and an electron farther apart have higher potential energy than a proton and an electron close together.

d. An HOCl molecule or an OH molecule and a chlorine atom formed by breaking the chlorine-oxygen bond in the HOCl molecule. (The conversion of HOCl into Cl and OH takes place in the stratosphere.)

> Separated atoms are less stable than atoms in a chemical bond, so the potential energy of OH and Cl is higher than HOCl. Energy is required to separate the oxygen atom and the chlorine atom being held together by mutual attraction in a chemical bond. The energy supplied goes to an increased potential energy of the separate OH and Cl compared to HOCl. If the bond is re-formed, the potential energy is converted into a form of energy that could be used to do work.

e. Two separate chlorine atoms in the stratosphere or the chlorine, Cl_2, molecule that can form when they collide.

> Separated atoms are less stable than atoms in a chemical bond, so the potential energy of two separate Cl atoms is greater than one Cl_2 molecule. When two Cl atoms are converted into a Cl_2 molecule, a new bond is formed, and some of the potential energy of the Cl atoms is released. The energy could be used to do some work.

$$2Cl(g) \rightarrow Cl_2(g)$$

f. Water in the liquid form or water in the gaseous form. (Assume that the two systems are at the same temperature.)

> Gaseous water has higher potential energy than liquid water. When water evaporates, the hydrogen bonds that link the water molecules together are broken. The energy that the water must absorb to break these attractions goes to an increased potential energy of the water vapor. If the water returns to the liquid form, hydrogen bonds are reformed, and the potential energy is converted into a form of energy that could be used to do work.

46. Energy is the capacity to do work. With reference to this definition, describe how you would demonstrate that each of the following has potential energy. (There is no one correct answer in these cases. There are many ways to demonstrate that a system has potential energy.)

a. A brick on the top of a tall building

> If you nudge the brick off the top of the building, its potential energy will be converted into kinetic energy as it falls. If it hits the roof of a parked car, it will move the metal of the roof down, making a dent. When an object is moved, work is done.

b. A stretched rubber band

> You can shoot the rubber band across the room at a paper airplane. When you release the rubber band, its potential energy is converted into kinetic energy, which is used to do the work of moving the airplane.

c. Alcohol molecules added to gasoline

It is possible to run a car on pure alcohol. When the alcohol is burned, its potential energy is converted into energy that does the work of moving the car.

48. For each of the following changes, describe whether (1) kinetic energy is being converted into potential energy, (2) potential energy is being converted into kinetic energy, (3) kinetic energy is transferred from one object to another. (More than one of these changes may be occurring.)

a. An archer pulls back a bow with the arrow in place.

Some of the **kinetic energy** of the moving hand is **transferred** to the string to set it moving and to the tips of the bow as it bends. Some of this **kinetic energy is converted into potential energy** of the stretched string and bow.

b. The archer releases the arrow, and it speeds toward the target.

Some of the **potential energy** of the stretched string and bow are **converted into kinetic energy** of the moving arrow.

51. Methyl bromide is an agricultural soil fumigant that can make its way into the stratosphere, where bromine atoms are stripped away by radiant energy from the sun. The bromine atoms react with ozone molecules (thus diminishing the earth's protective ozone layer) to produce BrO, which in turn reacts with nitrogen dioxide, NO_2, to form $BrONO_2$. For each of these reactions, indicate whether energy is absorbed or released. Explain why. Describe how energy is conserved in each reaction. *(Objs #7 & 8)*

a. $CH_3Br(g) \rightarrow CH_3(g) + Br(g)$

Because a chemical bond is broken in this reaction, energy would need to be absorbed to supply the energy necessary to move to the higher potential energy products. In the stratosphere, this energy comes from the radiant energy of a photon.

b. $BrO(g) + NO_2(g) \rightarrow BrONO_2(g)$

Because a chemical bond is made in this reaction, energy would be released as the system moves to the lower potential energy product. This potential energy could be converted into kinetic energy.

$$BrO(g) + NO_2(g) \rightarrow BrONO_2(g)$$

53. A silver bullet speeding toward a vampire's heart has both external kinetic energy and internal kinetic energy. Explain the difference between the two. *(Obj #11)*

A speeding bullet has a certain kinetic energy that is related to its overall mass and its velocity. This is its external kinetic energy. The bullet is also composed of silver atoms that, like all particles, are moving in a random way. The particles within the bullet are constantly moving, colliding with their neighbors, changing their direction of motion, and changing their velocities. The kinetic energy associated with this internal motion is the internal kinetic energy. The internal motion is independent of the overall motion of the bullet.

55. When a room-temperature thermometer is placed in a beaker of boiling water, heat is transferred from the hot water to the glass of the thermometer and then to the liquid mercury inside the thermometer. With reference to the motion of the particles in the water, glass, and mercury, describe the changes that are taking place during this heat transfer. What changes in thermal energy and average internal kinetic energy occur for each substance? Why do you think the mercury moves up the thermometer? *(Obj #14)*

A typical thermometer used in the chemical laboratory consists of a long cylindrical glass container with a bulb at the bottom that contains a reservoir of mercury and a thin tube running up the inside of the thermometer that the mercury can rise into as it expands.

When the thermometer is first placed in the hot water, the particles in the water have a greater average kinetic energy than the particles of the glass and mercury in the thermometer. When the more energetic water molecules collide with the particles in the glass, the particles of water slow down, and the particles in the glass speed up. The average kinetic energy of the water molecules decreases and the average kinetic energy of the particles of the glass increases. Thermal energy has been transferred from the water to the glass. The glass particles then collide with the less energetic mercury atoms, speeding them up and slowing down themselves. The average kinetic energy of the mercury increases, as thermal energy is transferred from the glass to the mercury. Thus, some of the kinetic energy of the water is transferred to the glass, which then transfers some of this energy to the mercury. This will continue until the particles of water, glass, and mercury all have the same average kinetic energy.

Try to picture the atoms of mercury in the liquid mercury. Now that they are moving faster, they collide with the particles around them with greater force. This pushes the particles farther apart and causes the liquid mercury to expand. Because the mercury has a larger volume, it moves farther up the thin column in the center of the thermometer.

56. With reference to both their particle nature and their wave nature, describe the similarities and differences between visible light and ultraviolet radiation. *(Objs #15-18)*

In the particle view, radiant energy is a stream of tiny, massless packets of energy called photons. Different forms of radiant energy differ with respect to the energy of each of their photons. The energies of the photons of visible light are lower than for ultraviolet radiation.

In the wave view, as radiant energy moves away from the source, it has an effect on the space around it that can be described as a wave consisting of an oscillating electric field perpendicular to an oscillating magnetic field (See Figure 7.11 on page 292 of the textbook.). Different forms of radiant energy differ with respect to the wavelengths and frequencies of these oscillating waves. The waves associated with visible light have longer wavelengths than the waves associated with ultraviolet radiation.

58. Consider the following forms of radiant energy: microwaves, infrared radiation, ultraviolet radiation, X rays, visible light, radio waves, and gamma rays. *(Obj #18)*

   a. List them in order of increasing energy.

   radio waves < microwaves < infrared radiation
   < visible light < ultraviolet radiation < X rays < gamma rays

   b. List them in order of increasing wavelength.

   gamma rays < X rays < ultraviolet radiation < visible light
   < infrared radiation < microwaves < radio waves

Section 7.2 Chemical Changes and Energy

60. Consider the following endergonic reaction. In general terms, explain why energy is absorbed in the process of this reaction. *(Obj #20)*

$$N_2(g) + O_2(g) \rightarrow 2NO(g)$$

The bonds in the products must be less stable and therefore higher potential energy than the bonds in the reactants. Energy is absorbed in the reaction to supply the energy necessary to increase the potential energy of the products compared to reactants.

more stable bonds + energy $\rightarrow$ less stable bonds

lower PE + energy $\rightarrow$ higher PE

62. Hydrazine, N_2H_4, is used as rocket fuel. Consider a system in which a sample of hydrazine is burned in a closed container, followed by heat transfer from the container to the surroundings. *(Objs #14 & 20)*

$$N_2H_4(g) + O_2(g) \rightarrow N_2(g) + 2H_2O(g)$$

   a. In general terms, explain why energy is released in the reaction.

   The bonds in the products must be more stable and therefore lower potential energy than the bonds in the reactants. The potential energy difference between reactants and products is released.

   b. Describe the average internal kinetic energy of the product particles, compared to the reactant particles, before heat energy is transferred to the surroundings. If the average internal kinetic energy for the product(s) is greater than that for the reactants, from where did this energy come? If the average internal kinetic energy for the product(s) is lower than that for the reactants, where did this energy go?

   Some of the potential energy of the reactants is converted into kinetic energy of the products, making the average kinetic energy of the products higher than the average kinetic energy of the reactants.

c. Describe the changes in particle motion that occur as heat is transferred from the products to the surroundings.

> The particles of the higher-temperature products collide with the particles of the lower-temperature container with greater average force than the particles of the container. Therefore, collisions between the particles of the products and the container speed up the particles of the container, increasing its internal kinetic energy or thermal energy, while slowing the particles of the products, decreasing their thermal energy. In this way, thermal energy is transferred from the products to the container. Likewise, the container, which is now at a higher temperature, transfers thermal energy to the lower-temperature surroundings. Heat has been transferred from the products to the container to the surroundings.

65. Classify each of the following changes as exothermic or endothermic.
    a. Leaves decaying in a compost heap   **exothermic**
    b. Dry ice (solid carbon dioxide) changing to carbon dioxide gas **endothermic**
    c. Dew forming on a lawn at night   **exothermic**

67. Explain why all chemical reactions either absorb or evolve energy. *(Obj #22)*

> See Figure 7.13 on page 295 in the textbook.

Section 7.3 Ozone: Pollutant and Protector

68. What characteristic of ozone makes it useful for some purposes and a problem in other situations? For what is it used in industry? What health problems does it cause? *(Obj #23)*

> Ozone is a very powerful oxidizing agent. This can be useful. For example, ozone mixed with oxygen can be used to sanitize hot tubs, and it is used in industry to bleach waxes, oils, and textiles. Conversely, when the levels in the air get too high, the highly reactive nature of ozone becomes a problem. For example, O_3 is a very strong respiratory irritant that can lead to shortness of breath, chest pain when inhaling, wheezing, and coughing. It also damages rubber and plastics, leading to premature deterioration of products made with these materials. According to the Agricultural Research Service of North Carolina State University, ozone damages plants more than all other pollutants combined.

70. Explain why UV radiation less than 400 nm in wavelength is able to break N–O bonds in NO_2 molecules, and explain why radiant energy greater than 400 nm in wavelength cannot break these bonds. *(Obj #26)*

> Shorter wavelengths of radiant energy are associated with higher energy photons. Radiant energy of wavelengths less than 400 nm has enough energy to break N–O bonds in NO_2 molecules, but radiant energy with wavelengths longer than 400 nm does not supply enough energy to separate the atoms.

72. Explain why UV-B radiation can be damaging to us and our environment if it reaches Earth in greater quantities than it does now. *(Obj #29)*

> The shorter wavelength UV-B radiation (from about 290 nm to 320 nm) has higher energy than UV-A radiation. Radiation in this portion of the spectrum has high enough energy so that excessive exposure can cause sunburn, premature skin aging, and skin cancer.

74. Explain how oxygen molecules, O_2, and ozone molecules, O_3, work together to protect us from high-energy ultraviolet radiation. *(Obj #31)*

> O_2 molecules absorb UV radiation with wavelengths less than 242 nm, and O_3 molecules absorb radiant energy with wavelengths from 240 nm to 320 nm. See Figure 7.16 on page 299 in the textbook.

Section 7.4   Chlorofluorocarbons: A Chemical Success Story Gone Wrong

76. Explain why CFCs eventually make their way into the stratosphere even though most chemicals released into the atmosphere do not. *(Obj #33)*

> Gases are removed from the lower atmosphere in 2 general ways. They either dissolve in the clouds and are rained out, or they react chemically to be converted into other substances. Neither of these mechanisms are important for CFCs. Chlorofluorocarbons are insoluble in water, and they are so stable that they can exist in the lower atmosphere for years. During this time, the CFC molecules wander around in the atmosphere, moving wherever the air currents take them. They can eventually make their way up into the stratosphere.

Additional Problems

79. Energy is the capacity to do work. With reference to this definition, describe how you would demonstrate that each of the following objects or substances has potential energy. (There is no one correct answer in these cases. There are many ways to demonstrate that a system has potential energy.)
    a.   Natural gas used to fuel a city bus

    > When the natural gas burns, some of its potential energy is used to do the work of moving the bus.

    b.   A compressed spring

    > If you put an object, such as a small ball of paper, on one end of the compressed spring and then release the spring, the potential energy stored in the compressed spring will be converted into kinetic energy of the moving spring as it stretches out. The kinetic energy of the moving spring will do the work of moving the object.

c.   A pinecone at the top of a tall tree

When a squirrel chews off the pinecone, allowing it to fall to the ground, the potential energy that it has because it is higher than the ground will be converted into kinetic energy as it falls. The falling pinecone can collide with another pinecone and do the work of knocking the second pinecone off its branch.

81. For each of the following changes, describe whether (1) kinetic energy is being converted into potential energy, (2) potential energy is being converted into kinetic energy, (3) kinetic energy is transferred from one object to another. (More than one of these changes may be occurring.)

a.   Using an elaborate system of ropes and pulleys, a piano mover hoists a piano up from the sidewalk to the outside of a large third floor window of a city apartment building.

The **kinetic energy** of the mover's hands is **transferred** to the kinetic energy of the moving ropes, which is transferred to the kinetic energy of the moving piano. Some of this **kinetic energy is converted into potential energy** as the piano rises.

b.   His hands slip and the piano drops 6 feet before he is able to stop the rope from unwinding.

Some of the **potential energy** of the piano is **converted into kinetic energy** as it falls.

83. The following changes combine to help move a car down the street. For each change, describe whether (1) kinetic energy is being converted into potential energy, (2) potential energy is being converted into kinetic energy, (3) kinetic energy is transferred from one object to another. (More than one of these changes may be occurring.)

a.   The combustion of gasoline in the cylinder releases heat energy, increasing the temperature of the gaseous products of the reaction.

Some of the **potential energy** of the gasoline molecules is **converted into the kinetic energy** of moving product particles. This is reflected in an increase in the temperature of the gaseous products.

b.   The hot gaseous products push the piston down in the cylinder.

Some of the **kinetic energy** of the moving gaseous product particles is **transferred into the kinetic energy** of the moving piston.

c.   The moving piston turns the crankshaft, which ultimately turns the wheels.

Some of the **kinetic energy** of the moving piston is **transferred into the kinetic energy** of the moving crankshaft and ultimately into the kinetic energy of the moving wheels.

85. Classify each of the following changes as exothermic or endothermic.

a.   The burning fuel in a camp stove   **exothermic**

b.   The melting of ice in a camp stove to provide water on a snow-camping trip **endothermic**

Chapter 8
Unit Conversions

$$? \text{ kg} = 4567.36 \ \mu g \left(\frac{1 \ g}{10^6 \ \mu g}\right)\left(\frac{1 \ kg}{10^3 \ g}\right) = \mathbf{4.56736 \times 10^{-6} \ kg}$$

Desired unit — Given value

Converts given metric unit to metric base unit

Converts metric base unit to desired metric unit

♦ Review Skills

8.1 Dimensional Analysis
- An Overview of the General Procedure
- Metric-Metric Unit Conversions
- English-Metric Unit Conversions

8.2 Rounding Off and Significant Figures
- Measurements, Calculations, and Uncertainty
- Rounding Off Answers Derived from Multiplication and Division
- Rounding Off Answers Derived from Addition and Subtraction

8.3 Density and Density Calculations
- Using Density as a Conversion Factor
- Determination of Mass Density

8.4 Percentage and Percentage Calculations

8.5 A Summary of the Dimensional Analysis Process

8.6 Temperature Conversions

♦ Chapter Glossary
  Internet: Glossary Quiz

♦ Chapter Objectives

Review Questions

Key Ideas

Chapter Problems

Section Goals and Introductions

Be sure that you can do the things listed in the Review Skills section before you spend too much time studying this chapter. They are especially important. You might also want to look at Appendices A, B, and C. Appendix A (*Measurement and Units*) provides tables that show units, their abbreviations, and relationships between units that lead to conversion factors. Appendix B (*Scientific Notation*) describes how to convert between regular decimal numbers and numbers expressed in scientific notation, and it shows how calculations using scientific notation are done. Appendix C (*Using a Scientific Calculator*) shows how calculations, such as those found in this chapter, can be done using common scientific calculators.

Section 8.1 Dimensional Analysis

Goals

- *To describe a procedure for making unit conversions called dimensional analysis.*
- *To describe metric-metric unit conversions.*
- *To describe English-metric unit conversions.*

Many chemical calculations include the conversion from a value expressed in one unit to the equivalent value expressed in a different unit. Dimensional analysis, which is described in this section, provides you with an organized format for making these unit conversions and gives you a logical thought process that will help you to reason through such calculations. It is extremely important that you master this technique. You'll be glad you have when you go on to Chapters 9 and 10, which describe chemical calculations that can be done using the dimensional analysis technique.

Section 8.2 Rounding Off and Significant Figures

Goal: To describe the procedures for rounding off answers to calculations.

When you use a calculator to complete your calculations, it's common that most of the numbers you see on the display at the end of the calculation are meaningless. This section describes why this is true and shows you simple techniques that you can use to round off your answers.

Section 8.3 Density and Density Calculations

Goal: To describe what density is, how it can be used as a conversion factor, and how density can be calculated.

Density calculations are common in chemistry. The examples in this section show you how these calculations are done, and perhaps more important, these density calculations provide more examples of the dimensional analysis techniques and the procedures for rounding.

Section 8.4 Percentage and Percentage Calculations

Goal: To show how percentages can be made into conversion factors and show how they are used in making unit conversions.

Some people have trouble with calculations using percentages. They multiply when they should divide or divide when they should multiply. This section shows you how to make conversion factors out of percentages and how these conversions can be used to do percent calculations with confidence.

Section 8.5 A Summary of the Dimensional Analysis Process

Goal: To summarize the dimensional analysis process.

This section summarizes the different types of unit conversions described in this chapter. It should help you organize your thought process for making unit conversions. Pay special attention to Figure 8.5.

Section 8.6 Temperature Conversions

Goal: To show how to convert among temperatures expressed in degrees Celsius, degrees Fahrenheit, and kelvins.

This section shows how to convert from one temperature unit to another. Pay close attention to the subtleties that arise in rounding off answers to temperature conversions.

Chapter 8 Map

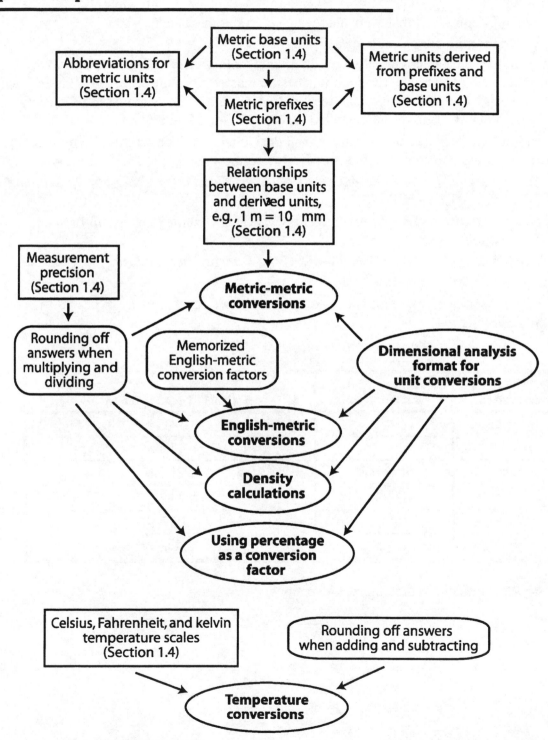

Chapter Checklist

☐ Read the Review Skills section. If there is any skill mentioned that you have not yet mastered, review the material on that topic before reading the present chapter.

☐ Read the chapter quickly before the lecture that describes it.

☐ Attend class meetings, take notes, and participate in class discussions.

☐ Work the Chapter Exercises, perhaps using the Chapter Examples as guides.

☐ Study the Chapter 8 Glossary and test yourself on our Web site:

   www.chemplace.com/college/

☐ Study all of the Chapter Objectives. You might want to write a description of how you will meet each objective. (Although it is best to master all of the objectives, the following objectives are especially important because they pertain to skills that you will need while studying other chapters of this text: 2, 3, 5, 8, 9, 11, 14, 15, and 16.)

☐ Reread the Study Sheets in this chapter and decide whether you will use them or some variation on them to complete the tasks they describe.

   Sample Study Sheet 8.1: Rounding Off Numbers Calculated Using Multiplication And Division

   Sample Study Sheet 8.2: Rounding Off Numbers Calculated Using Addition and Subtraction

   Sample Study Sheet 8.3: Calculations Using Dimensional Analysis

☐ Memorize the following.

   Be sure to check with your instructor to determine how much you are expected to know of the following.

   • English-metric conversion factors

Type of Measurement	Probably Most Useful to Know	Others Useful to Know		
Length	$\dfrac{2.54 \text{ cm}}{1 \text{ in.}}$ (exact)	$\dfrac{1.609 \text{ km}}{1 \text{ mi}}$	$\dfrac{39.37 \text{ in.}}{1 \text{ m}}$	$\dfrac{1.094 \text{ yd}}{1 \text{ m}}$
Mass	$\dfrac{453.6 \text{ g}}{1 \text{ lb}}$	$\dfrac{2.205 \text{ lb}}{1 \text{ kg}}$		
Volume	$\dfrac{3.785 \text{ L}}{1 \text{ gal}}$	$\dfrac{1.057 \text{ qt}}{1 \text{ L}}$		

   • Equations for temperature conversions

$$? \,°F = \text{number of } °C \left(\frac{1.8 \, °F}{1 \, °C} \right) + 32 \, °F$$

$$? \,°C = \left(\text{number of } °F - 32 \, °F \right) \left(\frac{1 \, °C}{1.8 \, °F} \right)$$

$$? \, K = \text{number of } °C + 273.15$$

$$? \,°C = \text{number of } K - 273.15$$

☐ To get a review of the most important topics in the chapter, fill in the blanks in the Key Ideas section.

☐ Work all of the selected problems at the end of the chapter, and check your answers with the solutions provided in this chapter of the study guide.

☐ Ask for help if you need it.

Web Resources   www.chemplace.com/college/

Glossary Quiz

Exercises Key

Exercise 8.1 - Conversion Factors: Write two conversion factors that relate the following pairs of metric units. Use positive exponents for each. *(Obj #2)*

a. joule and kilojoule $\dfrac{10^3\ J}{1\ kJ}$ and $\dfrac{1\ kJ}{10^3\ J}$

b. meter and centimeter $\dfrac{10^2\ cm}{1\ m}$ and $\dfrac{1\ m}{10^2\ cm}$

c. liter and gigaliter $\dfrac{10^9\ L}{1\ GL}$ and $\dfrac{1\ GL}{10^9\ L}$

d. gram and microgram $\dfrac{10^6\ \mu g}{1\ g}$ and $\dfrac{1\ g}{10^6\ \mu g}$

e. gram and megagram $\dfrac{10^6\ g}{1\ Mg}$ and $\dfrac{1\ Mg}{10^6\ g}$

Exercise 8.2 - Unit Conversions: Convert 4.352 micrograms to megagrams. *(Obj #3)*

$$? \ Mg \ = \ 4.352 \ \mu g \left(\frac{1\ g}{10^6\ \mu g} \right) \left(\frac{1\ Mg}{10^6\ g} \right) \ = \ \mathbf{4.352 \times 10^{-12}\ Mg}$$

Exercise 8.3 - Unit Conversions: The volume of the earth's oceans is estimated to be 1.5×10^{18} kiloliters. What is this volume in gallons? *(Obj #5)*

$$? \ gal \ = \ 1.5 \times 10^{18} \ kL \left(\frac{10^3\ L}{1\ kL} \right) \left(\frac{1\ gal}{3.785\ L} \right) \ = \ \mathbf{4.0 \times 10^{20}\ gal}$$

✍ **Exercise 8.4 - Rounding Off Answers Derived from Multiplication and Division:**   A first-class stamp allows you to send letters weighing up to 1 oz. (There are 16 ounces per pound.) You weigh a letter and find that it has a mass of 10.5 g. Can you mail this letter with one stamp? The dimensional analysis setup for converting 10.5 g to ounces follows. Identify whether each value in the setup is exact. Then determine the number of significant figures in each inexact value, calculate the answer, and report it to the correct number of significant figures. *(Obj #8)*

$$? \, oz = 10.5 \, g \left(\frac{1 \, lb}{453.6 \, g} \right) \left(\frac{16 \, oz}{1 \, lb} \right) = \textbf{0.370 oz}$$

The 10.5 comes from a measurement and has three significant figures. The 453.6 is calculated and rounded off. It has four significant figures. The 16 comes from a definition and is exact. We report three significant figures in our answer.

✍ **Exercise 8.5 - Rounding Off Answers Derived from Multiplication and Division:**   The re-entry speed of the Apollo 10 space capsule was 11.0 km/s. How many hours would it have taken for the capsule to fall through 25.0 miles of the stratosphere? The dimensional analysis setup for this calculation follows. Identify whether each value in the setup is exact. Then determine the number of significant figures in each inexact value, calculate the answer, and report it to the correct number of significant figures. *(Obj #8)*

$$? \, hr = 25.0 \, mi \left(\frac{5280 \, ft}{1 \, mi} \right) \left(\frac{12 \, in.}{1 \, ft} \right) \left(\frac{2.54 \, cm}{1 \, in.} \right) \left(\frac{1 \, m}{10^2 \, cm} \right) \left(\frac{1 \, km}{10^3 \, m} \right) \left(\frac{1 \, s}{11.0 \, km} \right) \left(\frac{1 \, min}{60 \, s} \right) \left(\frac{1 \, hr}{60 \, min} \right)$$

$$= \textbf{1.02} \times \textbf{10}^{-3} \, \textbf{hr}$$

The 25.0 comes from a measurement and has three significant figures. The 11.0 comes from a blend of measurement and calculations. It also has three significant figures. All of the other numbers come from definitions and are therefore exact. We report three significant figures in our answer.

✍ **Exercise 8.6 – Rounding Off Answers Derived from Addition and Subtraction:**   Report the answers to the following calculations to the correct number of decimal positions. Assume that each number is ±1 in the last decimal position reported. *(Obj #9)*

    a. 684 − 595.325 = **89**   b. 92.771 + 9.3 = **102.1**

✍ **Exercise 8.7 - Density Conversions:** *(Obj #11)*

    a. What is the mass in kilograms of 15.6 gallons of gasoline?

$$? \, kg = 15.6 \, gal \left(\frac{3.785 \, L}{1 \, gal} \right) \left(\frac{10^3 \, mL}{1 \, L} \right) \left(\frac{0.70 \, g \, gas.}{1 \, mL \, gas.} \right) \left(\frac{1 \, kg}{10^3 \, g} \right)$$

$$or \quad ? \, kg = 15.6 \, gal \left(\frac{3.785 \, L}{1 \, gal} \right) \left(\frac{0.70 \, kg \, gas.}{1 \, L \, gas.} \right) = \textbf{41 kg gasoline}$$

b. A shipment of iron to a steel-making plant has a mass of 242.6 metric tons. What is the volume in liters of this iron?

$$? \, L \; = \; 242.6 \text{ met. tons} \left(\frac{10^3 \text{ kg}}{1 \text{ met. ton}} \right) \left(\frac{10^3 \text{ g}}{1 \text{ kg}} \right) \left(\frac{1 \text{ mL Fe}}{7.86 \text{ g Fe}} \right) \left(\frac{1 \text{ L}}{10^3 \text{ mL}} \right)$$

$$\text{or} \quad ? \, L \; = \; 242.6 \text{ met. tons} \left(\frac{10^3 \text{ kg}}{1 \text{ met. ton}} \right) \left(\frac{1 \text{ L Fe}}{7.86 \text{ kg Fe}} \right)$$

$$= \mathbf{3.09 \times 10^4 \, L \; Fe}$$

Exercise 8.8 - Density Calculations: *(Obj #12)*

a. A graduated cylinder is weighed and found to have a mass of 48.737 g. A sample of hexane, C_6H_{14}, is added to the graduated cylinder, and the total mass is measured as 57.452 g. The volume of the hexane is 13.2 mL. What is the density of hexane?

$$\frac{? \, g}{mL} = \frac{(57.452 - 48.737) \, g}{13.2 \text{ mL}} = \mathbf{0.660 \; g/mL}$$

b. A tree trunk is found to have a mass of 1.2×10^4 kg and a volume of 2.4×10^4 L. What is the density of the tree trunk in g/mL?

$$\frac{? \, g}{1 \text{ mL}} = \frac{1.2 \times 10^4 \text{ kg}}{2.4 \times 10^4 \text{ L}} \left(\frac{10^3 \text{ g}}{1 \text{ kg}} \right) \left(\frac{1 \text{ L}}{10^3 \text{ mL}} \right) = \mathbf{0.50 \; g/mL}$$

Exercise 8.9 - Unit Conversions: *(Obj #14)*

a. The mass of the ocean is about 1.8×10^{21} kg. If the ocean contains 0.014% by mass hydrogen carbonate ions, HCO_3^-, how many pounds of HCO_3^- are in the ocean?

$$? \text{ lb } HCO_3^- = 1.8 \times 10^{21} \text{ kg ocean} \left(\frac{0.014 \text{ kg } HCO_3^-}{100 \text{ kg ocean}} \right) \left(\frac{2.205 \text{ lb}}{1 \text{ kg}} \right)$$

$$= \mathbf{5.6 \times 10^{17} \, lb \; HCO_3^-}$$

b. When you are doing heavy work, your muscles get about 75 to 80% by volume of your blood. If your body contains 5.2 liters of blood, how many liters of blood are in your muscles when you are working hard enough to send them 78% by volume of your blood?

$$? \text{ L blood to muscles} = 5.2 \text{ L blood total} \left(\frac{78 \text{ L blood to muscles}}{100 \text{ L blood total}} \right)$$

$$= \mathbf{4.1 \; L \; blood \; to \; muscles}$$

Exercise 8.10 - Unit Conversions: *(Obj #15)*

a. The diameter of a proton is 2×10^{-15} meter. What is this diameter in nanometers?

$$? \text{ nm} = 2 \times 10^{-15} \text{ m} \left(\frac{10^9 \text{ nm}}{1 \text{ m}} \right) = \mathbf{2 \times 10^{-6} \; nm}$$

b. The mass of an electron is $9.1093897 \times 10^{-31}$ kg. What is this mass in nanograms?

$$? \text{ lb} = 9.1093897 \times 10^{-31} \text{ kg} \left(\frac{10^3 \text{ g}}{1 \text{ kg}} \right) \left(\frac{10^9 \text{ ng}}{1 \text{ g}} \right) = \mathbf{9.1093897 \times 10^{-19} \; ng}$$

c. There are 4.070×10^6 lb of sulfuric acid used to make Jell-O each year. Convert this to kilograms.

$$? \text{ kg} = 4.070 \times 10^6 \text{ lb} \left(\frac{453.6 \text{ g}}{1 \text{ lb}}\right)\left(\frac{1 \text{ kg}}{10^3 \text{ g}}\right) = \mathbf{1.846 \times 10^6 \text{ kg}}$$

d. A piece of Styrofoam has a mass of 88.978 g and a volume of 2.9659 L. What is its density in g/mL?

$$\frac{? \text{ g}}{\text{mL}} = \frac{88.978 \text{ g}}{2.9659 \text{ L}}\left(\frac{1 \text{ L}}{10^3 \text{ mL}}\right) = \mathbf{0.030000 \text{ g/mL}}$$

e. The density of blood plasma is 1.03 g/mL. A typical adult has about 2.5 L of blood plasma. What is the mass in kilograms of the blood plasma in this person?

$$? \text{ kg} = 2.5 \text{ L}\left(\frac{10^3 \text{ mL}}{1 \text{ L}}\right)\left(\frac{1.03 \text{ g}}{1 \text{ mL}}\right)\left(\frac{1 \text{ kg}}{10^3 \text{ g}}\right)$$

$$\text{or} \quad ? \text{ kg} = 2.5 \text{ L}\left(\frac{1.03 \text{ kg}}{1 \text{ L}}\right) = \mathbf{2.6 \text{ kg}}$$

f. Pain signals are transferred through the nervous system at a speed between 12 and 30 meters per second. If a student drops a textbook on her toe, how long will it take for the signal, traveling at a velocity of 18 meters per second, to reach her brain 6.0 feet away?

$$? \text{ s} = 6.0 \text{ ft}\left(\frac{12 \text{ in.}}{1 \text{ ft}}\right)\left(\frac{2.54 \text{ cm}}{1 \text{ in.}}\right)\left(\frac{1 \text{ m}}{10^2 \text{ cm}}\right)\left(\frac{1 \text{ s}}{18 \text{ m}}\right) = \mathbf{0.10 \text{ s}}$$

g. An electron takes 6.2×10^{-9} second to travel across a TV set that is 22 inches wide. What is the velocity of the electron in km/hr?

$$\frac{? \text{ km}}{\text{hr}} = \frac{22 \text{ in.}}{6.2 \times 10^{-9} \text{ s}}\left(\frac{2.54 \text{ cm}}{1 \text{ in.}}\right)\left(\frac{1 \text{ m}}{10^2 \text{ cm}}\right)\left(\frac{1 \text{ km}}{10^3 \text{ m}}\right)\left(\frac{60 \text{ s}}{1 \text{ min}}\right)\left(\frac{60 \text{ min}}{1 \text{ hr}}\right)$$

$$= \mathbf{3.2 \times 10^8 \text{ km/hr}}$$

h. The mass of the ocean is about 1.8×10^{21} kg. If the ocean contains 0.041% by mass calcium ions, Ca^{2+}, how many tons of Ca^{2+} are in the ocean? (There are 2000 pounds per ton.)

$$? \text{ ton } Ca^{2+} = 1.8 \times 10^{21} \text{ kg ocean}\left(\frac{0.041 \text{ kg } Ca^{2+}}{100 \text{ kg ocean}}\right)\left(\frac{2.205 \text{ lb}}{1 \text{ kg}}\right)\left(\frac{1 \text{ ton}}{2000 \text{ lb}}\right)$$

$$= \mathbf{8.1 \times 10^{14} \text{ ton } Ca^{2+}}$$

i. When you are at rest, your heart pumps about 5.0 liters of blood per minute. Your brain gets about 15% by volume of your blood. What volume of blood, in liters, is pumped through your brain in 1.0 hour of rest?

$$? \text{ L to brain} = 1.0 \text{ hr}\left(\frac{60 \text{ min}}{1 \text{ hr}}\right)\left(\frac{5.0 \text{ L total}}{1 \text{ min}}\right)\left(\frac{15 \text{ L to brain}}{100 \text{ L total}}\right) = \mathbf{45 \text{ L}}$$

✍ **Exercise 8.11 - Temperature Conversions:** *(Obj #16)*

a. N,N-dimethylaniline, $C_6H_5N(CH_3)_2$, which is used to make dyes, melts at 2.5 °C. What is N,N-dimethylaniline's melting point in °F and K?

$$°F = 2.5°C\left(\frac{1.8°F}{1°C}\right) + 32°F = \textbf{36.5 °F}$$

$$K = 2.5°C + 273.15 = \textbf{275.7 K}$$

b. Benzenethiol, C_6H_5SH, a mosquito larvicide, melts at 5.4 °F. What is benzenethiol's melting point in °C and K?

$$°C = (5.4°F - 32°F)\frac{1°C}{1.8°F} = \textbf{–14.8 °C}$$

$$K = -14.8°C + 273.15 = \textbf{258.4 K}$$

c. The hottest part of the flame on a Bunsen burner is found to be 2.15×10^3 K. What is this temperature in °C and °F?

$$°C = 2.15 \times 10^3 K - 273.15 = \textbf{1.88} \times \textbf{10}^3 \textbf{°C}$$

$$°F = 1.88 \times 10^3 \, °C\left(\frac{1.8°F}{1°C}\right) + 32°F = \textbf{3.42} \times \textbf{10}^3 \textbf{°F}$$

or **3.41 × 10³ °F** if the unrounded answer to the first calculation is used in the second calculation.

Review Questions Key

The following questions give you a review of some of the important skills from earlier chapters that you will use in this chapter.

1. Write the metric base units and their abbreviations for length, mass, volume, energy, and gas pressure. (See Section 1.4.)

> Length, **meter (m)**, mass, **gram (g)**, volume, **liter (L)**, energy, **joule (J)**, and gas pressure, **pascal (Pa)**

2. Complete the following table by writing the type of measurement that the unit represents (mass, length, volume, or temperature) and either the name of a unit or the abbreviation for each unit. (See Section 1.4.)

Unit	Type of Measurement	Abbreviation	Unit	Type of Measurement	Abbreviation
milliliter	**volume**	**mL**	kilojoule	**energy**	kJ
microgram	**mass**	μg	kelvin	**temperature**	K

3. Complete the following relationships between units. (See Section 1.4.)
   - a. 10^{-6} m = 1 μm
   - b. 10^{6} g = 1 Mg
   - c. 10^{-3} L = 1 mL
   - d. 10^{-9} m = 1 nm
   - e. 1 cm^3 = 1 mL
   - f. 10^3 L = m^3
   - g. 10^3 kg = 1 t   (t = metric ton)
   - h. 1 Mg = 1 t   (t = metric ton)

4. An empty 2-L graduated cylinder is weighed on a balance and found to have a mass of 1124.2 g. Liquid methanol, CH_3OH, is added to the cylinder, and its volume is measured as 1.20 L. The total mass of the methanol and the cylinder is measured as 2073.9 g. On the basis of the way these data are reported, what do you assume is the range of possible values that each represents? (See Section 1.5.)

   We assume that each reported number is ± 1 in the last decimal position reported. Therefore, we assume the mass of the graduated cylinder **between 1124.1 g and 1124.3 g**, the volume of the methanol is between **1.19 L and 1.21 L**, and the total mass is between **2073.8 g and 2074.0 g**.

Key Ideas Answers

5. You will find that the stepwise thought process associated with the procedure called dimensional analysis not only guides you in figuring out how to set up **unit conversion** problems but also gives you confidence that your answers are **correct**.

7. Next, we multiply by one or more conversion factors that enable us to cancel the **unwanted** units and generate the **desired** units.

9. If you have used correct conversion factors in a dimensional analysis setup, and if your units **cancel** to yield the desired unit or units, you can be confident that you will arrive at the correct answer.

11. Unless we are told otherwise, we assume that values from measurements have an uncertainty of plus or minus **one** in the last decimal place reported.

13. When an answer is calculated by multiplying or dividing, we round it off to the same number of significant figures as the **inexact** value with the **fewest** significant figures.

15. Numbers that come from definitions and from **counting** are exact.

17. When adding or subtracting, round your answer to the same number of **decimal places** as the inexact value with the **fewest decimal places**.

19. The densities of liquids and solids are usually described in **grams per milliliter** or **grams per cubic centimeter**.

21. Because the density of a substance depends on the substance's **identity** and its temperature, it is possible to identify an unknown substance by comparing its density at a particular temperature to the **known** densities of substances at the same temperature.

23. Percentage by mass, the most common form of percentage used in chemical descriptions, is a value that tells us the number of mass units of the **part** for each 100 mass units of the **whole**.

25. Note that the numbers 1.8, 32, and 273.15 in the equations used for temperature conversions all come from **definitions**, so they are all exact.

Problems Key

Problems Relating to Appendices

If you have not yet read Appendix B, which describes scientific notation, and Appendix C, which contains pointers for using calculators, you might want to read them before working the problems that follow.

26. Convert the following ordinary decimal numbers to scientific notation. (See Appendix B at the end of the textbook if you need help with this.)
    a. 67,294 **6.7294 × 10^4**
    b. 438,763,102 **4.38763102 × 10^8**
    c. 0.000073 **7.3 × 10^{-5}**
    d. 0.0000000435 **4.35 × 10^{-8}**

28. Convert the following numbers expressed in scientific notation to ordinary decimal numbers. (See Appendix B at the end of the textbook if you need help with this.)
    a. 4.097 × 10^3 **4,097**
    b. 1.55412 × 10^4 **15,541.2**
    c. 2.34 × 10^{-5} **0.0000234**
    d. 1.2 × 10^{-8} **0.000000012**

30. Use your calculator to complete the following calculations. (See your calculator's instruction manual or Appendix C at the end of the textbook if you need help with this.)
    a. 34.25 × 84.00 = **2,877**
    b. 2607 ÷ 8.25 = **316**
    c. 425 ÷ 17 × 0.22 = **5.5**
    d. (27.001 − 12.866) ÷ 5.000 = **2.827**

32. Use your calculator to complete the following calculations. (See your calculator's instruction manual or Appendix C at the end of the textbook if you need help with this.)
    a. 10^9 × 10^3 = **10^{12}**
    b. 10^{12} ÷ 10^3 = **10^9**
    c. 10^3 × 10^6 ÷ 10^2 = **10^7**
    d. 10^9 × 10^{-4} = **10^5**
    e. 10^{23} ÷ 10^{-6} = **10^{29}**
    f. 10^{-4} × 10^2 ÷ 10^{-5} = **10^3**

34. Use your calculator to complete the following calculations. (See your calculator's instruction manual or Appendix C at the end of the textbook if you need help with this.)
    a. 9.5 × 10^5 • 8.0 × 10^9 = **7.6 × 10^{15}**
    b. 6.12 × 10^{19} ÷ 6.00 × 10^3 = **1.02 × 10^{16}**
    c. 2.75 × 10^4 × 6.00 × 10^7 ÷ 5.0 × 10^6 = **3.3 × 10^5**
    d. 8.50 × 10^{-7} • 2.20 × 10^3 = **1.87 × 10^{-3} or 0.00187**
    e. 8.203 × 10^9 ÷ 10^{-4} = **8.203 × 10^{13}**
    f. (7.679 × 10^{-4} − 3.457 × 10^{-4}) ÷ 2.000 × 10^{-8} = **2.111 × 10^4**

Section 8.1 Dimensional Analysis

36. Complete each of the following conversion factors by filling in the blank on the top of the ratio. *(Objs #2 & 4)*

a. $\left(\dfrac{10^3 \ m}{1 \ km}\right)$

d. $\left(\dfrac{1 \ cm^3}{1 \ mL}\right)$

b. $\left(\dfrac{10^2 \ cm}{1 \ m}\right)$

e. $\left(\dfrac{2.54 \ cm}{1 \ in.}\right)$

c. $\left(\dfrac{10^3 \ mm}{1 \ m}\right)$

f. $\left(\dfrac{453.6 \ g}{1 \ lb}\right)$

38. Complete each of the following conversion factors by filling in the blank on the top of the ratio. *(Objs #2 & 4)*

a. $\left(\dfrac{10^3 \ g}{1 \ kg}\right)$

c. $\left(\dfrac{1.094 \ yd}{1 \ m}\right)$

b. $\left(\dfrac{10^3 \ mg}{1 \ g}\right)$

d. $\left(\dfrac{2.205 \ lb}{1 \ kg}\right)$

40. The mass of an electron is $9.1093897 \times 10^{-31}$ kg. What is this mass in grams? *(Obj #3)*

$$? \ g = 9.1093897 \times 10^{-31} \ kg \left(\dfrac{10^3 \ g}{1 \ kg}\right) = \mathbf{9.1093897 \times 10^{-28} \ g}$$

42. The diameter of typical bacteria cells is 0.00032 centimeter. What is this diameter in micrometers? *(Obj #3)*

$$? \ \mu m = 0.00032 \ cm \left(\dfrac{1 \ m}{10^2 \ cm}\right)\left(\dfrac{10^6 \ \mu m}{1 \ m}\right) = \mathbf{3.2 \ \mu m}$$

44. The thyroid gland is the largest of the endocrine glands with a mass between 20 and 25 grams. What is the mass in pounds of a thyroid gland with a mass of 22.456 grams? *(Obj #5)*

$$? \ lb = 22.456 \ g \left(\dfrac{1 \ lb}{453.6 \ g}\right) = \mathbf{0.04951 \ lb}$$

$$\mathbf{or} \ \ ? \ lb = 22.456 \ g \left(\dfrac{1 \ lb}{453.59237 \ g}\right) = \mathbf{0.049507 \ lb}$$

46. The mass of a neutron is 1.674929×10^{-27} kg. Convert this to ounces. *(Obj #5)* (There are 16 oz/lb.)

$$? \ oz = 1.674929 \times 10^{-27} \ kg \left(\dfrac{10^3 \ g}{1 \ kg}\right)\left(\dfrac{1 \ lb}{453.6 \ g}\right)\left(\dfrac{16 \ oz}{1 \ lb}\right) = \mathbf{5.908 \times 10^{-26} \ oz}$$

$$? \ oz = 1.674929 \times 10^{-27} \ kg \left(\dfrac{10^3 \ g}{1 \ kg}\right)\left(\dfrac{1 \ lb}{453.59237 \ g}\right)\left(\dfrac{16 \ oz}{1 \ lb}\right) = \mathbf{5.908138 \times 10^{-26} \ oz}$$

48. A red blood cell is 8.7×10^{-5} inches thick. What is this thickness in micrometers? *(Obj #5)*

$$? \; \mu m = 8.7 \times 10^{-5} \; in. \left(\frac{2.54 \; cm}{1 \; in.} \right) \left(\frac{1 \; m}{10^2 \; cm} \right) \left(\frac{10^6 \; \mu m}{1 \; m} \right) = \textbf{2.2 } \boldsymbol{\mu}\textbf{m}$$

Section 8.2 Rounding Off and Significant Figures

50. Decide whether each of the numbers shown in boldface type is exact. If it is not exact, write the number of significant figures in it. *(Objs #6 and 7)*

   a.   The approximate volume of the ocean, $\textbf{1.5} \times \textbf{10}^{\textbf{21}}$ L.

   The value 1.5×10^{21} must have come from an estimate or a calculation, so it is **not exact** and has **two significant figures.**

   b.   A count of **24** instructors in the physical sciences division of a state college

   Counting leads to exact values. 24 is **exact.**

   c.   The **54%** of the instructors in the physical sciences division who are women (determined by counting 13 women in the total of 24 instructors and then calculating the percentage)

   Even if a value is calculated from exact values, if the answer is rounded off, the rounded answer is not exact. 13 divided by 24 and multiplied by 100 yields 54.166667 on a typical calculator. Thus, the 54% includes a value that was rounded off. The 54 is **not exact** and has **two significant figures.**

   d.   The **25%** of the instructors in the physical sciences division who are left-handed (determined by counting 6 left-handed instructors in the total of 24 and then calculating the percentage)

   Values calculated from exact values and not rounded off are exact. The 25 is **exact**.

   e. $\dfrac{16 \; oz}{1 \; lb}$

   English-English conversion factors with units of the same type of measurement top and bottom come from definitions, so the values in them are exact. 16 is **exact**.

   f. $\dfrac{10^6 \; mm}{1 \; m}$

   Metric-metric conversion factors with units of the same type of measurement top and bottom come from definitions, so the values in them are exact. 10^6 is **exact**.

g. $\dfrac{1.057 \text{ qt}}{1 \text{ L}}$

Except for $\dfrac{2.54 \text{ cm}}{1 \text{ in.}}$, the values in English-metric conversion factors that we will see are calculated and rounded off. They are **not exact**. 1.057 is **four significant figures**.

h. A measurement of **107.200** g water

Measurements never lead to exact values. Zeros in numbers that have decimal points are significant, and zeros between nonzero digits are significant. 107.200 is **not exact** and is **six significant figures**.

i. A mass of **0.2363** lb water [calculated from Part (h), using $\dfrac{453.6 \text{ g}}{1 \text{ lb}}$ as a conversion factor]

Measurements never lead to exact values. Thus, values calculated from measurements are not exact. 0.2363 is **not exact** and has **four significant figures**.

j. A mass of **1.182×10^{-4}** tons [calculated from the 0.2363 lb of the water described in Part (i).]

Values calculated from nonexact values are not exact. 1.182×10^{-4} is **not exact** and has **four significant figures**.

52. Assuming that each of the following are not exact, how many significant figures does each number have? *(Obj #7)*

a. 13.811   **5**

b. 0.0445   **3**

c. 505   **3**

d. 9.5004   **5**

e. 81.00   **4**

54. Assuming that each of the following are not exact, how many significant figures does each number have? *(Obj #7)*

a. 4.75×10^{23} **3**

b. 3.009×10^{-3} **4**

c. 4.000×10^{13} **4**

56. Convert each of the following to a number with 3 significant figures.

a. 34.579   **34.6**

b. 193.405   **193**

c. 23.995   **24.0**

d. 0.003882   **0.00388**

e. 0.023   **0.0230**

f. 2,846.5   **2.85×10^3**

g. 7.8354×10^4 **7.84×10^4**

58. Complete the following calculations and report your answers to the correct number of significant figures. The exponential factors, such as 10^3, are exact, and the 2.54 in Part (c) is exact. All other numbers are not exact. *(Obj #8)*

a. $$\frac{2.45 \times 10^{-5}\left(10^{12}\right)}{\left(10^3\right)237.00} = \textbf{103}$$

b. $$\frac{16.050\left(10^3\right)}{(24.8-19.4)(1.057)(453.6)} = \textbf{6.2}$$ Note that the subtraction yields 5.4, which limits the final number of significant figures to 2.

c. $$\frac{4.77 \times 10^{11}\left(2.54\right)^3\left(73.00\right)}{\left(10^3\right)} = \textbf{5.71} \times \textbf{10}^{\textbf{11}}$$

60. Report the answers to the following calculations to the correct number of decimal places. Assume that each number is ±1 in the last decimal place reported. *(Obj #9)*

a. $0.8995 + 99.24 = \textbf{100.14}$ b. $88 - 87.3 = \textbf{1}$

Section 8.3 Density and Density Calculations

62. A piece of balsa wood has a mass of 15.196 g and a volume of 0.1266 L. What is its density in g/mL?   *(Obj #12)*

$$\frac{?\,g}{mL} = \frac{15.196\,g}{0.1266\,L}\left(\frac{1\,L}{10^3\,mL}\right) = \textbf{0.1200 g/mL}$$

64. The density of water at 0 °C is 0.99987 g/mL. What is the mass in kilograms of 185.0 mL of water?   *(Obj #11)*

$$?\,kg = 185.0\,mL\left(\frac{0.99987\,g}{1\,mL}\right)\left(\frac{1\,kg}{10^3\,g}\right) = \textbf{0.1850 kg}$$

66. The density of a piece of ebony wood is 1.174 g/mL. What is the volume in quarts of a 2.1549-lb piece of this ebony wood?   *(Obj #11)*

$$?\,qt = 2.1549\,lb\left(\frac{453.6\,g}{1\,lb}\right)\left(\frac{1\,mL}{1.174\,g}\right)\left(\frac{1\,L}{10^3\,mL}\right)\left(\frac{1\,gal}{3.785\,L}\right)\left(\frac{4\,qt}{1\,gal}\right) = \textbf{0.8799 qt}$$

Section 8.4 Percentage and Percentage Calculations

68. The mass of the ocean is about 1.8×10^{21} kg. If the ocean contains 1.076% by mass sodium ions, Na^+, what is the mass in kilograms of Na^+ in the ocean?   *(Obj #14)*

$$?\,kg\,Na^+ = 1.8 \times 10^{21}\,kg\,ocean\left(\frac{1.076\,kg\,Na^+}{100\,kg\,ocean}\right) = \textbf{1.9} \times \textbf{10}^{\textbf{19}}\,\textbf{kg Na}^+$$

70. While you are doing heavy work, your heart pumps up to 25.0 L of blood per minute. Your brain gets about 3-4% by volume of your blood under these conditions. What volume of blood in liters is pumped through your brain in 125 minutes of work that causes your heart to pump 22.0 L per minute, 3.43% of which goes to your brain? *(Obj #14)*

$$? \text{ L to brain} = 125 \text{ min} \left(\frac{22.0 \text{ L total}}{1 \text{ min}} \right) \left(\frac{3.43 \text{ L to brain}}{100 \text{ L total}} \right) = \textbf{94.3 L to brain}$$

72. In chemical reactions that release energy, from 10^{-8}% to 10^{-7}% of the mass of the substances involved is converted into energy. Consider a chemical reaction for which 1.8×10^{-8}% of the mass is converted into energy. What mass in milligrams is converted into energy when 1.0×10^3 kilograms of substance reacts? *(Obj #14)*

$$? \text{ mg to energy} = 1.0 \times 10^3 \text{ kg reacts} \left(\frac{1.8 \times 10^{-8} \text{ kg to energy}}{100 \text{ kg reacts}} \right) \left(\frac{10^3 \text{ g}}{1 \text{ kg}} \right) \left(\frac{10^3 \text{ mg}}{1 \text{ g}} \right)$$

$$= \textbf{0.18 mg to energy}$$

Section 8.5 A Summary of the Dimensional Analysis Process

74. If an elevator moves 1340 ft to the 103rd floor of the Sears Tower in Chicago in 45 seconds, what is the velocity of the elevator in kilometers per hour? *(Obj #15)*

$$\frac{? \text{ km}}{\text{hr}} = \frac{1340 \text{ ft}}{45 \text{ s}} \left(\frac{12 \text{ in.}}{1 \text{ ft}} \right) \left(\frac{2.54 \text{ cm}}{1 \text{ in.}} \right) \left(\frac{1 \text{ m}}{10^2 \text{ cm}} \right) \left(\frac{1 \text{ km}}{10^3 \text{ m}} \right) \left(\frac{60 \text{ s}}{1 \text{ min}} \right) \left(\frac{60 \text{ min}}{1 \text{ hr}} \right)$$

$$= \textbf{33 km/hr}$$

76. Sound travels at a velocity of 333 m/s. How long does it take for sound to travel the length of a 100 yard football field? *(Obj #15)*

$$? \text{ s} = 100 \text{ yd} \left(\frac{3 \text{ ft}}{1 \text{ yd}} \right) \left(\frac{12 \text{ in.}}{1 \text{ ft}} \right) \left(\frac{2.54 \text{ cm}}{1 \text{ in.}} \right) \left(\frac{1 \text{ m}}{10^2 \text{ cm}} \right) \left(\frac{1 \text{ s}}{333 \text{ m}} \right) = \textbf{0.275 s}$$

78. A peanut butter sandwich provides about 1.4×10^3 kJ of energy. A typical adult uses about 95 kcal/hr of energy while sitting. If all of the energy in one peanut butter sandwich were to be burned off by sitting, how many hours would it be before this energy was used? (A kcal is a dietary calorie. There are 4.184 J/cal.) *(Obj #15)*

$$? \text{ hr} = 1.4 \times 10^3 \text{ kJ} \left(\frac{1 \text{ kcal}}{4.184 \text{ kJ}} \right) \left(\frac{1 \text{ hr}}{95 \text{ kcal}} \right) = \textbf{3.5 hr}$$

80. When one gram of hydrogen gas, $H_2(g)$, is burned, 141.8 kJ of heat is released. How much heat is released when 2.3456 kg of hydrogen gas is burned? *(Obj #15)*

$$? \text{ kJ} = 2.3456 \text{ kg } H_2 \left(\frac{10^3 \text{ g}}{1 \text{ kg}} \right) \left(\frac{141.8 \text{ kJ}}{1 \text{ g } H_2} \right) = \textbf{3.326} \times \textbf{10}^5 \textbf{ kJ}$$

82. When one gram of carbon in the graphite form is burned, 32.8 kJ of heat is released. How many kilograms of graphite must be burned to release 1.456×10^4 kJ of heat? *(Obj #15)*

$$? \text{ kg C} = 1.456 \times 10^4 \text{ kJ} \left(\frac{1 \text{ g C}}{32.8 \text{ kJ}} \right) \left(\frac{1 \text{ kg}}{10^3 \text{ g}} \right) = \textbf{0.444 kg C}$$

84. The average adult male needs about 58 g of protein in the diet each day. A can of vegetarian refried beans has 6.0 g of protein per serving. Each serving is 128 g of beans. If your only dietary source of protein were vegetarian refried beans, how many pounds of beans would you need to eat each day? *(Obj #15)*

$$\frac{?\text{ lb beans}}{\text{day}} = \frac{58 \text{ g protein}}{1 \text{ day}} \left(\frac{1 \text{ serving}}{6.0 \text{ g protein}} \right) \left(\frac{128 \text{ g beans}}{1 \text{ serving}} \right) \left(\frac{1 \text{ lb}}{453.6 \text{ g}} \right)$$

= 2.7 lb beans per day

86. About 6.0×10^5 tons of 30% by mass hydrochloric acid, HCl(*aq*), is used each year to remove metal oxides from metals to prepare them for painting or for the addition of a chrome covering. How many kilograms of pure HCl would be used to make this hydrochloric acid? (Assume that 30% has two significant figures. There are 2000 lb/ton.) *(Obj #15)*

$$?\text{ kg HCl} = 6.0 \times 10^5 \text{ ton HCl soln} \left(\frac{30 \text{ ton HCl}}{100 \text{ ton HCl soln}} \right) \left(\frac{2000 \text{ lb}}{1 \text{ ton}} \right) \left(\frac{1 \text{ kg}}{2.205 \text{ lb}} \right)$$

$$?\text{ kg HCl} = 6.0 \times 10^5 \text{ ton soln} \left(\frac{30 \text{ ton HCl}}{100 \text{ ton soln}} \right) \left(\frac{2000 \text{ lb}}{1 \text{ ton}} \right) \left(\frac{453.6 \text{ g}}{1 \text{ lb}} \right) \left(\frac{1 \text{ kg}}{10^3 \text{ g}} \right)$$

= 1.6×10^8 kg HCl

88. A typical nonobese male has about 11 kg of fat. Each gram of fat can provide the body with about 38 kJ of energy. If this person requires 8.0×10^3 kJ of energy per day to survive, how many days could he survive on his fat alone? *(Obj #15)*

$$?\text{ day} = 11 \text{ kg fat} \left(\frac{10^3 \text{ g}}{1 \text{ kg}} \right) \left(\frac{38 \text{ kJ}}{1 \text{ g}} \right) \left(\frac{1 \text{ day}}{8.0 \times 10^3 \text{ kJ}} \right) = \textbf{52 days}$$

90. During quiet breathing, a person breathes in about 6 L of air per minute. If a person breathes in an average of 6.814 L of air per minute, what volume of air in liters does this person breathe in 1 day? *(Obj #15)*

$$?\text{ L} = 1 \text{ day} \left(\frac{24 \text{ hr}}{1 \text{ day}} \right) \left(\frac{60 \text{ min}}{1 \text{ hr}} \right) \left(\frac{6.814 \text{ L air}}{1 \text{ min}} \right) = \textbf{9812 L air}$$

92. The kidneys of a normal adult female filter 115 mL of blood per minute. If this person has 5.345 quarts of blood, how many minutes will it take to filter all of her blood once? *(Obj #15)*

$$?\text{ min} = 5.345 \text{ qt blood} \left(\frac{1 \text{ gal}}{4 \text{ qt}} \right) \left(\frac{3.785 \text{ L}}{1 \text{ gal}} \right) \left(\frac{10^3 \text{ mL}}{1 \text{ L}} \right) \left(\frac{1 \text{ min}}{115 \text{ mL blood}} \right)$$

$$\text{or}\quad ?\text{ min} = 5.345 \text{ qt blood} \left(\frac{1 \text{ L}}{1.057 \text{ qt}} \right) \left(\frac{10^3 \text{ mL}}{1 \text{ L}} \right) \left(\frac{1 \text{ min}}{115 \text{ mL blood}} \right) = \textbf{44.0 minutes}$$

94. We lose between 0.2 and 1 liter of water from our skin and sweat glands each day. For a person who loses an average of 0.89 L H_2O/day in this manner, how many quarts of water are lost from the skin and sweat glands in 30 days? *(Obj #15)*

$$? \, qt = 30 \, day \left(\frac{0.89 \, L}{1 \, day} \right) \left(\frac{1 \, gal}{3.785 \, L} \right) \left(\frac{4 \, qt}{1 \, gal} \right)$$

$$or \quad ? \, qt = 30 \, day \left(\frac{0.89 \, L}{1 \, day} \right) \left(\frac{1.057 \, qt}{1 \, L} \right) = \textbf{28 qt}$$

96. The average heart rate is 75 beats/min. How many times does the average person's heart beat in a week? *(Obj #15)*

$$? \, beats = 1 \, week \left(\frac{7 \, day}{1 \, week} \right) \left(\frac{24 \, hr}{1 \, day} \right) \left(\frac{60 \, min}{1 \, hr} \right) \left(\frac{75 \, beats}{1 \, min} \right) = \textbf{7.6} \times \textbf{10}^{5} \textbf{ beats}$$

98. In optimum conditions, one molecule of the enzyme carbonic anhydrase can convert 3.6×10^5 molecules per minute of carbonic acid, H_2CO_3, into carbon dioxide, CO_2, and water, H_2O. How many molecules could be converted by one of these enzyme molecules in 1 week? *(Obj #15)*

$$? \, molecules = 1 \, week \left(\frac{7 \, day}{1 \, week} \right) \left(\frac{24 \, hr}{1 \, day} \right) \left(\frac{60 \, min}{1 \, hr} \right) \left(\frac{3.6 \times 10^5 \, molecules}{1 \, min} \right)$$

$$= \textbf{3.6} \times \textbf{10}^{9} \textbf{ molecules}$$

100. In optimum conditions, one molecule of the enzyme amylase can convert 1.0×10^5 molecules per minute of starch into the sugar maltose. How many days would it take one of these enzyme molecules to convert a billion (1.0×10^9) starch molecules? *(Obj #15)*

$$? \, days = 1.0 \times 10^9 \, molecules \left(\frac{1 \, min}{1.0 \times 10^5 \, molecules} \right) \left(\frac{1 \, hr}{60 \, min} \right) \left(\frac{1 \, day}{24 \, hr} \right) = \textbf{6.9 days}$$

102. When you sneeze, you close your eyes for about 1.00 second. If you are driving 65 miles per hour on the freeway and you sneeze, how many feet do you travel with your eyes closed? *(Obj #15)*

$$? \, ft = 1.00 \, s \left(\frac{1 \, min}{60 \, s} \right) \left(\frac{1 \, hr}{60 \, min} \right) \left(\frac{65 \, mi}{1 \, hr} \right) \left(\frac{5280 \, ft}{1 \, mi} \right) = \textbf{95 ft}$$

Section 8.6 Temperature Conversions

103. Butter melts at 31 °C. What is this temperature in °F? in K? *(Obj #16)*

$$°F = 31 \, °C \left(\frac{1.8 \, °F}{1 \, °C} \right) + 32 \, °F = \textbf{88 °F} \qquad K = 31 \, °C + 273.15 = \textbf{304 K}$$

105. A saturated salt solution boils at 226 °F. What is this temperature in °C? in K? *(Obj #16)*

$$°C = (226 \, °F - 32 \, °F) \left(\frac{1 \, °C}{1.8 \, °F} \right) = \textbf{108°C} \qquad K = 108 \, °C + 273.15 = \textbf{381 K}$$

107. Iron boils at 3023 K. What is this temperature in °C? in °F ? *(Obj #16)*

$$°C = 3023 \text{ K} - 273.15 = \mathbf{2.750 \times 10^3 \text{ °C}}$$

$$°F = 2.750 \times 10^3 \text{ °C} \left(\frac{1.8 \text{ °F}}{1 \text{ °C}} \right) + 32 \text{ °F} = \mathbf{4982 \text{ °F}}$$

109. The surface of the sun is 1.0×10^4 °F. What is this temperature in °C? in K? *(Obj #16)*

$$°C = \left(1.0 \times 10^4 \text{ °F} - 32 \text{ °F} \right) \left(\frac{1 \text{ °C}}{1.8 \text{ °F}} \right) = \mathbf{5.5 \times 10^3 \text{ °C}}$$

$$K = 5.5 \times 10^3 \text{ °C} + 273.15 = \mathbf{5.8 \times 10^3 \text{ K}}$$

Chapter 9
Chemical Calculations and Chemical Formulas

molar mass O = 15.9994 g/mol   molar mass H = 1.00794 g/mol

molar mass H_2O = 18.0153 g/mol

♦ Review Skills
9.1 A Typical Problem
9.2 Relating Mass to Number of Particles
- Atomic Mass and Counting Atoms by Weighing
- Molar Mass
9.3 Molar Mass and Chemical Compounds
- Molecular Mass and Molar Mass of Molecular Compounds
- Ionic Compounds, Formula Units, and Formula Mass
9.4 Relationships Between Masses of Elements and Compounds
  Internet: Percentage of an Element in a Compound
9.5 Determination of Empirical and Molecular Formulas
- Determining Empirical Formulas
- Converting Empirical Formulas Into Molecular Formulas
  Special Topic 9.1: Green Chemistry - Making Chemicals from Safer Reactants
  Internet: Combustion Analysis
  Special Topic 9.2: Safe and Effective?
♦ Chapter Glossary
  Internet: Glossary Quiz
♦ Chapter Objectives
Review Questions
Key Ideas
Chapter Problems

Section Goals and Introductions

Section 9.1 A Typical Problem

Goal: To introduce the chapter by describing a typical problem that you will be able to work after studying the chapter.

Sometimes a task becomes much easier if you know from the beginning where you are going to end up. It will help you to understand the importance of Sections 9.2 and 9.3 if you first know how the calculations described there can be used. This section shows a typical problem and gives you a sense of why it is important.

Section 9.2 Relating Mass to Number of Particles

Goals

- *To show how to do a procedure called counting by weighing.*
- *To introduce atomic mass and show how it can be used to convert between the mass of a sample of an element and the number of atoms that the sample contains.*

Even a tiny sample of an element contains a huge number of atoms. There's no way that you could ever count that high, even if you were able to count atoms one at a time (which you can't). So if you want to know the number of atoms in a sample of an element, you have to do it by an indirect technique called *counting by weighing*. This section introduces this technique and shows how it can be applied to the conversions between mass of a sample of an element and the number of atoms in the sample. An important unit called the mole is introduced in this section. It is very important that you understand what it is and how it is used.

Section 9.3 Molar Mass and Chemical Compounds

Goal: To introduce molecular mass and formula mass and show how they can be used to convert between the mass of a sample of a compound and the number of molecules or formula units that the sample contains.

This section shows how to calculate the number of molecules (expressed in moles) in a sample of a molecular compound from the mass of that sample and how to calculate the mass in a sample of a molecular compound from the moles of molecules it contains. The section also explains why ionic compounds do not contain molecules and how the term formula unit can be used to describe the units of ionic compounds that are like molecules of molecular compounds. Then you will see how to calculate the number of formula units in a sample of an ionic compound (expressed in moles) from the mass of that sample and how to calculate the mass of an ionic compound and the moles of formula units it contains. These calculations are very commonly done by chemists and chemistry students, so be sure you can do them quickly and correctly.

Section 9.4 Relationships Between Masses of Elements and Compounds

Goal: To show how you convert between mass of an element and mass of a compound that contains the element.

This section shows how you can use the skills you learned in Section 9.2 and 9.3 and some information derived from chemical formulas for compounds to convert between mass of an element and mass of a compound that contains the element.

The section on our Web site called *Percentage of an Element in a Compound* describes calculations that are related to this section.

www.chemplace.com/college/

Section 9.5 Determination of Empirical and Molecular Formulas

Goal: To describe what empirical and molecular formulas are and how they can be determined.

All compounds can be described with empirical formulas, and molecular compounds can also be described with molecular formulas. This section describes the information given by each type of formula and shows ways to determine them.

The section on our Web site called *Combustion Analysis* describes how empirical formulas can be determined.

www.chemplace.com/college/

Chapter 9 Map

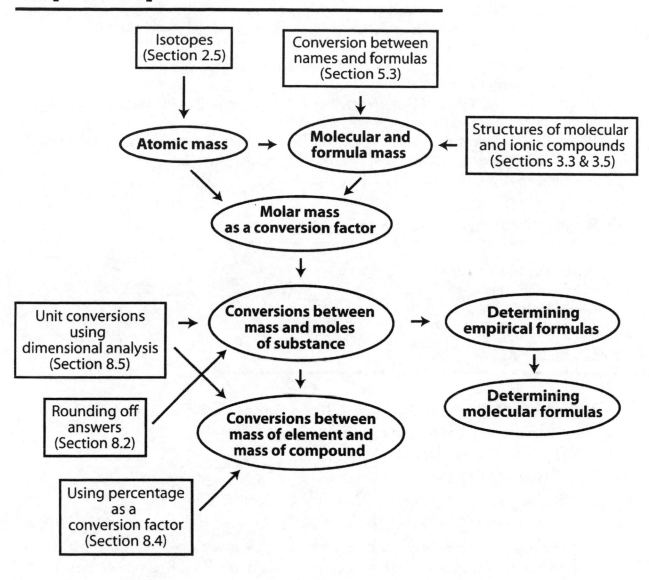

Chapter Checklist

- ☐ Read the Review Skills section. If there is any skill mentioned that you have not yet mastered, review the material on that topic before reading this chapter.
- ☐ Read the chapter quickly before the lecture that describes it.
- ☐ Attend class meetings, take notes, and participate in class discussions.
- ☐ Work the Chapter Exercises, perhaps using the Chapter Examples as guides.
- ☐ Study the Chapter Glossary and test yourself on our Web site:
     www.chemplace.com/college/
- ☐ Study all of the Chapter Objectives. You might want to write a description of how you will meet each objective. (Although it is best to master all of the objectives, the following objectives are especially important because they pertain to skills that you will need while studying other chapters of this text: 4, 5, 7, 10, and 12.)

☐ Reread the Study Sheets in this chapter and decide whether you will use them or some variation on them to complete the tasks they describe.

Sample Study Sheet 9.1: Converting Between Mass of Element and Mass of Compound Containing the Element

Sample Study Sheet 9.2: Calculating Empirical Formulas

Sample Study Sheet 9.3: Calculating Molecular Formulas

☐ To get a review of the most important topics in the chapter, fill in the blanks in the Key Ideas section.

☐ Work all of the selected problems at the end of the chapter, and check your answers with the solutions provided in this chapter of the study guide.

☐ Ask for help if you need it.

Web Resources   www.chemplace.com/college/

Percentage of an Element in a Compound
Combustion Analysis
Glossary Quiz

Exercises Key

Exercise 9.1 - Atomic Mass Calculations: Gold is often sold in units of troy ounces. There are 31.10 grams per troy ounce. *(Objs #3 & 4)*

a. What is the atomic mass of gold?

   196.9665 (from periodic table)

b. What is the mass in grams of 6.022×10^{23} gold atoms?

   196.9665 g (There are 6.022×10^{23} atoms per mole of atoms, and one mole of an element has a mass in grams equal to its atomic mass.)

c. Write the molar mass of gold as a conversion factor that can be used to convert between grams of gold and moles of gold.

   $$\left(\frac{196.9665 \text{ g Au}}{1 \text{ mol Au}} \right)$$

d. What is the mass in grams of 0.20443 mole of gold?

   $$? \text{ g Au} = 0.20443 \text{ mol Au} \left(\frac{196.9665 \text{ g Au}}{1 \text{ mol Au}} \right) = \textbf{40.266 g Au}$$

e. What is the mass in milligrams of 7.046×10^{-3} mole of gold?

   $$? \text{ mg Au} = 7.046 \times 10^{-3} \text{ mol Au} \left(\frac{196.9665 \text{ g Au}}{1 \text{ mol Au}} \right) \left(\frac{10^{3} \text{ mg}}{1 \text{ g}} \right) = \textbf{1388 mg Au}$$

f. How many moles of gold are in 1.00 troy ounce of pure gold?

   $$? \text{ mol Au} = 1.00 \text{ troy oz Au} \left(\frac{31.10 \text{ g}}{1 \text{ troy oz}} \right) \left(\frac{1 \text{ mol Au}}{196.9665 \text{ g Au}} \right) = \textbf{0.158 mol Au}$$

Exercise 9.2 - Molecular Mass Calculations: A typical glass of wine contains about 16 g of ethanol, C_2H_5OH.   *(Objs #5-7)*

    a.   What is the molecular mass of C_2H_5OH?

          $2(12.011) + 6(1.00794) + 1(15.9994) =$ **46.069**

    b.   What is the mass of 1 mole of C_2H_5OH?

          46.069 g *(One mole of a molecular compound has a mass in grams equal to its molecular mass.)*

    c.   Write a conversion factor that will convert between mass and moles of C_2H_5OH.

$$\left(\frac{46.069 \text{ g } C_2H_5OH}{1 \text{ mol } C_2H_5OH} \right)$$

    d.   How many moles of ethanol are there in 16 grams of C_2H_5OH?

$$? \text{ mol } C_2H_5OH = 16 \text{ g } C_2H_5OH \left(\frac{1 \text{ mol } C_2H_5OH}{46.069 \text{ g } C_2H_5OH} \right) = \textbf{0.35 mole } C_2H_5OH$$

    e.   What is the volume in milliliters of 1.0 mole of pure C_2H_5OH? (The density of ethanol is 0.7893 g/mL.)

$$? \text{ mL } C_2H_5OH = 1.0 \text{ mol } C_2H_5OH \left(\frac{46.069 \text{ g } C_2H_5OH}{1 \text{ mol } C_2H_5OH} \right) \left(\frac{1 \text{ mL } C_2H_5OH}{0.7893 \text{ g } C_2H_5OH} \right)$$

$$= \textbf{58 mL } C_2H_5OH$$

Exercise 9.3 - Formula Mass Calculations: A quarter teaspoon of a typical baking powder contains about 0.4 g of sodium hydrogen carbonate, $NaHCO_3$.

    a.   Calculate the formula mass of sodium hydrogen carbonate. *(Objs #10-12)*

          Formula Mass $= 1(22.9898) + 1(1.00794) + 1(12.011) + 3(15.9994)$

                    $=$ **84.007**

    b.   What is the mass in grams of 1 mole of $NaHCO_3$?

          84.007 g *(One mole of an ionic compound has a mass in grams equal to its formula mass.)*

    c.   Write a conversion factor to convert between mass and moles of $NaHCO_3$.

$$\left(\frac{84.007 \text{ g } NaHCO_3}{1 \text{ mol } NaHCO_3} \right)$$

    d.   How many moles of $NaHCO_3$ are in 0.4 g of $NaHCO_3$?

$$? \text{ mol } NaHCO_3 = 0.4 \text{ g } NaHCO_3 \left(\frac{1 \text{ mol } NaHCO_3}{84.007 \text{ g } NaHCO_3} \right) = \textbf{5} \times \textbf{10}^{-3} \textbf{ mol } NaHCO_3$$

✍ **Exercise 9.4 - Molar Ratios of Element to Compound:**   Find the requested conversion factors. *(Objs #13 & 14)*

a. Write a conversion factor that converts between moles of hydrogen and moles of C_2H_5OH.

$$\frac{6 \text{ mol H}}{1 \text{ mol C}_2\text{H}_5\text{OH}}$$

b. Write a conversion factor that converts between moles of oxygen and moles of $NaHCO_3$.

$$\frac{3 \text{ mol O}}{1 \text{ mol NaHCO}_3}$$

c. How many moles of hydrogen carbonate ions, HCO_3^-, are there in 1 mole of $NaHCO_3$?

There is 1 mole of HCO_3^- per 1 mole of $NaHCO_3$.

✍ **Exercise 9.5 - Molar Mass Calculations:**   Disulfur dichloride, S_2Cl_2, is used in vulcanizing rubber and in hardening soft woods. It can be made from the reaction of pure sulfur with chlorine gas. What is the mass of S_2Cl_2 that contains 123.8 g S?

$$? \text{ g } S_2Cl_2 = 123.8 \text{ g } S \left(\frac{1 \text{ mol } S}{32.066 \text{ g } S} \right) \left(\frac{1 \text{ mol } S_2Cl_2}{2 \text{ mol } S} \right) \left(\frac{135.037 \text{ g } S_2Cl_2}{1 \text{ mole } S_2Cl_2} \right) = \mathbf{260.7 \text{ g } S_2Cl_2}$$

✍ **Exercise 9.6 - Molar Mass Calculations:**   Vanadium metal, which is used as a component of steel and to catalyze various industrial reactions, is produced from the reaction of vanadium(V) oxide, V_2O_5, and calcium metal. What is the mass (in kilograms of vanadium) in 2.3 kilograms of V_2O_5?   *(Obj #15)*

$$? \text{ kg } V = 2.3 \text{ kg } V_2O_5 \left(\frac{10^3 \text{ g}}{1 \text{ kg}} \right) \left(\frac{1 \text{ mol } V_2O_5}{181.880 \text{ g } V_2O_5} \right) \left(\frac{2 \text{ mol } V}{1 \text{ mol } V_2O_5} \right) \left(\frac{50.9415 \text{ g } V}{1 \text{ mol } V} \right) \left(\frac{1 \text{ kg}}{10^3 \text{ g}} \right)$$

$$= \mathbf{1.3 \text{ kg } V}$$

✍ **Exercise 9.7 – Calculating an Empirical Formula:**   Bismuth ore, often called bismuth glance, contains an ionic compound that consists of the elements bismuth and sulfur. A sample of the pure compound is found to contain 32.516 g Bi and 7.484 g S. What is the empirical formula for this compound? What is its name?   *(Obj #16)*

$$? \text{ mol Bi} = 32.516 \text{ g Bi} \left(\frac{1 \text{ mol Bi}}{208.9804 \text{ g Bi}} \right) = 0.15559 \text{ mol Bi} \div 0.15559$$

$$= 1 \text{ mol Bi} \times 2 = 2 \text{ mol Bi}$$

$$? \text{ mol S} = 7.484 \text{ g S} \left(\frac{1 \text{ mol S}}{32.066 \text{ g S}} \right) = 0.2334 \text{ mol S} \div 0.15559$$

$$\cong 1\frac{1}{2} \text{ mol S} \times 2 = 3 \text{ mol S}$$

Our empirical formula is $\mathbf{Bi_2S_3}$ or bismuth(III) sulfide.

✍ Exercise 9.8 - Calculating an Empirical Formula: An ionic compound used in the brewing industry to clean casks and vats and in the wine industry to kill undesirable yeasts and bacteria is composed of 35.172% potassium, 28.846% sulfur, and 35.982% oxygen. What is the empirical formula for this compound? *(Obj #17)*

$$? \text{ mol K} = 35.172 \text{ g K} \left(\frac{1 \text{ mol K}}{39.0983 \text{ g K}} \right) = 0.89958 \text{ mol K} \div 0.89958$$

$$= 1 \text{ mol K} \times 2 = 2 \text{ mol K}$$

$$? \text{ mol S} = 28.846 \text{ g S} \left(\frac{1 \text{ mol S}}{32.066 \text{ g S}} \right) = 0.89958 \text{ mol S} \div 0.89958$$

$$= 1 \text{ mol S} \times 2 = 2 \text{ mol S}$$

$$? \text{ mol O} = 35.982 \text{ g O} \left(\frac{1 \text{ mol O}}{15.9994 \text{ g O}} \right) = 2.2490 \text{ mol O} \div 0.89958$$

$$\cong 2\frac{1}{2} \text{ mol O} \times 2 = 5 \text{ mol O}$$

Empirical Formula $K_2S_2O_5$

✍ Exercise 9.9 – Calculating a Molecular Formula Using the Percentage of Each Element in a Compound: Compounds called polychlorinated biphenyls (PCBs) have structures similar to chlorinated insecticides such as DDT. They have been used in the past for a variety of purposes, but because they have been identified as serious pollutants, their only legal use today is as insulating fluids in electrical transformers. This is a use for which no suitable substitute has been found. One PCB is 39.94% carbon, 1.12% hydrogen, and 58.94% chlorine and has a molecular mass of 360.88. What is its molecular formula? *(Obj #19)*

$$? \text{ mol C} = 39.94 \text{ g C} \left(\frac{1 \text{ mol C}}{12.011 \text{ g C}} \right) = 3.325 \text{ mol C} \div 1.11 = 3 \text{ mol C} \times 2 = 6 \text{ mol C}$$

$$? \text{ mol H} = 1.12 \text{ g H} \left(\frac{1 \text{ mol H}}{1.00794 \text{ g H}} \right) = 1.11 \text{ mol H} \div 1.11 = 1 \text{ mol H} \times 2 = 2 \text{ mol H}$$

$$? \text{ mol Cl} = 58.94 \text{ g Cl} \left(\frac{1 \text{ mol Cl}}{35.4527 \text{ g Cl}} \right) = 1.662 \text{ mol Cl} \div 1.11 \cong 1\frac{1}{2} \text{ mol Cl} \times 2 = 3 \text{ mol Cl}$$

Empirical Formula $C_6H_2Cl_3$ $n = \dfrac{\text{molecular mass}}{\text{empirical formula mass}} = \dfrac{360.88}{180.440} \cong 2$

Molecular Formula $= (C_6H_2Cl_3)_2$ or $\mathbf{C_{12}H_4Cl_6}$

Review Questions Key

1. Complete each of the following conversion factors by filling in the blank on the top of the ratio.

   a. $\left(\dfrac{10^3 \text{ g}}{1 \text{ kg}}\right)$ c. $\left(\dfrac{10^3 \text{ kg}}{1 \text{ metric ton}}\right)$

   b. $\left(\dfrac{10^3 \text{ mg}}{1 \text{ g}}\right)$ d. $\left(\dfrac{10^6 \text{ μg}}{1 \text{ g}}\right)$

2. Convert 3.45×10^4 kg into grams.

   $$? \text{ g} = 3.45 \times 10^4 \text{ kg} \left(\frac{10^3 \text{ g}}{1 \text{ kg}}\right) = \mathbf{3.45 \times 10^7 \text{ g}}$$

3. Convert 184.570 g into kilograms.

   $$? \text{ kg} = 184.570 \text{ g} \left(\frac{1 \text{ kg}}{10^3 \text{ g}}\right) = \mathbf{0.184570 \text{ kg}}$$

4. Convert 4.5000×10^6 g into megagrams.

   $$? \text{ Mg} = 4.5000 \times 10^6 \text{ g} \left(\frac{1 \text{ kg}}{10^3 \text{ g}}\right)\left(\frac{1 \text{ Mg}}{10^3 \text{ kg}}\right) = \mathbf{4.5000 \text{ Mg}}$$

5. Convert 871 Mg into grams.

   $$? \text{ g} = 871 \text{ Mg} \left(\frac{10^3 \text{ kg}}{1 \text{ Mg}}\right)\left(\frac{10^3 \text{ g}}{1 \text{ kg}}\right) = \mathbf{8.71 \times 10^8 \text{ g}}$$

6. Surinam bauxite is an ore that is 54-57% aluminum oxide, Al_2O_3. What is the mass (in kilograms) of Al_2O_3 in 1256 kg of Surinam bauxite that is 55.3% Al_2O_3?

   $$? \text{ kg } Al_2O_3 = 1256 \text{ kg Surinam bauxite} \left(\frac{55.3 \text{ kg } Al_2O_3}{100 \text{ kg Surinam bauxite}}\right) = \mathbf{695 \text{ kg } Al_2O_3}$$

Key Ideas Answers

7. Because of the size and number of carbon atoms in any normal sample of carbon, it is **impossible** to count the atoms directly.

9. The atomic mass of any element is the **weighted** average of the masses of the **naturally** occurring isotopes of the element.

11. The number of grams in the molar mass of an element is the same as the element's **atomic mass**.

13. In this text, the term **formula unit** will be used to describe ionic compounds in situations where molecule is used to describe molecular substances. It is the group represented by the substance's chemical formula, that is, a group containing the **kinds** and **numbers** of atoms or ions listed in the chemical formula.

15. Formula mass is the weighted average of the masses of the naturally occurring **formula units** of the substance.

17. When the subscripts in a chemical formula represent the **simplest** ratio of the kinds of atoms in the compound, the formula is called an empirical formula.

19. The subscripts in a molecular formula are always **whole-number** multiples of the subscripts in the empirical formula.

Problems Key

Section 9.2 Relating Mass to Numbers of Particles

21. What is the weighted average mass in atomic mass units (u) of each atom of the following elements?

The atomic mass for each element in the periodic table tells you the weighted average mass in atomic mass units (u) of each atom of that element.

    a. sodium   **22.9898 u**   b. oxygen   **15.9994 u**

23. What is the weighted average mass in grams of 6.022×10^{23} atoms of the following elements?

The atomic mass for each element in the periodic table tells you the mass, in grams, of 6.022×10^{23} atoms of that element.

    a. sulfur   **32.066 g**   b. fluorine   **18.9984 g**

25. What is the molar mass for each of the following elements?

The atomic mass for each element in the periodic table tells you the molar mass, in grams per mole, of that element.

    a. zinc   **65.39 g/mol**   b. aluminum   **26.9815 g/mol**

27. For each of the following elements, write a conversion factor that converts between mass in grams and moles of the substance.   *(Obj #3)*

    a. iron $\left(\dfrac{55.845 \text{ g Fe}}{1 \text{ mol Fe}} \right)$

    b. krypton $\left(\dfrac{83.80 \text{ g Kr}}{1 \text{ mol Kr}} \right)$

29. A vitamin supplement contains 50 micrograms of the element selenium in each tablet. How many moles of selenium does each tablet contain? *(Obj #4)*

$$? \text{ mol Se} = 50 \text{ μg Se} \left(\frac{1 \text{ g}}{10^6 \text{ μg}} \right) \left(\frac{1 \text{ mol Se}}{78.96 \text{ g Se}} \right) = \textbf{6.3} \times \textbf{10}^{-7} \textbf{ mol Se}$$

31. A multivitamin tablet contains 1.6×10^{-4} mole of iron per tablet. How many milligrams of iron does each tablet contain? *(Obj #4)*

$$? \text{ mg Fe} = 1.6 \times 10^{-4} \text{ mol Fe} \left(\frac{55.845 \text{ g Fe}}{1 \text{ mol Fe}} \right) \left(\frac{10^3 \text{ mg}}{1 \text{ g}} \right) = \textbf{8.9 mg Fe}$$

Section 9.3 Molar Mass and Chemical Compounds

33. For each of the following molecular substances, calculate its molecular mass and write a conversion factor that converts between mass in grams and moles of the substance. *(Objs #5 & 6)*

   a. H_3PO_2

   $$\text{molecular mass } H_3PO_2 = 3(1.00794) + 1(30.9738) + 2(15.9994)$$

   $$= 65.9964 \text{ leads to } \left(\frac{65.9964 \text{ g } H_3PO_2}{1 \text{ mol } H_3PO_2} \right)$$

   b. $C_6H_5NH_2$

   $$\text{molecular mass } C_6H_5NH_2 = 6(12.011) + 7(1.00794) + 1(14.0067)$$

   $$= 93.128 \text{ leads to } \left(\frac{93.128 \text{ g } C_6H_5NH_2}{1 \text{ mol } C_6H_5NH_2} \right)$$

35. Each dose of nighttime cold medicine contains 1000 mg of the analgesic acetaminophen. Acetaminophen, or N-acetyl-p-aminophenol, has the general formula C_8H_9NO. *(Obj #7)*

   a. How many moles of acetaminophen are in each dose?

   $$? \text{ mol } C_8H_9NO = 1000 \text{ mg } C_8H_9NO \left(\frac{1 \text{ g}}{10^3 \text{ mg}} \right) \left(\frac{1 \text{ mol } C_8H_9NO}{135.166 \text{ g } C_8H_9NO} \right)$$

   $$= \mathbf{7.398 \times 10^{-3} \text{ mol } C_8H_9NO}$$

   b. What is the mass in grams of 15.0 moles of acetaminophen?

   $$? \text{ g } C_8H_9NO = 15.0 \text{ mol } C_8H_9NO \left(\frac{135.166 \text{ g } C_8H_9NO}{1 \text{ mol } C_8H_9NO} \right) = \mathbf{2.03 \times 10^3 \text{ g } C_8H_9NO}$$

38. For each of the following examples, decide whether it would be better to use the term *molecule* or *formula unit*.   *(Obj #9)*

   a. Cl_2O **molecular compound - molecules**
   b. Na_2O **ionic compound – formula units**
   c. $(NH_4)_2SO_4$ **ionic compound – formula units**
   d. $HC_2H_3O_2$ **molecular compound - molecules**

40. For each of the following ionic substances, calculate its formula mass and write a conversion factor that converts between mass in grams and moles of the substance. *(Objs #10 & 11)*

   a. $BiBr_3$

   $$\text{formula mass } BiBr_3 = 1(208.9804) + 3(79.904)$$

   $$= 448.69 \text{ leads to } \left(\frac{448.69 \text{ g } BiBr_3}{1 \text{ mol } BiBr_3} \right)$$

   b. $Al_2(SO_4)_3$

   $$\text{formula mass } Al_2(SO_4)_3 = 2(26.9815) + 3(32.066) + 12(15.9994)$$

   $$= 342.154 \text{ leads to } \left(\frac{342.154 \text{ g } Al_2(SO_4)_3}{1 \text{ mol } Al_2(SO_4)_3} \right)$$

42. A common antacid tablet contains 500 mg of calcium carbonate, $CaCO_3$. *(Obj #12)*

    a. How many moles of $CaCO_3$ does each tablet contain?

$$? \text{ mol } CaCO_3 = 500 \text{ mg } CaCO_3 \left(\frac{1 g}{10^3 \text{ mg}} \right) \left(\frac{1 \text{ mol } CaCO_3}{100.087 \text{ g } CaCO_3} \right)$$

$$= \mathbf{5.00 \times 10^{-3} \text{ mol } CaCO_3}$$

    b. What is the mass in kilograms of 100.0 moles of calcium carbonate?

$$? \text{ kg } CaCO_3 = 100.0 \text{ mol } CaCO_3 \left(\frac{100.087 \text{ g } CaCO_3}{1 \text{ mol } CaCO_3} \right) \left(\frac{1 \text{ kg}}{10^3 \text{ g}} \right)$$

$$= \mathbf{10.01 \text{ kg } CaCO_3}$$

44. Rubies and other minerals in the durable corundum family are primarily composed of aluminum oxide, Al_2O_3, with trace impurities that lead to their different colors. For example, the red color in rubies comes from a small amount of chromium replacing some of the aluminum. If a 0.78-carat ruby were pure aluminum oxide, how many moles of Al_2O_3 would be in the stone? (There are exactly 5 carats per gram.) *(Obj #12)*

$$? \text{ mol } Al_2O_3 = 0.78 \text{ carat } Al_2O_3 \left(\frac{1 g}{5 \text{ carats}} \right) \left(\frac{1 \text{ mol } Al_2O_3}{101.9612 \text{ g } Al_2O_3} \right)$$

$$= \mathbf{1.5 \times 10^{-3} \text{ mol } Al_2O_3}$$

Section 9.4 Relationships Between Masses of Elements and Compounds

46. Write a conversion factor that converts between moles of nitrogen in nitrogen pentoxide, N_2O_5, and moles of N_2O_5. *(Obj #13)*

$$\frac{\mathbf{2 \text{ mol N}}}{\mathbf{1 \text{ mol } N_2O_5}}$$

48. The green granules on older asphalt roofing are chromium(III) oxide. Write a conversion factor that converts between moles of chromium atoms in chromium(III) oxide, Cr_2O_3, and moles of Cr_2O_3. *(Objs #13 & 14)*

$$\frac{\mathbf{2 \text{ mol Cr}}}{\mathbf{1 \text{ mol } Cr_2O_3}}$$

50. Ammonium oxalate is used for stain and rust removal. How many moles of ammonium ions are in 1 mole of ammonium oxalate, $(NH_4)_2C_2O_4$? *(Obj #14)*

    There are **2 moles of ammonium ions** in one mole of $(NH_4)_2C_2O_4$.

52. A nutritional supplement contains 0.405 g of $CaCO_3$. The recommended daily value of calcium is 1.000 g Ca. *(Objs #13 & 15)*

    a. Write a conversion factor that relates moles of calcium to moles of calcium carbonate.

$$\left(\frac{\mathbf{1 \text{ mol Ca}}}{\mathbf{1 \text{ mol } CaCO_3}} \right)$$

b. Calculate the mass in grams of calcium in 0.405 g of $CaCO_3$.

$$? \, g \, Ca = 0.405 \, g \, CaCO_3 \left(\frac{1 \, mol \, CaCO_3}{100.087 \, g \, CaCO_3} \right) \left(\frac{1 \, mol \, Ca}{1 \, mol \, CaCO_3} \right) \left(\frac{40.078 \, g \, Ca}{1 \, mol \, Ca} \right)$$

= 0.162 g Ca

c. What percentage of the daily value of calcium comes from this tablet?

$$\% \, Ca = \frac{0.162 \, g \, Ca \, in \, supp.}{1.000 \, g \, Ca \, total} \times 100 \quad \textbf{= 16.2\% of daily value Ca}$$

54. A multivitamin tablet contains 10 µg of vanadium in the form of sodium metavanadate, $NaVO_3$. How many micrograms of $NaVO_3$ does each tablet contain?   *(Obj #15)*

$$? \, \mu g \, NaVO_3 = 10 \, \mu g \, V \left(\frac{1 \, g}{10^6 \, \mu g} \right) \left(\frac{1 \, mol \, V}{50.9415 \, g \, V} \right) \left(\frac{1 \, mol \, NaVO_3}{1 \, mol \, V} \right) \left(\frac{121.9295 \, g \, NaVO_3}{1 \, mol \, NaVO_3} \right) \left(\frac{10^6 \, \mu g}{1 \, g} \right)$$

= 24 µg $NaVO_3$

56. There are several natural sources of the element titanium. One is the ore called rutile, which contains oxides of iron and titanium, FeO and TiO_2. Titanium metal can be made by first converting the TiO_2 in rutile to $TiCl_4$ by heating the ore to high temperature in the presence of carbon and chlorine. The titanium in $TiCl_4$ is then reduced from its +4 oxidation state to its zero oxidation state by reaction with a good reducing agent such as magnesium or sodium. What is the mass of titanium in kilograms in 0.401 Mg of $TiCl_4$? *(Obj #15)*

$$? \, kg \, Ti = 0.401 \, Mg \, TiCl_4 \left(\frac{10^6 \, g}{1 \, Mg} \right) \left(\frac{1 \, mol \, TiCl_4}{189.678 \, g \, TiCl_4} \right) \left(\frac{1 \, mol \, Ti}{1 \, mol \, TiCl_4} \right) \left(\frac{47.867 \, g \, Ti}{1 \, mol \, Ti} \right) \left(\frac{1 \, kg}{10^3 \, g} \right)$$

= 101 kg Ti

Section 9.5 Determination of Empirical and Molecular Formulas

59. An extremely explosive ionic compound is made from the reaction of silver compounds with ammonia. A sample of this compound is found to contain 17.261 g of silver and 0.743 g of nitrogen. What is the empirical formula for this compound? What is its chemical name? *(Obj #16)*

$$? \, mol \, Ag = 17.261 \, g \, Ag \left(\frac{1 \, mol \, Ag}{107.8682 \, g \, Ag} \right) = 0.16002 \, mol \, Ag \div 0.0530 \cong 3 \, mol \, Ag$$

$$? \, mol \, N = 0.743 \, g \, N \left(\frac{1 \, mol \, N}{14.0067 \, g \, N} \right) = 0.0530 \, mol \, N \div 0.0530 = 1 \, mol \, N$$

Empirical Formula   **Ag_3N silver nitride**

61. A sample of a compound used to polish dentures and as a nutrient and dietary supplement is analyzed and found to contain 9.2402 g of calcium, 7.2183 g of phosphorus, and 13.0512 g of oxygen. What is the empirical formula for this compound? *(Obj #16)*

$$? \text{ mol Ca } = 9.2402 \text{ g Ca} \left(\frac{1 \text{ mol Ca}}{40.078 \text{ g Ca}} \right) = 0.23056 \text{ mol Ca} \div 0.23056$$

$$= 1 \text{ mol Ca} \times 2 = 2 \text{ mol Ca}$$

$$? \text{ mol P} = 7.2183 \text{ g P} \left(\frac{1 \text{ mol P}}{30.9738 \text{ g P}} \right) = 0.23305 \text{ mol P} \div 0.23056$$

$$\cong 1 \text{ mol P} \times 2 = 2 \text{ mol P}$$

$$? \text{ mol O} = 13.0512 \text{ g O} \left(\frac{1 \text{ mol O}}{15.9994 \text{ g O}} \right) = 0.815731 \text{ mol O} \div 0.23056$$

$$\cong 3\frac{1}{2} \text{ mol O} \times 2 = 7 \text{ mol O}$$

Empirical Formula $Ca_2P_2O_7$

63. An ionic compound that is 38.791% nickel, 33.011% arsenic, and 28.198% oxygen is employed as a catalyst for hardening fats used to make soap. What is the empirical formula for this compound?   *(Obj #17)*

$$? \text{ mol Ni} = 38.791 \text{ g Ni} \left(\frac{1 \text{ mol Ni}}{58.6934 \text{ g Ni}} \right) = 0.66091 \text{ mol Ni} \div 0.44061$$

$$= 1\frac{1}{2} \text{ mol Ni} \times 2 = 3 \text{ mol Ni}$$

$$? \text{ mol As} = 33.011 \text{ g As} \left(\frac{1 \text{ mol As}}{74.9216 \text{ g As}} \right) = 0.44061 \text{ mol As} \div 0.44061$$

$$= 1 \text{ mol As} \times 2 = 2 \text{ mol As}$$

$$? \text{ mol O} = 28.198 \text{ g O} \left(\frac{1 \text{ mol O}}{15.9994 \text{ g O}} \right) = 1.7624 \text{ mol O} \div 0.44061$$

$$\cong 4 \text{ mol O} \times 2 = 8 \text{ mol O}$$

Empirical Formula $Ni_3As_2O_8$

65. An ionic compound that contains 10.279% calcium, 65.099% iodine, and 24.622% oxygen is used in deodorants and in mouthwashes. What is the empirical formula for this compound? *(Obj #17)*

$$? \text{ mol Ca} = 10.279 \text{ g Ca} \left(\frac{1 \text{ mol Ca}}{40.078 \text{ g Ca}} \right) = 0.25647 \text{ mol Ca} \div 0.25647 = 1 \text{ mol Ca}$$

$$? \text{ mol I} = 65.099 \text{ g I} \left(\frac{1 \text{ mol I}}{126.9045 \text{ g I}} \right) = 0.51298 \text{ mol I} \div 0.25647 \cong 2 \text{ mol I}$$

$$? \text{ mol O} = 24.622 \text{ g O} \left(\frac{1 \text{ mol O}}{15.9994 \text{ g O}} \right) = 1.5389 \text{ mol O} \div 0.25647 = 6 \text{ mol O}$$

Empirical Formula CaI_2O_6 or $Ca(IO_3)_2$ calcium iodate

67. In 1989 a controversy arose concerning the chemical daminozide, or Alar®, which was sprayed on apple trees to yield redder, firmer, and more shapely apples. Concerns about Alar's safety stemmed from the suspicion that one of its breakdown products, unsymmetrical dimethylhydrazine (UDMH), was carcinogenic. Alar is no longer sold for food uses. UDMH has the empirical formula of CNH_4 and has a molecular mass of 60.099. What is the molecular formula for UDMH?   *(Obj #17)*

Molecular Formula = (Empirical Formula)$_n$

$$n = \frac{\text{molecular mass}}{\text{empirical formula mass}} = \frac{60.099}{30.049} \cong 2$$

Molecular Formula = $(CNH_4)_2$ or $\mathbf{C_2N_2H_8}$

69. Lindane is one of the chlorinated pesticides the use of which is now restricted in the United States. It is 24.78% carbon, 2.08% hydrogen, and 73.14% chlorine and has a molecular mass of 290.830. What is lindane's molecular formula?   *(Obj #18)*

$$? \text{ mol C} = 24.78 \text{ g C} \left(\frac{1 \text{ mol C}}{12.011 \text{ g C}} \right) = 2.063 \text{ mol C} \div 2.063 = 1 \text{ mol C}$$

$$? \text{ mol H} = 2.08 \text{ g H} \left(\frac{1 \text{ mol H}}{1.00794 \text{ g H}} \right) = 2.064 \text{ mol H} \div 2.063 \cong 1 \text{ mol H}$$

$$? \text{ mol Cl} = 73.14 \text{ g Cl} \left(\frac{1 \text{ mol Cl}}{35.4527 \text{ g Cl}} \right) = 2.063 \text{ mol Cl} \div 2.063 = 1 \text{ mol Cl}$$

Empirical Formula   CHCl

$$n = \frac{\text{molecular mass}}{\text{empirical formula mass}} = \frac{290.830}{48.472} \cong 6$$

Molecular Formula = $(CHCl)_6$ or $\mathbf{C_6H_6Cl_6}$

71. Melamine is a compound used to make the melamine-formaldehyde resins in very hard surface materials such as Formica®. It is 28.57% carbon, 4.80% hydrogen, and 66.63% nitrogen and has a molecular mass of 126.121. What is melamine's molecular formula? *(Obj #18)*

$$? \text{ mol C} = 28.57 \text{ g C} \left(\frac{1 \text{ mol C}}{12.011 \text{ g C}} \right) = 2.379 \text{ mol C} \div 2.379 = 1 \text{ mol C}$$

$$? \text{ mol H} = 4.80 \text{ g H} \left(\frac{1 \text{ mol H}}{1.00794 \text{ g H}} \right) = 4.76 \text{ mol H} \div 2.379 \cong 2 \text{ mol H}$$

$$? \text{ mol N} = 66.63 \text{ g N} \left(\frac{1 \text{ mol N}}{14.0067 \text{ g N}} \right) = 4.757 \text{ mol N} \div 2.379 \cong 2 \text{ mol N}$$

Empirical Formula CH_2N_2

$$n = \frac{\text{molecular mass}}{\text{empirical formula mass}} = \frac{126.121}{42.040} \cong 3$$

Molecular Formula = $(CH_2N_2)_3$ or $\mathbf{C_3H_6N_6}$

Additional Problems

73. Your boss at the hardware store points you to a bin of screws and asks you to find out the approximate number of screws it contains. You weigh the screws and find that their total mass is 68 pounds. You take out 100 screws and weigh them individually, and you find that 7 screws weigh 2.65 g, 4 screws weigh 2.75 g, and 89 screws weigh 2.90 g. Calculate the weighted average mass of each screw. How many screws are in the bin? How many gross of screws are in the bin?

$$weighted\ average = 0.07(2.65\ g) + 0.04(2.75\ g) + 0.89(2.90\ g) = \textbf{2.88 g}$$

$$?\ screws = 68\ lb\ screws \left(\frac{453.6\ g}{1\ lb}\right)\left(\frac{1\ screw}{2.88\ g}\right) = \textbf{1.1} \times \textbf{10}^{\textbf{4}}\ \textbf{screws}$$

$$?\ screws = 68\ lb\ screws \left(\frac{453.6\ g}{1\ lb}\right)\left(\frac{1\ screw}{2.88\ g}\right)\left(\frac{1\ gross\ screws}{144\ screws}\right) = \textbf{74 gross screws}$$

75. As a member of the corundum family of minerals, sapphire (the September birthstone) consists primarily of aluminum oxide, Al_2O_3. Small amounts of iron and titanium give it its rich dark blue color. Gem cutter Norman Maness carved a giant sapphire into the likeness of Abraham Lincoln. If this 2302-carat sapphire were pure aluminum oxide, how many moles of Al_2O_3 would it contain? (There are exactly 5 carats per gram.)

$$?\ mol\ Al_2O_3 = 2302\ carat\ Al_2O_3 \left(\frac{1\ g}{5\ carats}\right)\left(\frac{1\ mol\ Al_2O_3}{101.9612\ g\ Al_2O_3}\right)$$

$$= \textbf{4.515 mol Al}_2\textbf{O}_3$$

77. Aquamarine (the March birthstone) is a light blue member of the beryl family, which is made up of natural silicates of beryllium and aluminum that have the general formula $Be_3Al_2(SiO_3)_6$. Aquamarine's bluish color is caused by trace amounts of iron(II) ions. A 43-pound aquamarine mined in Brazil in 1910 remains the largest gem-quality crystal ever found. If this stone were pure $Be_3Al_2(SiO_3)_6$, how many moles of beryllium would it contain?

$$?\ mol\ Be_3Al_2(SiO_3)_6 = 43\ lb\ Be_3Al_2(SiO_3)_6 \left(\frac{453.6\ g}{1\ lb}\right)\left(\frac{1\ mol\ Be_3Al_2(SiO_3)_6}{537.502\ g\ Be_3Al_2(SiO_3)_6}\right)\left(\frac{3\ mol\ Be}{1\ mol\ Be_3Al_2(SiO_3)_6}\right)$$

$$= \textbf{1.1} \times \textbf{10}^{\textbf{2}}\ \textbf{mol Be}$$

79. November's birthstone is citrine, a yellow member of the quartz family. It is primarily silicon dioxide, but small amounts of iron(III) ions give it its yellow color. A high-quality citrine containing about 0.040 moles of SiO_2 costs around $225. If this stone were pure SiO_2, how many carats would it weigh? (There are exactly 5 carats per gram.)

$$?\ carat\ SiO_2 = 0.040\ mol\ SiO_2 \left(\frac{60.0843\ g\ SiO_2}{1\ mol\ SiO_2}\right)\left(\frac{5\ carat}{1\ g}\right) = \textbf{12 carats}$$

81. A common throat lozenge contains 29 mg of phenol, C_6H_5OH.

    a. How many moles of C_6H_5OH are there in 5.0 mg of phenol?

$$?\ mol\ C_6H_5OH = 5.0\ mg\ C_6H_5OH \left(\frac{1\ g}{10^3\ mg}\right)\left(\frac{1\ mol\ C_6H_5OH}{94.113\ g\ C_6H_5OH}\right)$$

$$= \textbf{5.3} \times \textbf{10}^{\textbf{-5}}\ \textbf{mol C}_6\textbf{H}_5\textbf{OH}$$

b.   What is the mass in kilograms of 0.9265 mole of phenol?

$$? \text{ kg } C_6H_5OH = 0.9265 \text{ mol } C_6H_5OH \left(\frac{94.113 \text{ g } C_6H_5OH}{1 \text{ mol } C_6H_5OH} \right) \left(\frac{1 \text{ kg}}{10^3 \text{ g}} \right)$$

$$= \mathbf{0.08720 \text{ kg } C_6H_5OH}$$

83. Beryl, $Be_3Al_2(SiO_3)_6$, is a natural source of beryllium, a known carcinogen. What is the mass in kilograms of beryllium in 1.006 Mg of $Be_3Al_2(SiO_3)_6$?

$$? \text{ kg Be} = 1.006 \text{ Mg } Be_3Al_2(SiO_3)_6 \left(\frac{10^6 \text{ g}}{1 \text{ Mg}} \right) \left(\frac{1 \text{ mol } Be_3Al_2(SiO_3)_6}{537.502 \text{ g } Be_3Al_2(SiO_3)_6} \right) \left(\frac{3 \text{ mol Be}}{1 \text{ mol } Be_3Al_2(SiO_3)_6} \right) \left(\frac{9.0122 \text{ g Be}}{1 \text{ mol Be}} \right) \left(\frac{1 \text{ kg}}{10^3 \text{ g}} \right)$$

$$= \mathbf{50.60 \text{ kg Be}}$$

85. Cermets (for *cer*amic plus *met*al) are synthetic substances with both ceramic and metallic components. They combine the strength and toughness of metal with the resistance to heat and oxidation that ceramics offer. One cermet containing molybdenum and silicon is used to coat molybdenum engine parts on space vehicles. A sample of this compound is analyzed and found to contain 14.212 g of molybdenum and 8.321 g of silicon. What is the empirical formula for this compound?

$$? \text{ mol Mo} = 14.212 \text{ g Mo} \left(\frac{1 \text{ mol Mo}}{95.94 \text{ g Mo}} \right) = 0.1481 \text{ mol Mo} \div 0.1481 = 1 \text{ mol Mo}$$

$$? \text{ mol Si} = 8.321 \text{ g Si} \left(\frac{1 \text{ mol Si}}{28.0855 \text{ g Si}} \right) = 0.2963 \text{ mol Si} \div 0.1481 \cong 2 \text{ mol Si}$$

Empirical Formula $MoSi_2$

87. A compound that is sometimes called sorrel salt can be used to remove ink stains or to clean wood. It is 30.52% potassium, 0.787% hydrogen, 18.75% carbon, and 49.95% oxygen. What is the empirical formula for this compound?

$$? \text{ mol K} = 30.52 \text{ g K} \left(\frac{1 \text{ mol K}}{39.0983 \text{ g K}} \right) = 0.7806 \text{ mol K} \div 0.7806 = 1 \text{ mol K}$$

$$? \text{ mol H} = 0.787 \text{ g H} \left(\frac{1 \text{ mol H}}{1.00794 \text{ g H}} \right) = 0.781 \text{ mol H} \div 0.7806 \cong 1 \text{ mol H}$$

$$? \text{ mol C} = 18.75 \text{ g C} \left(\frac{1 \text{ mol C}}{12.011 \text{ g C}} \right) = 1.561 \text{ mol C} \div 0.7806 \cong 2 \text{ mol C}$$

$$? \text{ mol O} = 49.95 \text{ g O} \left(\frac{1 \text{ mol O}}{15.9994 \text{ g O}} \right) = 3.122 \text{ mol O} \div 0.7806 \cong 4 \text{ mol O}$$

Empirical Formula KHC_2O_4

89. An ionic compound that is 24.186% sodium, 33.734% sulfur, and 42.080% oxygen is used as a food preservative. What is its empirical formula?

$$? \, mol \, Na \; = \; 24.186 \, g \, Na \left(\frac{1 \, mol \, Na}{22.9898 \, g \, Na} \right) \; = \; 1.0520 \, mol \, Na \; \div \; 1.0520$$

$$= \; 1 \, mol \, Na \; \times \; 2 = 2 \, mol \, Na$$

$$? \, mol \, S \; = \; 33.734 \, g \, S \left(\frac{1 \, mol \, S}{32.066 \, g \, S} \right) \; = \; 1.0520 \, mol \, S \; \div \; 1.0520$$

$$\cong \; 1 \, mol \, S \; \times \; 2 = 2 \, mol \, S$$

$$? \, mol \, O \; = \; 42.080 \, g \, O \left(\frac{1 \, mol \, O}{15.9994 \, g \, O} \right) \; = \; 2.6301 \, mol \, O \; \div \; 1.0520$$

$$= \; 2\frac{1}{2} \, mol \, O \; \times \; 2 = 5 \, mol \, O$$

Empirical Formula $Na_2S_2O_5$

91. An ionic compound 22.071% manganese, 1.620% hydrogen, 24.887% phosphorus, and 51.422% oxygen is used as a food additive and dietary supplement. What is the empirical formula for this compound? What do you think its chemical name is? (Consider the possibility that this compound contains more than one polyatomic ion.)

$$? \, mol \, Mn \; = \; 22.071 \, g \, Mn \left(\frac{1 \, mol \, Mn}{54.9380 \, g \, Mn} \right) \; = \; 0.40174 \, mol \, Mn \; \div \; 0.40174 \; = \; 1 \, mol \, Mn$$

$$? \, mol \, H \; = \; 1.620 \, g \, H \left(\frac{1 \, mol \, H}{1.00794 \, g \, H} \right) \; = \; 1.607 \, mol \, H \; \div \; 0.40174 \; \cong \; 4 \, mol \, H$$

$$? \, mol \, P \; = \; 24.887 \, g \, P \left(\frac{1 \, mol \, P}{30.9738 \, g \, P} \right) \; = \; 0.80349 \, mol \, P \; \div \; 0.40174 \; \cong \; 2 \, mol \, P$$

$$? \, mol \, O \; = \; 51.422 \, g \, O \left(\frac{1 \, mol \, O}{15.9994 \, g \, O} \right) \; = \; 3.2140 \, mol \, O \; \div \; 0.40174 \; = \; 8 \, mol \, O$$

Empirical Formula $MnH_4P_2O_8$ or $Mn(H_2PO_4)_2$ manganese(II) dihydrogen phosphate

93. Thalidomide was used as a tranquilizer and flu medicine for pregnant women in Europe until it was found to cause birth defects. (The horrible effects of this drug played a significant role in the passage of the Kefauver-Harris Amendment to the Food and Drug Act, requiring that drugs be proved safe before they are put on the market.) Thalidomide is 60.47% carbon, 3.90% hydrogen, 24.78% oxygen, and 10.85% nitrogen and has a molecular mass of 258.23. What is the molecular formula for thalidomide?

$$? \text{ mol C} = 60.47 \text{ g C} \left(\frac{1 \text{ mol C}}{12.011 \text{ g C}} \right) = 5.035 \text{ mol C} \div 0.7746 = 6.5 \text{ mol C} \times 2 = 13 \text{ mol C}$$

$$? \text{ mol H} = 3.90 \text{ g H} \left(\frac{1 \text{ mol H}}{1.00794 \text{ g H}} \right) = 3.87 \text{ mol H} \div 0.7746 \cong 5 \text{ mol H} \times 2 = 10 \text{ mol H}$$

$$? \text{ mol O} = 24.78 \text{ g O} \left(\frac{1 \text{ mol O}}{15.9994 \text{ g O}} \right) = 1.549 \text{ mol O} \div 0.7746 \cong 2 \text{ mol O} \times 2 = 4 \text{ mol O}$$

$$? \text{ mol N} = 10.85 \text{ g N} \left(\frac{1 \text{ mol N}}{14.0067 \text{ g N}} \right) = 0.7746 \text{ mol N} \div 0.7746 = 1 \text{ mol N} \times 2 = 2 \text{ mol N}$$

Empirical Formula $C_{13}H_{10}O_4N_2$

$$n = \frac{\text{molecular mass}}{\text{empirical formula mass}} = \frac{258.23}{258.23} = 1$$

Molecular Formula = **$C_{13}H_{10}O_4N_2$**

Challenge Problems

95. Calamine is a naturally occurring zinc silicate that contains the equivalent of 67.5% zinc oxide, ZnO. (The term *calamine* also refers to a substance used to make calamine lotion.) What is the mass, in kilograms, of zinc in 1.347×10^4 kg of natural calamine that is 67.5% ZnO?

$$? \text{ kg Zn} = 1.347 \times 10^4 \text{ kg calamine} \left(\frac{67.5 \text{ kg ZnO}}{100 \text{ kg calamine}} \right)\left(\frac{10^3 \text{ g}}{1 \text{ kg}} \right)\left(\frac{1 \text{ mol ZnO}}{81.39 \text{ g ZnO}} \right)\left(\frac{1 \text{ mol Zn}}{1 \text{ mol ZnO}} \right)\left(\frac{65.39 \text{ g Zn}}{1 \text{ mol Zn}} \right)\left(\frac{1 \text{ kg}}{10^3 \text{ g}} \right)$$

   $= 7.30 \times 10^3$ kg Zn

97. Flue dust from the smelting of copper and lead contains As_2O_3. (Smelting is the heating of a metal ore until it melts, so that its metallic components can be separated.) When this flue dust is collected, it contains 90% to 95% As_2O_3. The arsenic in As_2O_3 can be reduced to the element arsenic by reaction with charcoal. What is the maximum mass, in kilograms, of arsenic that can be formed from 67.3 kg of flue dust that is 93% As_2O_3?

$$? \text{ kg As} = 67.3 \text{ kg flue dust} \left(\frac{93 \text{ kg As}_2\text{O}_3}{100 \text{ kg flue dust}} \right)\left(\frac{10^3 \text{ g}}{1 \text{ kg}} \right)\left(\frac{1 \text{ mol As}_2\text{O}_3}{197.841 \text{ g As}_2\text{O}_3} \right)\left(\frac{2 \text{ mol As}}{1 \text{ mol As}_2\text{O}_3} \right)\left(\frac{74.9216 \text{ g As}}{1 \text{ mol As}} \right)\left(\frac{1 \text{ kg}}{10^3 \text{ g}} \right)$$

$$\text{or} \quad ? \text{ kg As} = 67.3 \text{ kg flue dust} \left(\frac{93 \text{ kg As}_2\text{O}_3}{100 \text{ kg flue dust}} \right)\left(\frac{2 \times 74.9216 \text{ g As}}{197.841 \text{ g As}_2\text{O}_3} \right) = \textbf{47 kg As}$$

99. Magnesium metal, which is used to make die-cast auto parts, missiles, and space vehicles, is obtained by the electrolysis of magnesium chloride. Magnesium hydroxide forms magnesium chloride when it reacts with hydrochloric acid. There are two common sources of magnesium hydroxide.

    a. Magnesium ions can be precipitated from seawater as magnesium hydroxide, $Mg(OH)_2$. Each kiloliter of seawater yields about 3.0 kg of the compound. How many metric tons of magnesium metal can be made from the magnesium hydroxide derived from 1.0×10^5 kL of seawater?

$$? \text{ t Mg} = 1.0 \times 10^5 \text{ kL seawater} \left(\frac{3.0 \text{ kg Mg(OH)}_2}{1 \text{ kL seawater}} \right) \left(\frac{10^3 \text{ g}}{1 \text{ kg}} \right) \left(\frac{1 \text{ mol Mg(OH)}_2}{58.3197 \text{ g Mg(OH)}_2} \right) \left(\frac{1 \text{ mol Mg}}{1 \text{ mol Mg(OH)}_2} \right) \left(\frac{24.3050 \text{ g Mg}}{1 \text{ mol Mg}} \right) \left(\frac{1 \text{ t}}{1 \text{ Mg}} \right)$$

$$= \mathbf{1.3 \times 10^2 \text{ t Mg}}$$

    b. Brucite is a natural form of magnesium hydroxide. A typical crude ore containing brucite is 29% $Mg(OH)_2$. What minimum mass, in metric tons, of this crude ore is necessary to make 34.78 metric tons of magnesium metal?

$$? \text{ t ore} = 34.78 \text{ t Mg} \left(\frac{10^6 \text{ g}}{1 \text{ t}} \right) \left(\frac{1 \text{ mol Mg}}{24.3050 \text{ g Mg}} \right) \left(\frac{1 \text{ mol Mg(OH)}_2}{1 \text{ mol Mg}} \right) \left(\frac{58.3197 \text{ g Mg(OH)}_2}{1 \text{ mol Mg(OH)}_2} \right) \left(\frac{1 \text{ t}}{10^6 \text{ g}} \right) \left(\frac{100 \text{ t ore}}{29 \text{ t Mg(OH)}_2} \right)$$

$$= \mathbf{2.9 \times 10^2 \text{ t ore}}$$

101. The element fluorine can be obtained by the electrolysis of combinations of hydrofluoric acid and potassium fluoride. These compounds can be made from the calcium fluoride, CaF_2, found in nature as the mineral fluorite. Fluorite's commercial name is fluorspar. Crude ores containing fluorite have 15% to 90% CaF_2. What minimum mass, in metric tons, of crude ore is necessary to make 2.4 metric tons of fluorine if the ore is 72% CaF_2?

$$? \text{ t ore} = 2.4 \text{ t F} \left(\frac{10^6 \text{ g}}{1 \text{ t}} \right) \left(\frac{1 \text{ mol F}}{18.9984 \text{ g F}} \right) \left(\frac{1 \text{ mol CaF}_2}{2 \text{ mol F}} \right) \left(\frac{78.075 \text{ g CaF}_2}{1 \text{ mol CaF}_2} \right) \left(\frac{1 \text{ t}}{10^6 \text{ g}} \right) \left(\frac{100 \text{ t ore}}{72 \text{ t CaF}_2} \right)$$

$$= \mathbf{6.8 \text{ t ore}}$$

103. What mass of baking powder that is 36% $NaHCO_3$ contains 1.0 mole of sodium hydrogen carbonate?

$$? \text{ g baking powder} = 1.0 \text{ mol NaHCO}_3 \left(\frac{84.007 \text{ g NaHCO}_3}{1 \text{ mol NaHCO}_3} \right) \left(\frac{100 \text{ g baking powder}}{36 \text{ g NaHCO}_3} \right)$$

$$= \mathbf{2.3 \times 10^2 \text{ g baking powder}}$$

105. Hafnium metal is used to make control rods in water-cooled nuclear reactors and to make filaments in light bulbs. The hafnium is found with zirconium in zircon sand, which is about 1% hafnium(IV) oxide, HfO_2. What minimum mass, in metric tons, of zircon sand is necessary to make 120.5 kg of hafnium metal if the sand is 1.3% HfO_2?

$$? \text{ t sand} = 120.5 \text{ kg Hf} \left(\frac{10^3 \text{ g}}{1 \text{ kg}} \right) \left(\frac{1 \text{ mol Hf}}{178.49 \text{ g Hf}} \right) \left(\frac{1 \text{ mol HfO}_2}{1 \text{ mol Hf}} \right) \left(\frac{210.49 \text{ g HfO}_2}{1 \text{ mol HfO}_2} \right) \left(\frac{1 \text{ kg}}{10^3 \text{ g}} \right) \left(\frac{100 \text{ kg sand}}{1.3 \text{ kg HfO}_2} \right) \left(\frac{1 \text{ t}}{10^3 \text{ kg}} \right)$$

$$= \mathbf{11 \text{ t zircon sand}}$$

Chapter 10
Chemical Calculations and Chemical Equations

Converts given unit to liters.

Coefficients from balanced equation convert moles of one substance to moles of another substance.

Converts liters to desired unit.

$$? \text{ mL AgNO}_3 \text{ soln} = 25.00 \text{ mL Na}_3\text{PO}_4 \text{ soln} \left(\frac{1 \text{ L}}{10^3 \text{ mL}} \right) \left(\frac{0.500 \text{ mol Na}_3\text{PO}_4}{1 \text{ L Na}_3\text{PO}_4 \text{ soln}} \right) \left(\frac{3 \text{ mol AgNO}_3}{1 \text{ mol Na}_3\text{PO}_4} \right) \left(\frac{1 \text{ L AgNO}_3 \text{ soln}}{1.00 \text{ mol AgNO}_3} \right) \left(\frac{10^3 \text{ mL}}{1 \text{ L}} \right)$$

Molarity as a conversion factor converts liters to moles.

Molarity as a conversion factor converts moles to liters.

◆ Review Skills

10.1 Equation Stoichiometry

  Internet: Equation Stoichiometry Problems with Mixtures

10.2 Real-World Applications of Equation Stoichiometry

  • Limiting Reactants

  • Percent Yield

  Special Topic 10.1: Big Problems Require Bold Solutions - Global Warming and Limiting Reactants

  Internet: Global Warming, Greenhouse Gases, the Ocean, and Limiting Reactants

10.3 Molarity and Equation Stoichiometry

  • Reactions in Solution and Molarity

  • Equation Stoichiometry and Reactions in Solution

  Internet: Acid-Base Titrations

  Internet: Making Solutions of a Certain Molarity

◆ Chapter Glossary

  Internet: Glossary Quiz

◆ Chapter Objectives

Review Questions

Key Ideas

Chapter Problems

Section Goals and Introductions

Section 10.1 Equation Stoichiometry

Goal: To show how the coefficients in a balanced chemical equation can be used to convert from mass of one substance in a given chemical reaction to the corresponding mass of another substance participating in the same reaction.

It's common that chemists and chemistry students are asked to convert from an amount of one substance in a given chemical reaction to the corresponding amount of another substance participating in the same reaction. This type of calculation, which uses the coefficients in a balanced equation to convert from moles of one substance to moles of another, is called equation stoichiometry. This section shows how to do equation stoichiometry problems for which you are asked to convert from *mass* of one substance in a given chemical reaction to the corresponding *mass* of another substance participating in the same reaction. For a related section, see *Equation Stoichiometry Problems with Mixtures* on our Web site.

> **www.chemplace.com/college/**

Section 10.2 Real-World Applications of Equation Stoichiometry
Goals

- *To explain why chemists sometimes deliberately use a limited amount of one reactant (called the limiting reactant) and excessive amounts of others.*

- *To show how to determine which reactant in a chemical reaction is the limiting reactant.*

- *To show how to calculate the maximum amount of product that they can form from given amounts of two or more reactants in a chemical reaction.*

- *To explain why the actual yield in a reaction might be less than the maximum possible yield (called the theoretical yield).*

- *To explain what percent yield is and to show how to calculate the percent yield given the actual yield and enough information to determine the theoretical yield.*

Chemistry in the real world is sometimes more complicated than we make it seem in introductory chemistry texts. You will see in this section that reactions run in real laboratories never have exactly the right amounts of reactants for each to react completely, and even if they did, it is unlikely that all of the reactants would form the desired products. This section shows you why this is true and shows you how to do calculations that reflect these realities. For a related topic, see *Global Warming, Greenhouse Gases, the Ocean, and Limiting Reactants* on our Web site.

> **www.chemplace.com/college/**

Section 10.3 Molarity and Equation Stoichiometry
Goals

- *To show how the concentration of solute in solution can be described with molarity, which is moles of solute per liter of solution.*

- *To show how to calculate molarity.*

- *To show how the molarity of a solution can be translated into a conversion factor that converts between moles of solute and volume of solution.*

Section 10.1 shows the general equation stoichiometry steps as

> measurable property #1 → moles #1 → moles #2 → measurable property #2

When the reactants and products of a reaction are pure solids and pure liquids, mass is the conveniently measurable property, but many chemical changes take place in either the gas phase or in solution. The masses of gases or of solutes in solution cannot be measured directly. For reactions run in solution, it's more convenient to measure the volume of the solution that contains the solute reactants and products. Therefore, to complete equation stoichiometry problems for reactions done in solution, we need a conversion factor that converts between volume of solution and moles of solute. This section defines that conversion factor, called molarity (moles of solute per liter of solution), shows you how it can be determined, and shows you how molarity can be translated into a conversion factor that allows you to convert between the measurable property of volume of solution and moles of solute.

The section ends with a summary of equation stoichiometry problems and shows how the skills developed in Section 10.1 can be mixed with the new skills developed in this section. Section 13.3 completes our process of describing equation stoichiometry problems by showing how to combine the information found in this chapter with calculations that convert between volume of gas and moles of gas.

See the two topics on our Web site that relate to this section: *Acid-Base Titrations* and *Making Solutions of a Certain Molarity*.

www.chemplace.com/college/

Chapter 10 Map

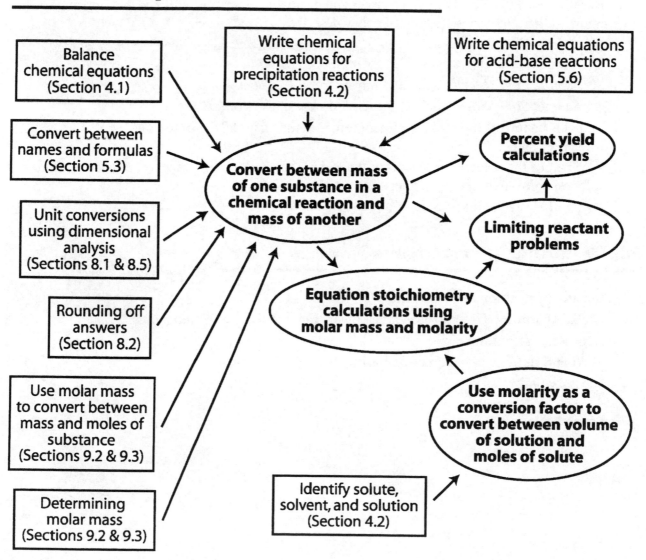

Chapter Checklist

☐ Read the Review Skills section. If there is any skill mentioned that you have not yet mastered, review the material on that topic before reading this chapter.

☐ Read the chapter quickly before the lecture that describes it.

☐ Attend class meetings, take notes, and participate in class discussions.

☐ Work the Chapter Exercises, perhaps using the Chapter Examples as guides.

☐ Study the Chapter Glossary and test yourself on our Web site:

        www.chemplace.com/college/

☐ Study all of the Chapter Objectives. You might want to write a description of how you will meet each objective. (Although it is best to master all of the objectives, the following objectives are especially important because they pertain to skills that you will need while studying other chapters of this text: 4, 9, 12, and 13.)

☐ Reread the Study Sheets in this chapter and decide whether you will use them or some variation on them to complete the tasks they describe.

    Sample Study Sheet 10.1: Basic Equation Stoichiometry - Converting Mass of One Compound in a Reaction to Mass of Another

    Sample Study Sheet 10.2: Limiting Reactant Problems

    Sample Study Sheet 10.3: Equation Stoichiometry Problems

☐ To get a review of the most important topics in the chapter, fill in the blanks in the Key Ideas section.

☐ Work all of the selected problems at the end of the chapter, and check your answers with the solutions provided in this chapter of the study guide.

☐ Ask for help if you need it.

Web Resources   www.chemplace.com/college/

    Equation Stoichiometry Problems with Mixtures

    Global Warming, Greenhouse Gases, the Ocean, and Limiting Reactants

    Acid-Base Titrations

    Making Solutions of a Certain Molarity

    Glossary Quiz

Exercises Key

✍ **Exercise 10.1 - Equation Stoichiometry:** Tetrachloroethene, C_2Cl_4, often called perchloroethylene (perc), is a colorless liquid used in dry cleaning. It can be formed in several steps from the reaction of dichloroethane, chlorine gas, and oxygen gas. The equation for the net reaction is

$$8C_2H_4Cl_2(l) + 6Cl_2(g) + 7O_2(g) \rightarrow 4C_2HCl_3(l) + 4C_2Cl_4(l) + 14H_2O(l)$$

a. Fifteen different conversion factors for relating moles of one reactant or product to moles of another reactant or product can be derived from this equation. Write five of them.

All fifteen possibilities are below.

$$\left(\frac{8 \text{ mol } C_2H_4Cl_2}{6 \text{ mol } Cl_2}\right) \text{ or } \left(\frac{8 \text{ mol } C_2H_4Cl_2}{7 \text{ mol } O_2}\right) \text{ or } \left(\frac{8 \text{ mol } C_2H_4Cl_2}{4 \text{ mol } C_2HCl_3}\right) \text{ or } \left(\frac{8 \text{ mol } C_2H_4Cl_2}{4 \text{ mol } C_2Cl_4}\right)$$

$$\text{or } \left(\frac{8 \text{ mol } C_2H_4Cl_2}{14 \text{ mol } H_2O}\right) \text{ or } \left(\frac{6 \text{ mol } Cl_2}{7 \text{ mol } O_2}\right) \text{ or } \left(\frac{6 \text{ mol } Cl_2}{4 \text{ mol } C_2HCl_3}\right) \text{ or } \left(\frac{6 \text{ mol } Cl_2}{4 \text{ mol } C_2Cl_4}\right)$$

$$\text{or } \left(\frac{6 \text{ mol } Cl_2}{14 \text{ mol } H_2O}\right) \text{ or } \left(\frac{7 \text{ mol } O_2}{4 \text{ mol } C_2HCl_3}\right) \text{ or } \left(\frac{7 \text{ mol } O_2}{4 \text{ mol } C_2Cl_4}\right) \text{ or } \left(\frac{7 \text{ mol } O_2}{14 \text{ mol } H_2O}\right)$$

$$\text{or } \left(\frac{4 \text{ mol } C_2HCl_3}{4 \text{ mol } C_2Cl_4}\right) \text{ or } \left(\frac{4 \text{ mol } C_2HCl_3}{14 \text{ mol } H_2O}\right) \text{ or } \left(\frac{4 \text{ mol } C_2Cl_4}{14 \text{ mol } H_2O}\right)$$

b. How many grams of water form when 362.47 grams of tetrachloroethene, C_2Cl_4, are made in the reaction above?

$$? \text{ g } H_2O = 362.47 \text{ g } C_2Cl_4 \left(\frac{1 \text{ mol } C_2Cl_4}{165.833 \text{ g } C_2Cl_4}\right)\left(\frac{14 \text{ mol } H_2O}{4 \text{ mol } C_2Cl_4}\right)\left(\frac{18.0153 \text{ g } H_2O}{1 \text{ mol } H_2O}\right)$$

$$\text{or } ? \text{ g } H_2O = 362.47 \text{ g } C_2Cl_4 \left(\frac{14 \times 18.0153 \text{ g } H_2O}{4 \times 165.833 \text{ g } C_2Cl_4}\right) = \textbf{137.82 g } \mathbf{H_2O}$$

c. What is the maximum mass of perchloroethylene, C_2Cl_4, that can be formed from 23.75 kilograms of dichloroethane, $C_2H_4Cl_2$?

$$? \text{ kg } C_2Cl_4 = 23.75 \text{ kg } C_2H_4Cl_2 \left(\frac{10^3 \text{ g}}{1 \text{ kg}}\right)\left(\frac{1 \text{ mol } C_2H_4Cl_2}{98.959 \text{ g } C_2H_4Cl_2}\right)\left(\frac{4 \text{ mol } C_2Cl_4}{8 \text{ mol } C_2H_4Cl_2}\right)\left(\frac{165.833 \text{ g } C_2Cl_4}{1 \text{ mol } C_2Cl_4}\right)\left(\frac{1 \text{ kg}}{10^3 \text{ g}}\right)$$

$$\text{or } ? \text{ kg } C_2Cl_4 = 23.75 \text{ kg } C_2H_4Cl_2 \left(\frac{4 \times 165.833 \text{ kg } C_2Cl_4}{8 \times 98.959 \text{ kg } C_2H_4Cl_2}\right) = \textbf{19.90 kg } \mathbf{C_2Cl_4}$$

Exercise 10.2 - Limiting Reactant: The uranium(IV) oxide, UO_2, which is used as fuel in nuclear power plants, has a higher percentage of the fissionable isotope uranium-235 than is present in the UO_2 found in nature. To make fuel-grade UO_2, chemists first convert uranium oxides to uranium hexafluoride, UF_6, whose concentration of uranium-235 can be increased by a process called gas diffusion. The enriched UF_6 is then converted back to UO_2 in a series of reactions, beginning with

$$UF_6 + 2H_2O \rightarrow UO_2F_2 + 4HF$$

a. How many megagrams of UO_2F_2 can be formed from the reaction of 24.543 Mg UF_6 with 8.0 Mg of water?

$$? \text{ Mg } UO_2F_2 = 24.543 \text{ Mg } UF_6 \left(\frac{10^6 \text{ g}}{1 \text{ Mg}} \right) \left(\frac{1 \text{ mol } UF_6}{352.019 \text{ g } UF_6} \right) \left(\frac{1 \text{ mol } UO_2F_2}{1 \text{ mol } UF_6} \right) \left(\frac{308.0245 \text{ g } UO_2F_2}{1 \text{ mol } UO_2F_2} \right) \left(\frac{1 \text{ Mg}}{10^6 \text{ g}} \right)$$

or $? \text{ Mg } UO_2F_2 = 24.543 \text{ Mg } UF_6 \left(\dfrac{1 \times 308.0245 \text{ Mg } UO_2F_2}{1 \times 352.019 \text{ Mg } UF_6} \right) = \mathbf{21.476 \text{ Mg } UO_2F_2}$

$$? \text{ Mg } UO_2F_2 = 8.0 \text{ Mg } H_2O \left(\frac{10^6 \text{ g}}{1 \text{ Mg}} \right) \left(\frac{1 \text{ mol } H_2O}{18.0153 \text{ g } H_2O} \right) \left(\frac{1 \text{ mol } UO_2F_2}{2 \text{ mol } H_2O} \right) \left(\frac{308.0245 \text{ g } UO_2F_2}{1 \text{ mol } UO_2F_2} \right) \left(\frac{1 \text{ Mg}}{10^6 \text{ g}} \right)$$

or $? \text{ Mg } UO_2F_2 = 8.0 \text{ Mg } H_2O \left(\dfrac{1 \times 308.0245 \text{ Mg } UO_2F_2}{2 \times 18.0153 \text{ Mg } H_2O} \right) = 68 \text{ Mg } UO_2F_2$

b. Why do you think the reactant in excess was chosen to be in excess?

Water is much less toxic and less expensive than the radioactive and rare uranium compound. Water in the form of either liquid or steam is also very easy to separate from the solid product mixture.

Exercise 10.3 - Percent yield: The raw material used as a source of chromium and chromium compounds is a chromium-iron ore called chromite. For example, sodium chromate, Na_2CrO_4, is made by roasting chromite with sodium carbonate, Na_2CO_3. (Roasting means heating in the presence of air or oxygen.) A simplified version of the net reaction is

$$4FeCr_2O_4 + 8Na_2CO_3 + 7O_2 \rightarrow 8Na_2CrO_4 + 2Fe_2O_3 + 8CO_2$$

What is the percent yield if 1.2 kg of Na_2CrO_4 is produced from ore that contains 1.0 kg of $FeCr_2O_4$?

$$? \text{ kg } Na_2CrO_4 = 1.0 \text{ kg } FeCr_2O_4 \left(\frac{10^3 \text{ g}}{1 \text{ kg}} \right) \left(\frac{1 \text{ mol } FeCr_2O_4}{223.835 \text{ g } FeCr_2O_4} \right) \left(\frac{8 \text{ mol } Na_2CrO_4}{4 \text{ mol } FeCr_2O_4} \right) \left(\frac{161.9733 \text{ g } Na_2CrO_4}{1 \text{ mol } Na_2CrO_4} \right) \left(\frac{1 \text{ kg}}{10^3 \text{ g}} \right)$$

or $? \text{ kg } Na_2CrO_4 = 1.0 \text{ kg } FeCr_2O_4 \left(\dfrac{8 \times 161.9733 \text{ kg } Na_2CrO_4}{4 \times 223.835 \text{ kg } FeCr_2O_4} \right) = 1.4 \text{ kg } Na_2CrO_4$

$$\text{Percent Yield} = \frac{\text{actual yield}}{\text{theoretical yield}} \times 100 = \frac{1.2 \text{ kg } Na_2CrO_4}{1.4 \text{ kg } Na_2CrO_4} \times 100 = \mathbf{86\% \text{ yield}}$$

✍ **Exercise 10.4 - Calculating a Solution's Molarity:**   A silver perchlorate solution was made by dissolving 29.993 g of pure $AgClO_4$ in water and then diluting the mixture with additional water to achieve a total volume of 50.00 mL. What is the solution's molarity?

$$\text{Molarity} = \frac{?\ \text{mol } AgClO_4}{1\ \text{L } AgClO_4\ \text{soln}} = \frac{29.993\ \text{g } AgClO_4}{50.0\ \text{mL } AgClO_4\ \text{soln}} \left(\frac{1\ \text{mol } AgClO_4}{207.3185\ \text{g } AgClO_4} \right) \left(\frac{10^3\ \text{mL}}{1\ \text{L}} \right)$$

$$= \frac{2.893\ \text{mol } AgClO_4}{1\ \text{L } AgClO_4\ \text{soln}} = \textbf{2.893 M } AgClO_4$$

✍ **Exercise 10.5 - Molarity and Equation Stoichiometry:**   How many milliliters of 6.00 M HNO_3 are necessary to neutralize the carbonate in 75.0 mL of 0.250 M Na_2CO_3?

$$2HNO_3(aq) + Na_2CO_3(aq) \rightarrow H_2O(l) + CO_2(g) + 2NaNO_3(aq)$$

$$?\ \text{mL } HNO_3\ \text{soln} = 75.0\ \text{mL } Na_2CO_3 \left(\frac{0.250\ \text{mol } Na_2CO_3}{10^3\ \text{mL } Na_2CO_3} \right) \left(\frac{2\ \text{mol } HNO_3}{1\ \text{mol } Na_2CO_3} \right) \left(\frac{10^3\ \text{mL } HNO_3\ \text{soln}}{6.00\ \text{mol } HNO_3} \right)$$

$$= \textbf{6.25 mL } HNO_3 \textbf{ soln}$$

✍ **Exercise 10.6 - Molarity and Equation Stoichiometry:**   What is the maximum number of grams of silver chloride that will precipitate from a solution made by mixing 25.00 mL of 0.050 M $MgCl_2$ with an excess of $AgNO_3$ solution?

$$2AgNO_3(aq) + MgCl_2(aq) \rightarrow 2AgCl(s) + Mg(NO_3)_2(aq)$$

$$?\ \text{g } AgCl = 25.00\ \text{mL } MgCl_2 \left(\frac{0.050\ \text{mol } MgCl_2}{10^3\ \text{mL } MgCl_2} \right) \left(\frac{2\ \text{mol } AgCl}{1\ \text{mol } MgCl_2} \right) \left(\frac{143.3209\ \text{g } AgCl}{1\ \text{mol } AgCl} \right)$$

$$= \textbf{0.36 g } AgCl$$

Review Questions Key

This chapter requires many of the same skills that were listed as review skills for Chapter 9. Thus you should make sure that you can do the Review Questions at the end of Chapter 9 before continuing with the questions that follow.

1. Write balanced equations for the following reactions. You do *not* need to include the substance's states.

   a. Hydrofluoric acid reacts with silicon dioxide to form silicon tetrafluoride and water.

   $$\textbf{4HF + SiO}_2 \rightarrow \textbf{SiF}_4 \textbf{ + 2H}_2\textbf{O}$$

   b. Ammonia reacts with oxygen gas to form nitrogen monoxide and water.

   $$\textbf{4NH}_3 \textbf{ + 5O}_2 \rightarrow \textbf{4NO + 6H}_2\textbf{O}$$

   c. Water solutions of nickel(II) acetate and sodium phosphate react to form solid nickel(II) phosphate and aqueous sodium acetate.

   $$\textbf{3Ni(C}_2\textbf{H}_3\textbf{O}_2\textbf{)}_2 \textbf{ + 2Na}_3\textbf{PO}_4 \rightarrow \textbf{Ni}_3\textbf{(PO}_4\textbf{)}_2 \textbf{ + 6NaC}_2\textbf{H}_3\textbf{O}_2$$

    d. Phosphoric acid reacts with potassium hydroxide to form water and potassium phosphate.

$$H_3PO_4 + 3KOH \rightarrow 3H_2O + K_3PO_4$$

2. Write complete equations, including states, for the precipitation reaction that takes place between the reactants in Part (a) and the neutralization reaction that takes place in Part (b).

    a. $Ca(NO_3)_2(aq) + Na_2CO_3(aq) \rightarrow CaCO_3(s) + 2NaNO_3(aq)$

    b. $3HNO_3(aq) + Al(OH)_3(s) \rightarrow 3H_2O(l) + Al(NO_3)_3(aq)$

3. How many moles of phosphorous acid are there in 68.785 g of phosphorous acid?

Molecular mass of $H_3PO_3 = 3(1.00794) + 30.9738 + 3(15.9994) = 81.9958$

$$? \text{ mol } H_3PO_3 = 68.785 \text{ g } H_3PO_3 \left(\frac{1 \text{ mol } H_3PO_3}{81.9958 \text{ g } H_3PO_3} \right) = \textbf{0.83888 mol } \mathbf{H_3PO_3}$$

4. What is the mass in kilograms of 0.8459 mole of sodium sulfate?

Molecular mass of $Na_2SO_4 = 2(22.9898) + 32.066 + 4(15.9994) = 142.043$

$$? \text{ mol } Na_2SO_4 = 0.8459 \text{ mol } Na_2SO_4 \left(\frac{142.043 \text{ g } Na_2SO_4}{1 \text{ mol } Na_2SO_4} \right) \left(\frac{1 \text{ kg}}{10^3 \text{ g}} \right)$$

$$= \textbf{0.1202 kg } \mathbf{Na_2SO_4}$$

Key Ideas Answers

5. If a calculation calls for you to convert from an amount of one substance in a given chemical reaction to the corresponding amount of another substance participating in the same reaction, it is an equation **stoichiometry** problem.

7. For some chemical reactions, chemists want to mix reactants in amounts that are as close as possible to the ratio that would lead to the complete reaction of each. This ratio is sometimes called the **stoichiometric** ratio.

9. Sometimes one product is more important than others are, and the amounts of reactants are chosen to **optimize** its production.

11. Because some of the reactant that was added in excess is likely to be mixed in with the product, chemists would prefer that the substance in excess be a substance that is **easy to separate** from the primary product.

13. The tip-off for limiting reactant problems is that you are given **two or more amounts** of reactants in a chemical reaction, and you are asked to calculate the maximum **amount of product** that they can form.

15. There are many reasons why the actual yield in a reaction might be less than the theoretical yield. One key reason is that many chemical reactions are significantly **reversible**.

17. Another factor that affects the actual yield is a reaction's rate. Sometimes a reaction is so **slow** that it has not reached the maximum yield by the time the product is isolated.

19. When two solutions are mixed to start a reaction, it is more convenient to measure their **volumes** than their masses.

21. Conversion factors constructed from molarities can be used in stoichiometric calculations in very much the same way conversion factors from **molar mass** are used. When a substance is

pure, its molar mass can be used to convert back and forth between the measurable property of **mass** and moles. When a substance is in solution, its molarity can be used to convert between the measurable property of **volume of solution** and moles of solute.

Problems Key

Section 10.1 Equation Stoichiometry and Section 10.2 Real-World Applications of Equation Stoichiometry

22. Because the bond between fluorine atoms in F_2 is relatively weak, and because the bonds between fluorine atoms and atoms of other elements are relatively strong, it is difficult to make diatomic fluorine, F_2. One way it can be made is to run an electric current through liquid hydrogen fluoride, HF. This reaction yields hydrogen gas, H_2, and fluorine gas, F_2.

    a.   Write a complete balanced equation, including states, for this reaction.

$$\overset{\textbf{electric current}}{2HF(l) \quad\longrightarrow\quad H_2(g) + F_2(g)}$$

    b.   Redraw your equation, substituting rough drawings of space-filling models for the coefficients and formulas for the reactants and products. Fluorine atoms have a little more than twice the diameter of hydrogen atoms.

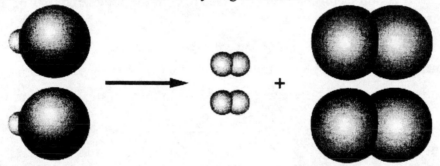

    c.   Write a conversion factor that could be used to convert between moles of HF and moles of F_2.

$$\left(\frac{1\ \textbf{mol F}_2}{2\ \textbf{mol HF}}\right) \quad \text{or} \quad \left(\frac{2\ \textbf{mol HF}}{1\ \textbf{mol F}_2}\right)$$

    d.   How many moles of F_2 form when one mole of HF reacts completely?

The number of moles of F_2 that form is one-half the number of moles of HF that react, so one mole of HF forms **one-half mole of F_2.**

$$? \text{ mol } F_2 = 1 \text{ mol HF} \left(\frac{1 \text{ mol } F_2}{2 \text{ mol HF}}\right) = \textbf{0.5 mol F}_2$$

    e.   How many moles of HF react to yield 3.452 moles of H_2?

$$? \text{ mol HF} = 3.452 \text{ mol } H_2 \left(\frac{2 \text{ mol HF}}{1 \text{ mol } H_2}\right) = \textbf{6.904 mol HF}$$

24. The bond between nitrogen atoms in N_2 molecules is very strong, making N_2 very unreactive. Because of this, magnesium is one of the few metals that react with nitrogen gas directly. This reaction yields solid magnesium nitride.

   a. Write a complete balanced equation, without including states, for the reaction between magnesium and nitrogen to form magnesium nitride.

   $$3Mg + N_2 \rightarrow Mg_3N_2$$

   b. Write a conversion factor that could be used to convert between moles of magnesium and moles of magnesium nitride.

   $$\left(\frac{3 \text{ mol Mg}}{1 \text{ mol Mg}_3\text{N}_2} \right) \quad \text{or} \quad \left(\frac{1 \text{ mol Mg}_3\text{N}_2}{3 \text{ mol Mg}} \right)$$

   c. How many moles of magnesium nitride form when 1.0 mole of magnesium reacts completely?

   The number of moles of Mg_3N_2 that form is one-third the number of moles of Mg that react, so one mole of Mg forms **one-third mole of Mg_3N_2.**

   $$? \text{ mol Mg}_3\text{N}_2 = 1 \text{ mol Mg} \left(\frac{1 \text{ mol Mg}_3\text{N}_2}{3 \text{ mol Mg}} \right) = \textbf{0.33 mol Mg}_3\textbf{N}_2$$

   d. Write a conversion factor that could be used to convert between moles of nitrogen and moles of magnesium nitride.

   $$\left(\frac{1 \text{ mol N}_2}{1 \text{ mol Mg}_3\text{N}_2} \right) \quad \text{or} \quad \left(\frac{1 \text{ mol Mg}_3\text{N}_2}{1 \text{ mol N}_2} \right)$$

   e. How many moles of nitrogen react to yield 3.452 moles of magnesium nitride?

   $$? \text{ mol N}_2 = 3.452 \text{ mol Mg}_3\text{N}_2 \left(\frac{1 \text{ mol N}_2}{1 \text{ mol Mg}_3\text{N}_2} \right) = \textbf{3.452 mol N}_2$$

26. For many years, it was thought that the formation of sodium perbromate was impossible. But the production of xenon difluoride, XeF_2, which was also thought to be impossible to make, led to the discovery of the following reaction that yields the illusive sodium perbromate.

   $$NaBrO_3 + XeF_2 + H_2O \rightarrow NaBrO_4 + 2HF + Xe$$

   a. Write a conversion factor that could be used to convert between moles of xenon difluoride, XeF_2, and moles of hydrogen fluoride, HF.

   $$\left(\frac{2 \text{ mol HF}}{1 \text{ mol XeF}_2} \right) \quad \text{or} \quad \left(\frac{1 \text{ mol XeF}_2}{2 \text{ mol HF}} \right)$$

   b. How many moles of XeF_2 are necessary to form 16 moles of hydrogen fluoride?

   The number of moles of HF that form is 2 times the number of moles of XeF_2 that react, so **8 moles of XeF_2** form 16 moles of HF.

   $$? \text{ mol XeF}_2 = 16 \text{ mol HF} \left(\frac{1 \text{ mol XeF}_2}{2 \text{ mol HF}} \right) = \textbf{8.0 mol XeF}_2$$

c. What is the maximum number of moles of $NaBrO_4$ that could form in the combination of 2 moles of $NaBrO_3$ and 3 moles of XeF_2?

> The 2 moles of $NaBrO_3$ form a maximum of 2 moles of $NaBrO_4$. It only takes 2 moles of XeF_2 to make 2 moles of $NaBrO_4$, so the XeF_2 is in excess. A maximum of **2 moles of $NaBrO_4$** can form.

$$? \text{ mol } NaBrO_4 = 2 \text{ mol } NaBrO_3 \left(\frac{1 \text{ mol } NaBrO_4}{1 \text{ mol } NaBrO_3} \right) = \textbf{2 mol } NaBrO_4$$

$$? \text{ mol } NaBrO_4 = 3 \text{ mol } XeF_2 \left(\frac{1 \text{ mol } NaBrO_4}{1 \text{ mol } XeF_2} \right) = 3 \text{ mol } NaBrO_4$$

> The 2 moles of $NaBrO_3$ run out first, so it is the limiting reactant.

d. What is the maximum number of moles of $NaBrO_4$ that could form in the combination of 2 moles of $NaBrO_3$ and 3 million moles of XeF_2?

> The XeF_2 is in excess, so no matter how much extra XeF_2 is added, the maximum yield is **2 moles of $NaBrO_4$**.

e. Write a conversion factor that could be used to convert between moles of sodium perbromate, $NaBrO_4$, and moles of hydrogen fluoride, HF.

$$\left(\frac{\textbf{2 mol HF}}{\textbf{1 mol } NaBrO_4} \right) \quad \textbf{or} \quad \left(\frac{\textbf{1 mol } NaBrO_4}{\textbf{2 mol HF}} \right)$$

f. How many moles of HF form along with 5.822 moles of sodium perbromate, $NaBrO_4$?

$$? \text{ mol HF} = 5.822 \text{ mol } NaBrO_4 \left(\frac{2 \text{ mol HF}}{1 \text{ mol } NaBrO_4} \right) = \textbf{11.64 mol HF}$$

28. In Chapter 3, you were told that you can expect halogen atoms to form one covalent bond, but there are many compounds that contain halogen atoms with more than one bond. For example, bromine pentafluoride, which is used as an oxidizing agent in rocket propellants, has bromine atoms with five covalent bonds. Liquid bromine pentafluoride is the only product in the reaction of gaseous bromine monofluoride with fluorine gas.

a. Write a complete balanced equation, including states, for this reaction.

> $$\textbf{BrF(g) + 2F}_2\textbf{(g)} \rightarrow \textbf{BrF}_5\textbf{(l)}$$

b. Write a conversion factor that could be used to convert between moles of fluorine and moles of bromine pentafluoride.

$$\left(\frac{\textbf{1 mol BrF}_5}{\textbf{2 mol F}_2} \right) \quad \textbf{or} \quad \left(\frac{\textbf{2 mol F}_2}{\textbf{1 mol BrF}_5} \right)$$

c. How many moles of bromine pentafluoride form when 6 moles of fluorine react completely?

The number of moles of BrF_5 that form is one-half the number of moles of F_2 that react, so **3 moles of BrF_5** form from 6 moles of F_2.

$$? \text{ mol } BrF_5 = 6 \text{ mol } BrF \left(\frac{1 \text{ mol } BrF_5}{1 \text{ mol } BrF} \right) = \textbf{3 mol } BrF_5$$

d. What is the maximum number of moles of bromine pentafluoride that could form in the combination of 8 moles of bromine monofluoride with 12 moles of fluorine?

The 8 moles of BrF form a maximum of 8 moles of BrF_5. The 12 moles of F_2 form a maximum of 6 moles of BrF_5. Therefore, the F_2 is limiting, and the BrF is in excess. A maximum of **6 moles of BrF_5** can form.

$$? \text{ mol } BrF_5 = 8 \text{ mol } BrF \left(\frac{1 \text{ mol } BrF_5}{1 \text{ mol } BrF} \right) = 8 \text{ mol } BrF_5$$

$$? \text{ mol } BrF_5 = 12 \text{ mol } F_2 \left(\frac{1 \text{ mol } BrF_5}{2 \text{ mol } F_2} \right) = \textbf{6 mol } BrF_5$$

The 12 moles of F_2 runs out first, so it is the limiting reactant.

e. Write a conversion factor that could be used to convert between moles of bromine monofluoride and moles of bromine pentafluoride.

$$\left(\frac{\textbf{1 mol } BrF_5}{\textbf{1 mol } BrF} \right) \quad \textbf{or} \quad \left(\frac{\textbf{1 mol } BrF}{\textbf{1 mol } BrF_5} \right)$$

f. How many moles of bromine monofluoride must react to yield 0.78 mole of bromine pentafluoride?

$$? \text{ mol } BrF = 0.78 \text{ mol } BrF_5 \left(\frac{1 \text{ mol } BrF}{1 \text{ mol } BrF_5} \right) = \textbf{0.78 mol } BrF$$

31. Aniline, $C_6H_5NH_2$, is used to make many different chemicals, including dyes, photographic chemicals, antioxidants, explosives, and herbicides. It can be formed from nitrobenzene, $C_6H_5NO_2$, in the following reaction with iron(II) chloride as a catalyst. *(Objs #2-4, & 9)*

$$4C_6H_5NO_2 + 9Fe + 4H_2O \xrightarrow{\text{FeCl}_2} 4C_6H_5NH_2 + 3Fe_3O_4$$

a. Write a conversion factor that could be used to convert between moles of iron and moles of nitrobenzene.

$$\left(\frac{\textbf{9 mol Fe}}{\textbf{4 mol } C_6H_5NO_2} \right) \quad \textbf{or} \quad \left(\frac{\textbf{4 mol } C_6H_5NO_2}{\textbf{9 mol Fe}} \right)$$

b. What is the minimum mass of iron that would be necessary to react completely with 810.5 g of nitrobenzene, $C_6H_5NO_2$?

$$? \, g \, Fe \; = \; 810.5 \, g \, C_6H_5NO_2 \left(\frac{1 \, mol \, C_6H_5NO_2}{123.111 \, g \, C_6H_5NO_2} \right) \left(\frac{9 \, mol \, Fe}{4 \, mol \, C_6H_5NO_2} \right) \left(\frac{55.845 \, g \, Fe}{1 \, mol \, Fe} \right)$$

or $\quad ? \, g \, Fe \; = \; 810.5 \, g \, C_6H_5NO_2 \left(\dfrac{9 \times 55.845 \, g \, Fe}{4 \times 123.111 \, g \, C_6H_5NO_2} \right) = $ **827.2 g Fe**

c. Write a conversion factor that could be used to convert between moles of aniline and moles of nitrobenzene.

$$\left(\frac{4 \, mol \, C_6H_5NH_2}{4 \, mol \, C_6H_5NO_2} \right) \quad or \quad \left(\frac{1 \, mol \, C_6H_5NH_2}{1 \, mol \, C_6H_5NO_2} \right)$$

$$or \quad \left(\frac{4 \, mol \, C_6H_5NO_2}{4 \, mol \, C_6H_5NH_2} \right) \quad or \quad \left(\frac{1 \, mol \, C_6H_5NO_2}{1 \, mol \, C_6H_5NH_2} \right)$$

d. What is the maximum mass of aniline, $C_6H_5NH_2$, that can be formed from 810.5 g of nitrobenzene, $C_6H_5NO_2$, with excess iron and water?

$$? \, g \, C_6H_5NH_2 \; = \; 810.5 \, g \, C_6H_5NO_2 \left(\frac{1 \, mol \, C_6H_5NO_2}{123.111 \, g \, C_6H_5NO_2} \right) \left(\frac{4 \, mol \, C_6H_5NH_2}{4 \, mol \, C_6H_5NO_2} \right) \left(\frac{93.128 \, g \, C_6H_5NH_2}{1 \, mol \, C_6H_5NH_2} \right)$$

or $\quad ? \, g \, C_6H_5NH_2 \; = \; 810.5 \, g \, C_6H_5NO_2 \left(\dfrac{4 \times 93.128 \, g \, C_6H_5NH_2}{4 \times 123.111 \, g \, C_6H_5NO_2} \right) = $ **613.1 g C$_6$H$_5$NH$_2$**

e. Write a conversion factor that could be used to convert between moles of Fe_3O_4 and moles of aniline.

$$\left(\frac{3 \, mol \, Fe_3O_4}{4 \, mol \, C_6H_5NH_2} \right) \quad or \quad \left(\frac{4 \, mol \, C_6H_5NH_2}{3 \, mol \, Fe_3O_4} \right)$$

f. What is the mass of Fe_3O_4 formed with the amount of aniline, $C_6H_5NH_2$, calculated in Part (d)?

$$? \, g \, Fe_3O_4 \; = \; 613.1 \, g \, C_6H_5NH_2 \left(\frac{1 \, mol \, C_6H_5NH_2}{93.128 \, g \, C_6H_5NH_2} \right) \left(\frac{3 \, mol \, Fe_3O_4}{4 \, mol \, C_6H_5NH_2} \right) \left(\frac{231.53 \, g \, Fe_3O_4}{1 \, mol \, Fe_3O_4} \right)$$

or $\quad ? \, g \, Fe_3O_4 \; = \; 613.1 \, g \, C_6H_5NH_2 \left(\dfrac{3 \times 231.53 \, g \, Fe_3O_4}{4 \times 93.128 \, g \, C_6H_5NH_2} \right)$

$$= \textbf{1143 g Fe}_3\textbf{O}_4 \textbf{ or 1.143 kg Fe}_3\textbf{O}_4$$

g. If 478.2 g of aniline, $C_6H_5NH_2$, are formed from the reaction of 810.5 g of nitrobenzene, $C_6H_5NO_2$, with excess iron and water, what is the percent yield?

$$\% \, yield = \frac{actual \; yield}{theoretical \; yield} \times 100 = \frac{478.2 \, g \, C_6H_5NH_2}{613.1 \, g \, C_6H_5NH_2} \times 100 = \textbf{78.00\% yield}$$

33. Because of its red-orange color, sodium dichromate, $Na_2Cr_2O_7$, is used in the manufacture of pigments. It can be made by reacting sodium chromate, Na_2CrO_4, with sulfuric acid. The products other than sodium dichromate are sodium sulfate and water. *(Objs #2-4)*

   a. Write a balanced equation for this reaction. (You do not need to write the states.)

$$2Na_2CrO_4 + H_2SO_4 \rightarrow Na_2Cr_2O_7 + Na_2SO_4 + H_2O$$

   b. How many kilograms of sodium chromate, Na_2CrO_4, are necessary to produce 84.72 kg of sodium dichromate, $Na_2Cr_2O_7$?

$$? \text{ kg } Na_2CrO_4 = 84.72 \text{ kg } Na_2Cr_2O_7 \left(\frac{10^3 \text{ g}}{1 \text{ kg}}\right)\left(\frac{1 \text{ mol } Na_2Cr_2O_7}{261.968 \text{ g } Na_2Cr_2O_7}\right)\left(\frac{2 \text{ mol } Na_2CrO_4}{1 \text{ mol } Na_2Cr_2O_7}\right)\left(\frac{161.9733 \text{ g } Na_2CrO_4}{1 \text{ mol } Na_2CrO_4}\right)\left(\frac{1 \text{ kg}}{10^3 \text{ g}}\right)$$

or $$? \text{ kg } Na_2CrO_4 = 84.72 \text{ kg } Na_2Cr_2O_7 \left(\frac{2 \times 161.9733 \text{ kg } Na_2CrO_4}{1 \times 261.968 \text{ kg } Na_2Cr_2O_7}\right) = \textbf{104.8 kg } Na_2CrO_4$$

   c. How many kilograms of sodium sulfate are formed with 84.72 kg of $Na_2Cr_2O_7$?

$$? \text{ kg } Na_2SO_4 = 84.72 \text{ kg } Na_2Cr_2O_7 \left(\frac{10^3 \text{ g}}{1 \text{ kg}}\right)\left(\frac{1 \text{ mol } Na_2Cr_2O_7}{261.968 \text{ g } Na_2Cr_2O_7}\right)\left(\frac{1 \text{ mol } Na_2SO_4}{1 \text{ mol } Na_2Cr_2O_7}\right)\left(\frac{142.043 \text{ g } Na_2SO_4}{1 \text{ mol } Na_2SO_4}\right)\left(\frac{1 \text{ kg}}{10^3 \text{ g}}\right)$$

or $$? \text{ kg } Na_2SO_4 = 84.72 \text{ kg } Na_2Cr_2O_7 \left(\frac{1 \times 142.043 \text{ kg } Na_2SO_4}{1 \times 261.968 \text{ kg } Na_2Cr_2O_7}\right) = \textbf{45.94 kg } Na_2SO_4$$

35. The tanning agent, $Cr(OH)SO_4$, is formed in the reaction of sodium dichromate ($Na_2Cr_2O_7$), sulfur dioxide, and water. Tanning protects animal hides from bacterial attack, reduces swelling, and prevents the fibers from sticking together when the hides dry. This leads to a softer, more flexible leather. *(Objs #2-4)*

$$Na_2Cr_2O_7 + 3SO_2 + H_2O \rightarrow 2Cr(OH)SO_4 + Na_2SO_4$$

   a. How many kilograms of sodium dichromate, $Na_2Cr_2O_7$, are necessary to produce 2.50 kg of $Cr(OH)SO_4$?

$$? \text{ kg } Na_2Cr_2O_7 = 2.50 \text{ kg } Cr(OH)SO_4 \left(\frac{10^3 \text{ g}}{1 \text{ kg}}\right)\left(\frac{1 \text{ mol } Cr(OH)SO_4}{165.067 \text{ g } Cr(OH)SO_4}\right)\left(\frac{1 \text{ mol } Na_2Cr_2O_7}{2 \text{ mol } Cr(OH)SO_4}\right)\left(\frac{261.968 \text{ g } Na_2Cr_2O_7}{1 \text{ mol } Na_2Cr_2O_7}\right)\left(\frac{1 \text{ kg}}{10^3 \text{ g}}\right)$$

or $$? \text{ kg } Na_2Cr_2O_7 = 2.50 \text{ kg } Cr(OH)SO_4 \left(\frac{1 \times 261.968 \text{ kg } Na_2Cr_2O_7}{2 \times 165.067 \text{ kg } Cr(OH)SO_4}\right) = \textbf{1.98 kg } Na_2Cr_2O_7$$

   b. How many megagrams of sodium sulfate are formed with 2.50 Mg of $Cr(OH)SO_4$?

$$? \text{ Mg } Na_2SO_4 = 2.50 \text{ Mg } Cr(OH)SO_4 \left(\frac{10^6 \text{ g}}{1 \text{ Mg}}\right)\left(\frac{1 \text{ mol } Cr(OH)SO_4}{165.067 \text{ g } Cr(OH)SO_4}\right)\left(\frac{1 \text{ mol } Na_2SO_4}{2 \text{ mol } Cr(OH)SO_4}\right)\left(\frac{142.043 \text{ g } Na_2SO_4}{1 \text{ mol } Na_2SO_4}\right)\left(\frac{1 \text{ Mg}}{10^6 \text{ g}}\right)$$

or $$? \text{ Mg } Na_2SO_4 = 2.50 \text{ Mg } Cr(OH)SO_4 \left(\frac{1 \times 142.043 \text{ Mg } Na_2SO_4}{2 \times 165.067 \text{ Mg } Cr(OH)SO_4}\right) = \textbf{1.08 Mg } Na_2SO_4$$

38. Chromium(III) oxide can be made from the reaction of sodium dichromate and ammonium chloride. What is the maximum mass, in grams, of chromium(III) oxide that can be produced from the complete reaction of 123.5 g of sodium dichromate, $Na_2Cr_2O_7$, with 59.5 g of ammonium chloride? The other products are sodium chloride, nitrogen gas, and water. *(Obj #6)*

$$Na_2Cr_2O_7 + 2NH_4Cl \rightarrow 2NaCl + Cr_2O_3 + 4H_2O + N_2$$

$$? \text{ g } Cr_2O_3 = 123.5 \text{ g } Na_2Cr_2O_7 \left(\frac{1 \times 151.990 \text{ g } Cr_2O_3}{1 \times 261.968 \text{ g } Na_2Cr_2O_7} \right) = \mathbf{71.65 \text{ g } Cr_2O_3}$$

$$? \text{ g } Cr_2O_3 = 59.5 \text{ g } NH_4Cl \left(\frac{1 \times 151.990 \text{ g } Cr_2O_3}{2 \times 53.4912 \text{ g } NH_4Cl} \right) = 84.5 \text{ g } Cr_2O_3$$

40. Tetraboron carbide, B_4C, which is used as a protective material in nuclear reactors, can be made from boric acid, H_3BO_3. *(Objs #6 & 7)*

$$4H_3BO_3 + 7C \xrightarrow{2400 \text{ °C}} B_4C + 6CO + 6H_2O$$

a. What is the maximum mass, in kilograms, of B_4C formed in the reaction of 30.0 kg of carbon with 54.785 kg of H_3BO_3?

$$? \text{ kg } B_4C = 30.0 \text{ kg } C \left(\frac{1 \times 55.255 \text{ kg } B_4C}{7 \times 12.011 \text{ kg } C} \right) = 19.7 \text{ kg } B_4C$$

$$? \text{ kg } B_4C = 54.785 \text{ kg } H_3BO_3 \left(\frac{1 \times 55.255 \text{ kg } B_4C}{4 \times 61.833 \text{ kg } H_3BO_3} \right) = \mathbf{12.239 \text{ kg } B_4C}$$

b. Explain why one of the substances in Part (a) is in excess and one is limiting.

There are two reasons why we are not surprised that the carbon is in excess. (1) We would expect carbon to be less expensive than the less common boric acid, and (2) the excess carbon can be separated easily from the solid B_4C by converting it to gaseous carbon dioxide or carbon monoxide.

42. Aniline, $C_6H_5NH_2$, which is used to make antioxidants, can be formed from nitrobenzene, $C_6H_5NO_2$, in the following reaction. *(Objs #6 & 7)*

$$4C_6H_5NO_2 + 9Fe + 4H_2O \xrightarrow{FeCl_2} 4C_6H_5NH_2 + 3Fe_3O_4$$

a. What is the maximum mass of aniline, $C_6H_5NH_2$, formed in the reaction of 810.5 g of nitrobenzene, $C_6H_5NO_2$, with 985.0 g of Fe and 250 g of H_2O?

$$? \text{ g } C_6H_5NH_2 = 810.5 \text{ g } C_6H_5NO_2 \left(\frac{4 \times 93.128 \text{ g } C_6H_5NH_2}{4 \times 123.111 \text{ g } C_6H_5NO_2} \right) = \mathbf{613.1 \text{ g } C_6H_5NH_2}$$

$$? \text{ g } C_6H_5NH_2 = 985.0 \text{ g } Fe \left(\frac{4 \times 93.128 \text{ g } C_6H_5NH_2}{9 \times 55.845 \text{ g } Fe} \right) = 730.0 \text{ g } C_6H_5NH_2$$

$$? \text{ g } C_6H_5NH_2 = 250 \text{ g } H_2O \left(\frac{4 \times 93.128 \text{ g } C_6H_5NH_2}{4 \times 18.0153 \text{ g } H_2O} \right) = 1.29 \times 10^3 \text{ g } C_6H_5NH_2$$

b. Explain why two of these substances are in excess and one is limiting.

> Both iron and water would be less expensive than nitrobenzene. They would also be expected to be less toxic than nitrobenzene.

44. Calcium carbide, CaC_2, is formed in the reaction between calcium oxide and carbon. The other product is carbon monoxide.   *(Obj #6)*

a. Write a balanced equation for this reaction. (You do not need to write the states.)

$$CaO + 3C \rightarrow CaC_2 + CO$$

b. If you were designing the procedure for producing calcium carbide from calcium oxide and carbon, which of the reactants would you have as the limiting reactant? Why?

> The carbon is probably best to have in excess. We would expect carbon to be less expensive than the calcium oxide, and the excess carbon can be separated easily from the solid CaC_2 by converting it to gaseous carbon dioxide or carbon monoxide. Thus, the CaO would be limiting.

c. Assuming 100% yield from the limiting reactant, what are the approximate amounts of CaO and carbon that you would combine to form 860.5 g of CaC_2?

$$? \text{ g } CaO = 860.5 \text{ g } CaC_2 \left(\frac{1 \times 56.077 \text{ g } CaO}{1 \times 64.100 \text{ g } CaC_2} \right) = \textbf{752.8 g CaO}$$

$$? \text{ g } C = 860.5 \text{ g } CaC_2 \left(\frac{3 \times 12.011 \text{ g } C}{1 \times 64.100 \text{ g } CaC_2} \right) = \textbf{483.7 g C}$$

> We would add 752.8 g CaO and well over 483.7 g C.

46. Give four reasons why the actual yield in a chemical reaction is less than the theoretical yield.   *(Obj #8)*

> (1) Many chemical reactions are significantly reversible. Because there is a constant conversion of reactants to products and products to reactants, the reaction never proceeds completely to products. (2) It is common, especially in reactions involving organic compounds, to have side reactions. These reactions form products other than the desired product. (3) Sometimes a reaction is so slow that it has not reached the maximum yield by the time the product is isolated. (4) Even if 100% of the limiting reactant proceeds to products, the product still usually needs to be separated from the other components in the product mixture. (The other components include excess reactants, products of side reactions, and other impurities.) This separation generally involves some loss of product.

48. Does the reactant in excess affect the *actual* yield for a reaction? If it does, explain how.

> Although the maximum (or theoretical) yield of a reaction is determined by the limiting reactant rather than reactants in excess, reactants that are in excess can affect the actual yield of an experiment. Sometimes the actual yield is less than the theoretical yield because the reaction is reversible. Adding a large excess of one of the reactants

ensures that the limiting reactant reacts as completely as possible (by speeding up the forward rate in the reversible reaction and driving the reaction toward a greater actual yield of products).

Section 10.3 Molarity and Equation Stoichiometry

50. What is the molarity of a solution made by dissolving 37.452 g of aluminum sulfate, $Al_2(SO_4)_3$, in water and diluting with water to 250.0 mL total?   *(Obj #10)*

$$\text{Molarity} = \frac{?\ \text{mol}\ Al_2(SO_4)_3}{\text{L solution}} = \frac{37.452\ g\ Al_2(SO_4)_3}{250.0\ \text{mL soln}} \left(\frac{1\ \text{mol}\ Al_2(SO_4)_3}{342.154\ g\ Al_2(SO_4)_3} \right) \left(\frac{10^3\ \text{mL}}{1\ L} \right)$$

$$= \textbf{0.4378 M } Al_2(SO_4)_3$$

52. The following equation represents the first step in the conversion of UO_3, found in uranium ore, into the uranium compounds called "yellow cake."

$$UO_3 + H_2SO_4 \rightarrow UO_2SO_4 + H_2O$$

a. How many milliliters of 18.0 M H_2SO_4 are necessary to react completely with 249.6 g of UO_3?   *(Objs #11-13)*

$$?\ \text{mL}\ H_2SO_4\ \text{soln} = 249.6\ g\ UO_3 \left(\frac{1\ \text{mol}\ UO_3}{286.0271\ g\ UO_3} \right) \left(\frac{1\ \text{mol}\ H_2SO_4}{1\ \text{mol}\ UO_3} \right) \left(\frac{10^3\ \text{mL}\ H_2SO_4\ \text{soln}}{18.0\ \text{mol}\ H_2SO_4} \right)$$

$$= \textbf{48.5 mL } H_2SO_4 \textbf{ soln}$$

b. What is the maximum mass, in grams, of UO_2SO_4 that forms from the complete reaction of 125 mL of 18.0 M H_2SO_4?   *(Objs #11-13)*

$$?\ g\ UO_2SO_4 = 125\ \text{mL}\ H_2SO_4\ \text{soln} \left(\frac{18.0\ \text{mol}\ H_2SO_4}{10^3\ \text{mL}\ H_2SO_4\ \text{soln}} \right) \left(\frac{1\ \text{mol}\ UO_2SO_4}{1\ \text{mol}\ H_2SO_4} \right) \left(\frac{366.091\ g\ UO_2SO_4}{1\ \text{mol}\ UO_2SO_4} \right)$$

$$= \textbf{824 g } UO_2SO_4$$

54. When a water solution of sodium sulfite, Na_2SO_3, is added to a water solution of iron(II) chloride, $FeCl_2$, iron(II) sulfite, $FeSO_3$, precipitates from the solution. *(Objs #11-13)*

a. Write a balanced equation for this reaction.

$$Na_2SO_3(aq) + FeCl_2(aq) \rightarrow 2NaCl(aq) + FeSO_3(s)$$

b. What is the maximum mass of iron(II) sulfite that will precipitate from a solution prepared by adding an excess of an Na_2SO_3 solution to 25.00 mL of 1.009 M $FeCl_2$?

$$?\ g\ FeSO_3 = 25.00\ \text{mL}\ FeCl_2\ \text{soln} \left(\frac{1.009\ \text{mol}\ FeCl_2}{10^3\ \text{mL}\ FeCl_2\ \text{soln}} \right) \left(\frac{1\ \text{mol}\ FeSO_3}{1\ \text{mol}\ FeCl_2} \right) \left(\frac{135.909\ g\ FeSO_3}{1\ \text{mol}\ FeSO_3} \right)$$

$$= \textbf{3.428 g } FeSO_3$$

56. Consider the neutralization reaction that takes place when nitric acid reacts with aqueous potassium hydroxide. *(Objs #11-13)*

a. Write a conversion factor that relates moles of HNO_3 to moles of KOH for this reaction.

$$\left(\frac{1\ \textbf{mol } HNO_3}{1\ \textbf{mol } KOH} \right)$$

b. What is the minimum volume of 1.50 M HNO_3 necessary to neutralize completely the hydroxide in 125.0 mL of 0.501 M KOH?

$$? \text{ mL } HNO_3 \text{ soln} = 125.0 \text{ mL KOH soln} \left(\frac{0.501 \text{ mol KOH}}{10^3 \text{ mL KOH soln}} \right) \left(\frac{1 \text{ mol } HNO_3}{1 \text{ mol KOH}} \right) \left(\frac{10^3 \text{ mL } HNO_3 \text{ soln}}{1.50 \text{ mol } HNO_3} \right)$$

= 41.8 mL HNO_3 soln

58. Consider the neutralization reaction that takes place when sulfuric acid reacts with aqueous sodium hydroxide. (*Objs #11-13*)

a. Write a conversion factor that relates moles of H_2SO_4 to moles of NaOH for this reaction.

$$\left(\frac{\textbf{1 mol } H_2SO_4}{\textbf{2 mol NaOH}} \right)$$

b. What is the minimum volume of 6.02 M H_2SO_4 necessary to neutralize completely the hydroxide in 47.5 mL of 2.5 M NaOH?

$$? \text{ mL } H_2SO_4 \text{ soln} = 47.5 \text{ mL NaOH soln} \left(\frac{2.5 \text{ mol NaOH}}{10^3 \text{ mL NaOH soln}} \right) \left(\frac{1 \text{ mol } H_2SO_4}{2 \text{ mol NaOH}} \right) \left(\frac{10^3 \text{ mL } H_2SO_4 \text{ soln}}{6.02 \text{ mol } H_2SO_4} \right)$$

= 9.9 mL H_2SO_4 soln

60. Consider the neutralization reaction that takes place when hydrochloric acid reacts with solid cobalt(II) hydroxide. (*Objs #11-13*)

a. Write a conversion factor that relates moles of HCl to moles of $Co(OH)_2$ for this reaction.

$$\left(\frac{\textbf{2 mol HCl}}{\textbf{1 mol Co(OH)}_2} \right)$$

b. What is the minimum volume of 6.14 M HCl necessary to react completely with 2.53 kg of solid cobalt(II) hydroxide, $Co(OH)_2$?

$$? \text{ L HCl soln} = 2.53 \text{ kg } Co(OH)_2 \left(\frac{10^3 \text{ g}}{1 \text{ kg}} \right) \left(\frac{1 \text{ mol } Co(OH)_2}{92.9479 \text{ g } Co(OH)_2} \right) \left(\frac{2 \text{ mol HCl}}{1 \text{ mol } Co(OH)_2} \right) \left(\frac{1 \text{ L HCl soln}}{6.14 \text{ mol HCl}} \right)$$

= 8.87 L HCl soln

62. Consider the neutralization reaction that takes place when nitric acid reacts with solid chromium(III) hydroxide. (*Objs #19-21*)

a. Write a conversion factor that relates moles of HNO_3 to moles of $Cr(OH)_3$ for this reaction.

$$\left(\frac{\textbf{3 mol } HNO_3}{\textbf{1 mol Cr(OH)}_3} \right)$$

a. What is the minimum volume of 2.005 M HNO_3 necessary to react completely with 0.5187 kg of solid chromium(III) hydroxide, $Cr(OH)_3$?

$$? \text{ L } HNO_3 \text{ soln} = 0.5187 \text{ kg } Cr(OH)_3 \left(\frac{10^3 \text{ g}}{1 \text{ kg}} \right) \left(\frac{1 \text{ mol } Cr(OH)_3}{103.0181 \text{ g } Cr(OH)_3} \right) \left(\frac{3 \text{ mol } HNO_3}{1 \text{ mol } Cr(OH)_3} \right) \left(\frac{1 \text{ L } HNO_3 \text{ soln}}{2.005 \text{ mol } HNO_3} \right)$$

= 7.534 L HNO_3 soln

Additional Problems

64. Because nitrogen and phosphorus are both nonmetallic elements in group 15 on the periodic table, we expect them to react with other elements in similar ways. This is true, but there are also distinct differences between them. For example, nitrogen atoms form stable triple bonds to carbon atoms in substances such as hydrogen cyanide (often called hydrocyanic acid), HCN. Phosphorus atoms also form triple bonds to carbon atoms in substances such as HCP, but the substances that form are much less stable. The compound HCP can be formed in the following reaction.

$$CH_4 + PH_3 \xrightarrow{\text{electric arc}} HCP + 3H_2$$

a. Write a conversion factor that could be used to convert between moles of HCP and moles of H_2.

$$\left(\frac{3 \text{ mol } H_2}{1 \text{ mol HCP}}\right) \quad \text{or} \quad \left(\frac{1 \text{ mol HCP}}{3 \text{ mol } H_2}\right)$$

b. How many moles of HCP form along with 9 moles of H_2?

The number of moles of HCP that form is one-third the number of moles of H_2 that forms, so **3 moles of HCP** form with 9 moles of H_2.

$$? \text{ mol HCP} = 9 \text{ mol } H_2 \left(\frac{1 \text{ mol HCP}}{3 \text{ mol } H_2}\right) = \textbf{3 mol HCP}$$

c. Write a conversion factor that could be used to convert between moles of methane, CH_4, and moles of hydrogen, H_2.

$$\left(\frac{3 \text{ mol } H_2}{1 \text{ mol } CH_4}\right) \quad \text{or} \quad \left(\frac{1 \text{ mol } CH_4}{3 \text{ mol } H_2}\right)$$

d. How many moles of hydrogen gas form when 1.8834 moles of CH_4 react with an excess of PH_3?

$$? \text{ mol } H_2 = 1.8834 \text{ mol } CH_4 \left(\frac{3 \text{ mol } H_2}{1 \text{ mol } CH_4}\right) = \textbf{5.6502 mol } \mathbf{H_2}$$

66. Iodine pentafluoride is an incendiary agent, which is a substance that ignites combustible materials. Iodine pentafluoride is usually made by passing fluorine gas over solid iodine, but it also forms when iodine monofluoride changes into the element iodine and iodine pentafluoride.

a. Write a balanced equation, without including states, for the conversion of iodine monofluoride into iodine and iodine pentafluoride.

$$\mathbf{5IF} \rightarrow \mathbf{2I_2} + \mathbf{IF_5}$$

b. How many moles of the element iodine form when 15 moles of iodine monofluoride react completely?

The number of moles of I_2 that form is two-fifths the number of moles of IF that react, so **6.0 moles of I_2** form from 15 moles of IF.

$$? \text{ mol } I_2 = 15 \text{ mol IF} \left(\frac{2 \text{ mol } I_2}{5 \text{ mol IF}}\right) = \textbf{6.0 mol } \mathbf{I_2}$$

c. How many moles of iodine pentafluoride form when 7.939 moles of iodine monofluoride react completely?

$$? \text{ mol IF}_5 = 7.939 \text{ mol IF} \left(\frac{1 \text{ mol IF}_5}{5 \text{ mol IF}} \right) = \textbf{1.588 mol IF}_5$$

68. Xenon hexafluoride is a better fluorinating agent than the xenon difluoride described in the previous problem, but it must be carefully isolated from any moisture. This is because xenon hexafluoride reacts with water to form the dangerously explosive xenon trioxide and hydrogen fluoride (hydrogen monofluoride).

a. Write a balanced equation, without including states, for the reaction of xenon hexafluoride and water to form xenon trioxide and hydrogen fluoride.

$$\textbf{XeF}_6 + \textbf{3H}_2\textbf{O} \rightarrow \textbf{XeO}_3 + \textbf{6HF}$$

b. How many moles of hydrogen fluoride form when 0.50 mole of xenon hexafluoride reacts completely?

The number of moles of HF that forms is 6 times the number of moles of XeF_6 that reacts, so **3.0 moles of HF** form when 0.50 mole of XeF_6 react.

$$? \text{ mol HF} = 0.50 \text{ mol XeF}_6 \left(\frac{6 \text{ mol HF}}{1 \text{ mol XeF}_6} \right) = \textbf{3.0 mol HF}$$

c. What is the maximum number of moles of xenon trioxide that can form in the combination of 7 moles of xenon hexafluoride and 18 moles of water?

A maximum of 7 moles of XeO_3 forms from 7 moles of XeF_6. A maximum of 6 moles of XeO_3 forms from 18 moles of H_2O. Therefore, the H_2O is limiting, and the XeF_6 is in excess. A maximum of **6 moles of XeO₃** can form.

$$? \text{ mol XeO}_3 = 7 \text{ mol XeF}_6 \left(\frac{1 \text{ mol XeO}_3}{1 \text{ mol XeF}_6} \right) = 7 \text{ mol XeO}_3$$

$$? \text{ mol XeO}_3 = 18 \text{ mol H}_2\text{O} \left(\frac{1 \text{ mol XeO}_3}{3 \text{ mol H}_2\text{O}} \right) = \textbf{6 mol XeO}_3$$

70. Hydriodic acid is produced industrially by the reaction of hydrazine, N_2H_4, with iodine, I_2. HI(*aq*) is used to make iodine salts such as AgI, which are used to seed clouds to promote rain. What is the minimum mass of iodine, I_2, necessary to react completely with 87.0 g of hydrazine, N_2H_4?

$$N_2H_4 + 2I_2 \rightarrow 4HI + N_2$$

$$? \text{ g I}_2 = 87.0 \text{ g N}_2\text{H}_4 \left(\frac{1 \text{ mol N}_2\text{H}_4}{32.0452 \text{ g N}_2\text{H}_4} \right) \left(\frac{2 \text{ mol I}_2}{1 \text{ mol N}_2\text{H}_4} \right) \left(\frac{253.8090 \text{ g I}_2}{1 \text{ mol I}_2} \right)$$

$$\text{or } ? \text{ g I}_2 = 87.0 \text{ g N}_2\text{H}_4 \left(\frac{2 \times 253.8090 \text{ g I}_2}{1 \times 32.0452 \text{ g N}_2\text{H}_4} \right) = \textbf{1.38} \times \textbf{10}^3 \textbf{ g I}_2 \text{ or } \textbf{1.38 kg I}_2$$

72. Because plants need nitrogen compounds, potassium compounds, and phosphorus compounds to grow, these are often added to the soil as fertilizers. Potassium sulfate, which is used to make fertilizers, is made industrially by reacting potassium chloride with sulfur dioxide gas, oxygen gas, and water. Hydrochloric acid is formed with the potassium sulfate.

   a. Write a balanced equation for this reaction. (You do not need to include states.)

$$2KCl + SO_2 + \tfrac{1}{2}O_2 + H_2O \rightarrow K_2SO_4 + 2HCl$$

$$or \quad 4KCl + 2SO_2 + O_2 + 2H_2O \rightarrow 2K_2SO_4 + 4HCl$$

   b. What is the maximum mass, in kilograms, of potassium sulfate that can be formed from 2.76×10^5 kg of potassium chloride with excess sulfur dioxide, oxygen, and water?

$$? \text{ kg } K_2SO_4 = 2.76 \times 10^5 \text{ kg KCl} \left(\frac{10^3 \text{ g}}{1 \text{ kg}}\right)\left(\frac{1 \text{ mol KCl}}{74.5510 \text{ g KCl}}\right)\left(\frac{2 \text{ mol } K_2SO_4}{4 \text{ mol KCl}}\right)\left(\frac{174.260 \text{ g } K_2SO_4}{1 \text{ mol } K_2SO_4}\right)\left(\frac{1 \text{ kg}}{10^3 \text{ g}}\right)$$

$$or \quad ? \text{ kg } K_2SO_4 = 2.76 \times 10^5 \text{ kg KCl} \left(\frac{2 \times 174.260 \text{ kg } K_2SO_4}{4 \times 74.5510 \text{ kg KCl}}\right) = \textbf{3.23} \times \textbf{10}^{\textbf{5}} \textbf{ kg } \textbf{K}_2\textbf{SO}_4$$

   c. If 2.94×10^5 kg of potassium sulfate is isolated from the reaction of 2.76×10^5 kg of potassium chloride, what is the percent yield?

$$\% \text{ yield} = \frac{\text{actual yield}}{\text{theoretical yield}} \times 100 = \frac{2.94 \times 10^5 \text{ kg } K_2SO_4}{3.23 \times 10^5 \text{ kg } K_2SO_4} \times 100 = \textbf{91.0\% yield}$$

74. The element phosphorus can be made by reacting carbon in the form of coke with calcium phosphate, $Ca_3(PO_4)_2$, which is found in phosphate rock.

$$Ca_3(PO_4)_2 + 5C \rightarrow 3CaO + 5CO + 2P$$

   a. What is the minimum mass of carbon, C, necessary to react completely with 67.45 Mg of $Ca_3(PO_4)_2$?

$$? \text{ Mg C} = 67.45 \text{ Mg } Ca_3(PO_4)_2 \left(\frac{10^6 \text{ g}}{1 \text{ Mg}}\right)\left(\frac{1 \text{ mol } Ca_3(PO_4)_2}{310.18 \text{ g } Ca_3(PO_4)_2}\right)\left(\frac{5 \text{ mol C}}{1 \text{ mol } Ca_3(PO_4)_2}\right)\left(\frac{12.011 \text{ g C}}{1 \text{ mol C}}\right)\left(\frac{1 \text{ Mg}}{10^6 \text{ g}}\right)$$

$$or \quad ? \text{ Mg C} = 67.45 \text{ Mg } Ca_3(PO_4)_2 \left(\frac{5 \times 12.011 \text{ Mg C}}{1 \times 310.18 \text{ Mg } Ca_3(PO_4)_2}\right) = \textbf{13.06 Mg C}$$

   b. What is the maximum mass of phosphorus produced from the reaction of 67.45 Mg of $Ca_3(PO_4)_2$ with an excess of carbon?

$$? \text{ Mg P} = 67.45 \text{ Mg } Ca_3(PO_4)_2 \left(\frac{10^6 \text{ g}}{1 \text{ Mg}}\right)\left(\frac{1 \text{ mol } Ca_3(PO_4)_2}{310.18 \text{ g } Ca_3(PO_4)_2}\right)\left(\frac{2 \text{ mol P}}{1 \text{ mol } Ca_3(PO_4)_2}\right)\left(\frac{30.9738 \text{ g P}}{1 \text{ mol P}}\right)\left(\frac{1 \text{ Mg}}{10^6 \text{ g}}\right)$$

$$or \quad ? \text{ Mg P} = 67.45 \text{ Mg } Ca_3(PO_4)_2 \left(\frac{2 \times 30.9738 \text{ Mg P}}{1 \times 310.18 \text{ Mg } Ca_3(PO_4)_2}\right) = \textbf{13.47 Mg P}$$

c. What mass of calcium oxide, CaO, is formed with the mass of phosphorus calculated in Part (b)?

$$? \text{ Mg CaO} = 13.47 \text{ Mg P} \left(\frac{10^6 \text{ g}}{1 \text{ Mg}} \right) \left(\frac{1 \text{ mol P}}{30.9738 \text{ g P}} \right) \left(\frac{3 \text{ mol CaO}}{2 \text{ mol P}} \right) \left(\frac{56.077 \text{ g CaO}}{1 \text{ mol CaO}} \right) \left(\frac{1 \text{ Mg}}{10^6 \text{ g}} \right)$$

or $? \text{ Mg CaO} = 13.47 \text{ Mg P} \left(\dfrac{3 \times 56.077 \text{ Mg CaO}}{2 \times 30.9738 \text{ Mg P}} \right) = \textbf{36.58 Mg CaO}$

d. If 11.13 Mg of phosphorus is formed in the reaction of 67.45 Mg of $Ca_3(PO_4)_2$ with an excess of carbon, what is the percent yield?

$$\% \text{ yield} = \frac{\text{actual yield}}{\text{theoretical yield}} \times 100 = \frac{11.13 \text{ Mg P}}{13.47 \text{ Mg P}} \times 100 = \textbf{82.63\% yield}$$

76. Thionyl chloride, $SOCl_2$, is a widely used source of chlorine in the formation of pesticides, pharmaceuticals, dyes, and pigments. It can be formed from disulfur dichloride in the following reaction.

$$2SO_2 + S_2Cl_2 + 3Cl_2 \rightarrow 4SOCl_2$$

If 1.140 kg of thionyl chloride is isolated from the reaction of 457.6 grams of disulfur dichloride, S_2Cl_2, with excess sulfur dioxide and chlorine gas, what is the percent yield?

$$? \text{ kg SOCl}_2 = 457.6 \text{ g S}_2\text{Cl}_2 \left(\frac{1 \text{ mol S}_2\text{Cl}_2}{135.037 \text{ g S}_2\text{Cl}_2} \right) \left(\frac{4 \text{ mol SOCl}_2}{1 \text{ mol S}_2\text{Cl}_2} \right) \left(\frac{118.971 \text{ g SOCl}_2}{1 \text{ mol SOCl}_2} \right) \left(\frac{1 \text{ kg}}{10^3 \text{ g}} \right)$$

or $? \text{ kg SOCl}_2 = 457.6 \text{ g S}_2\text{Cl}_2 \left(\dfrac{4 \times 118.971 \text{ g SOCl}_2}{1 \times 135.037 \text{ g S}_2\text{Cl}_2} \right) \left(\dfrac{1 \text{ kg}}{10^3 \text{ g}} \right) = 1.613 \text{ kg SOCl}_2$

$$\% \text{ yield} = \frac{\text{actual yield}}{\text{theoretical yield}} \times 100 = \frac{1.140 \text{ kg SOCl}_2}{1.613 \text{ kg SOCl}_2} \times 100 = \textbf{70.68\% yield}$$

78. Sodium dichromate, $Na_2Cr_2O_7$, is converted to chromium(III) sulfate, which is used in the tanning of animal hides. Sodium dichromate can be made by reacting sodium chromate, Na_2CrO_4, with water and carbon dioxide.

$$2Na_2CrO_4 + H_2O + 2CO_2 \rightleftharpoons Na_2Cr_2O_7 + 2NaHCO_3$$

a. Show that the sodium chromate is the limiting reactant when 87.625 g of Na_2CrO_4 reacts with 10.008 g of water and excess carbon dioxide.

$$? \text{ g Na}_2\text{Cr}_2\text{O}_7 = 87.625 \text{ g Na}_2\text{CrO}_4 \left(\frac{1 \text{ mol Na}_2\text{CrO}_4}{161.9733 \text{ g Na}_2\text{CrO}_4} \right) \left(\frac{1 \text{ mol Na}_2\text{Cr}_2\text{O}_7}{2 \text{ mol Na}_2\text{CrO}_4} \right) \left(\frac{261.968 \text{ g Na}_2\text{Cr}_2\text{O}_7}{1 \text{ mol Na}_2\text{Cr}_2\text{O}_7} \right)$$

or $? \text{ g Na}_2\text{Cr}_2\text{O}_7 = 87.625 \text{ g Na}_2\text{CrO}_4 \left(\dfrac{1 \times 261.968 \text{ g Na}_2\text{Cr}_2\text{O}_7}{2 \times 161.9733 \text{ g Na}_2\text{CrO}_4} \right) = 70.860 \text{ g Na}_2\text{Cr}_2\text{O}_7$

$$? \text{ g Na}_2\text{Cr}_2\text{O}_7 = 10.008 \text{ g H}_2\text{O} \left(\frac{1 \text{ mol H}_2\text{O}}{18.0153 \text{ g H}_2\text{O}} \right) \left(\frac{1 \text{ mol Na}_2\text{Cr}_2\text{O}_7}{1 \text{ mol H}_2\text{O}} \right) \left(\frac{261.968 \text{ g Na}_2\text{Cr}_2\text{O}_7}{1 \text{ mol Na}_2\text{Cr}_2\text{O}_7} \right)$$

or $? \text{ g Na}_2\text{Cr}_2\text{O}_7 = 10.008 \text{ g H}_2\text{O} \left(\dfrac{1 \times 261.968 \text{ g Na}_2\text{Cr}_2\text{O}_7}{1 \times 18.0153 \text{ g H}_2\text{O}} \right) = 145.53 \text{ g Na}_2\text{Cr}_2\text{O}_7$

Since the Na_2CrO_4 forms the least product, it is the limiting reactant.

b. Explain why the carbon dioxide and water are in excess and sodium chromate is limiting.

> Both water and carbon dioxide are very inexpensive and nontoxic. Since CO_2 is a gas and since water can be easily converted to steam, they are also very easily separated from solid products. Adding an excess of these substances drives the reversible reaction toward products and yields a more complete conversion of Na_2CrO_4 to $Na_2Cr_2O_7$.

80. What is the molarity of a solution made by dissolving 100.065 g of $SnBr_2$ in water and diluting with water to 1.00 L total?

$$\frac{?\ mol\ SnBr_2}{1\ L\ soln} = \frac{100.065\ g\ SnBr_2}{1.00\ L\ soln}\left(\frac{1\ mol\ SnBr_2}{278.52\ g\ SnBr_2}\right) = \textbf{0.359 M SnBr}_2$$

82. A precipitation reaction takes place when a water solution of sodium carbonate, Na_2CO_3, is added to a water solution of chromium(III) nitrate, $Cr(NO_3)_3$.

a. Write a balanced equation for this reaction.

$$3Na_2CO_3(aq) + 2Cr(NO_3)_3(aq) \rightarrow Cr_2(CO_3)_3(s) + 6NaNO_3(aq)$$

b. What is the maximum mass of chromium(III) carbonate that will precipitate from a solution prepared by adding an excess of an Na_2CO_3 solution to 10.00 mL of 0.100 M $Cr(NO_3)_3$?

$$?\ g\ Cr_2(CO_3)_3 = 10.00\ mL\ Cr(NO_3)_3\ soln\left(\frac{0.100\ mol\ Cr(NO_3)_3}{10^3\ mL\ Cr(NO_3)_3\ soln}\right)\left(\frac{1\ mol\ Cr_2(CO_3)_3}{2\ mol\ Cr(NO_3)_3}\right)\left(\frac{284.020\ g\ Cr_2(CO_3)_3}{1\ mol\ Cr_2(CO_3)_3}\right)$$

$$= \textbf{0.142 g Cr}_2\textbf{(CO}_3\textbf{)}_3$$

84. Consider the neutralization reaction between nitric acid and aqueous barium hydroxide.

a. Write a conversion factor that shows the ratio of moles of nitric acid to moles of barium hydroxide.

$$\left(\frac{\textbf{2 mol HNO}_3}{\textbf{1 mol Ba(OH)}_2}\right)$$

b. What volume of 1.09 M nitric acid would be necessary to neutralize the hydroxide in 25.00 mL of 0.159 M barium hydroxide?

$$?\ mL\ HNO_3\ soln = 25.00\ mL\ Ba(OH)_2\ soln\left(\frac{0.159\ mol\ Ba(OH)_2}{10^3\ mL\ Ba(OH)_2}\right)\left(\frac{2\ mol\ HNO_3}{1\ mol\ Ba(OH)_2}\right)\left(\frac{10^3\ mL\ HNO_3\ soln}{1.09\ mol\ HNO_3}\right)$$

$$= \textbf{7.29 mL HNO}_3\textbf{ soln}$$

86. Consider the neutralization reaction between hydrochloric acid and solid zinc carbonate.

a. Write a conversion factor that shows the ratio of moles of hydrochloric acid to moles of zinc carbonate.

$$\left(\frac{\textbf{2 mol HCl}}{\textbf{1 mol ZnCO}_3}\right)$$

b. What volume of 0.500 M hydrochloric acid would be necessary to neutralize and dissolve 562 milligrams of solid zinc carbonate?

$$? \text{ mL HCl soln} = 562 \text{ mg } ZnCO_3 \left(\frac{1 \text{ g}}{10^3 \text{ mg}}\right) \left(\frac{1 \text{ mol } ZnCO_3}{125.40 \text{ g } ZnCO_3}\right) \left(\frac{2 \text{ mol HCl}}{1 \text{ mol } ZnCO_3}\right) \left(\frac{10^3 \text{ mL HCl soln}}{0.500 \text{ mol HCl}}\right)$$

= 17.9 mL HCl soln

Challenge Problems

Some of problems 88-98 include conversions between masses of pure substances and masses of mixtures that contain the pure substances, using percentages as conversion factors. If you need some help with them, visit the Web address below:

www.chemplace.com/college/

88. A solution is made by adding 22.609 g of a solid that is 96.3% NaOH to a beaker of water. What volume of 2.00 M H_2SO_4 is necessary to neutralize the NaOH in this solution?

$$H_2SO_4(aq) + 2NaOH(aq) \rightarrow 2H_2O(l) + Na_2SO_4(aq)$$

$$? \text{ mL } H_2SO_4 = 22.609 \text{ g solid} \left(\frac{96.3 \text{ g NaOH}}{100 \text{ g solid}}\right) \left(\frac{1 \text{ mol NaOH}}{39.9971 \text{ g NaOH}}\right) \left(\frac{1 \text{ mol } H_2SO_4}{2 \text{ mol NaOH}}\right) \left(\frac{10^3 \text{ mL } H_2SO_4}{2.00 \text{ mol } H_2SO_4}\right)$$

= 136 mL H_2SO_4 solution

90. Aluminum sulfate, often called alum, is used in paper making to increase the paper's stiffness and smoothness and to help keep the ink from running. It is made from the reaction of sulfuric acid with the aluminum oxide found in bauxite ore. The products are aluminum sulfate and water. Bauxite ore is 30% to 75% aluminum oxide.

a. Write a balanced equation for this reaction. (You do not need to write the states.)

$$3H_2SO_4 + Al_2O_3 \rightarrow Al_2(SO_4)_3 + 3H_2O$$

b. What is the maximum mass, in kilograms, of aluminum sulfate that could be formed from 2.3×10^3 kilograms of bauxite ore that is 62% aluminum oxide?

$$? \text{ kg } Al_2(SO_4)_3 = 2.3 \times 10^3 \text{ kg ore} \left(\frac{62 \text{ kg } Al_2O_3}{100 \text{ kg ore}}\right) \left(\frac{1 \times 342.154 \text{ kg } Al_2(SO_4)_3}{1 \times 101.9612 \text{ kg } Al_2O_3}\right)$$

= 4.8×10^3 kg Al₂(SO₄)₃

92. Sodium tripolyphosphate, or STPP, $Na_5P_3O_{10}$, is used in detergents. It is made by combining phosphoric acid with sodium carbonate at 300 to 500 °C. What is the minimum mass, in kilograms, of sodium carbonate that would be necessary to react with excess phosphoric acid to make enough STPP to produce 1.025×10^5 kg of a detergent that is 32% $Na_5P_3O_{10}$?

$$6H_3PO_4 + 5Na_2CO_3 \rightarrow 2Na_5P_3O_{10} + 9H_2O + 5CO_2$$

$$? \text{ kg } Na_2CO_3 = 1.025 \times 10^5 \text{ kg det.} \left(\frac{32 \text{ kg } Na_5P_3O_{10}}{100 \text{ kg det.}}\right) \left(\frac{5 \times 105.989 \text{ kg } Na_2CO_3}{2 \times 367.864 \text{ kg } Na_5P_3O_{10}}\right)$$

= 2.4×10^4 kg Na₂CO₃

94. Urea, NH_2CONH_2, is a common nitrogen source used in fertilizers. When urea is made industrially, its temperature must be carefully controlled because heat turns urea into biuret, $NH_2CONHCONH_2$, a compound that is harmful to plants. Consider a pure sample of urea that has a mass of 92.6 kg. If 0.5% of the urea in this sample decomposes to form biuret, what mass, in grams, of $NH_2CONHCONH_2$ will it contain?

$$2NH_2CONH_2 \rightarrow NH_2CONHCONH_2 + NH_3$$
$$\text{biuret}$$

$$? \text{ kg } NH_2CONHCONH_2 = 92.6 \text{ kg } NH_2CONH_2 \text{ original} \left(\frac{0.5 \text{ kg } NH_2CONH_2 \text{ decomposed}}{100 \text{ kg } NH_2CONH_2 \text{ original}} \right)$$

$$\left(\frac{1 \times 103.081 \text{ kg } NH_2CONHCONH_2}{2 \times 60.056 \text{ kg } NH_2CONH_2} \right) \left(\frac{10^3 \text{ g}}{1 \text{ kg}} \right)$$

$$= \mathbf{4 \times 10^2 \text{ g } NH_2CONHCONH_2}$$

96. The white pigment titanium(IV) oxide (often called titanium dioxide), TiO_2, is made from rutile ore that is about 95% TiO_2. Before the TiO_2 can be used, it must be purified. The equation that follows represents the first step in this purification.

$$\overset{900\ °C}{3TiO_2(s) + 4C(s) + 6Cl_2(g) \rightarrow 3TiCl_4(l) + 2CO(g) + 2CO_2(g)}$$

a. How many pounds of $TiCl_4$ can be made from the reaction of 1.250×10^5 pounds of rutile ore that is 95% TiO_2 with 5.0×10^4 pounds of carbon?

$$? \text{ lb } TiCl_4 = 1.250 \times 10^5 \text{ lb ore} \left(\frac{95 \text{ lb } TiO_2}{100 \text{ lb ore}} \right) \left(\frac{3 \times 189.678 \text{ lb } TiCl_4}{3 \times 79.866 \text{ lb } TiO_2} \right)$$

$$= \mathbf{2.8 \times 10^5 \text{ lb } TiCl_4}$$

$$? \text{ lb } TiCl_4 = 5.0 \times 10^4 \text{ lb C} \left(\frac{3 \times 189.678 \text{ lb } TiCl_4}{4 \times 12.011 \text{ lb C}} \right) = 5.9 \times 10^5 \text{ lb } TiCl_4$$

b. Explain why two of these substances are in excess and one is limiting.

Carbon is inexpensive, nontoxic, and easy to convert to gaseous CO or CO_2, which are easy to separate from solid and liquid products. Although chlorine gas is a more dangerous substance, it is inexpensive and easy to separate from the product mixture. Because the ultimate goal is to convert the titanium in TiO_2 into $TiCl_4$, the TiO_2 is the more important reactant, so it is limiting.

98. What is the maximum mass of calcium hydrogen phosphate, $CaHPO_4$, that can form in the mixture of 12.50 kg of a solution that contains 84.0% H_3PO_4, 25.00 kg of $Ca(NO_3)_2$, 25.00 L of 14.8 M NH_3, and an excess of CO_2 and H_2O?

$$3H_3PO_4 + 5Ca(NO_3)_2 + 10NH_3 + 2CO_2 + 2H_2O \rightarrow 10NH_4NO_3 + 2CaCO_3 + 3CaHPO_4$$

$$? \text{ kg } CaHPO_4 = 12.50 \text{ kg } H_3PO_4 \text{ soln} \left(\frac{84.0 \text{ kg } H_3PO_4}{100 \text{ kg } H_3PO_4 \text{ soln}} \right) \left(\frac{3 \times 136.057 \text{ kg } CaHPO_4}{3 \times 97.9952 \text{ kg } H_3PO_4} \right)$$

$$= 14.6 \text{ kg } CaHPO_4$$

$$? \text{ kg } CaHPO_4 = 25.00 \text{ kg } Ca(NO_3)_2 \left(\frac{3 \times 136.057 \text{ kg } CaHPO_4}{5 \times 164.088 \text{ kg } Ca(NO_3)_2} \right) = \textbf{12.44 kg CaHPO}_\textbf{4}$$

$$? \text{ kg } CaHPO_4 = 25.00 \text{ L } NH_3 \text{ soln} \left(\frac{14.8 \text{ mol } NH_3}{1 \text{ L } NH_3 \text{ soln}} \right) \left(\frac{3 \text{ mol } CaHPO_4}{10 \text{ mol } NH_3} \right) \left(\frac{136.057 \text{ g } CaHPO_4}{1 \text{ mol } CaHPO_4} \right) \left(\frac{1 \text{ kg}}{10^3 \text{ g}} \right)$$

$$= 15.1 \text{ kg } CaHPO_4$$

Chapter 11
Modern Atomic Theory

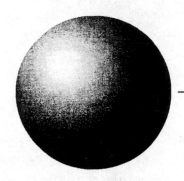

_____ Sphere enclosing almost all of the electron's negative charge

♦ Review Skills

11.1 The Mysterious Electron

- Standing Waves and Guitar Strings
- Electrons as Standing Waves
- Waveforms for Hydrogen Atoms
- Particle Interpretation of the Wave Character of the Electron
- Other Important Waveforms
- Overall Organization of Principal Energy Levels, Sublevels, and Orbitals

11.2 Multi-Electron Atoms

- Helium and Electron Spin
- The Second-Period Elements
- The Periodic Table and the Modern Model of the Atom
- Abbreviated Electron Configurations

    Special Topic 11.1: Why Does Matter Exist, and Why Should We Care About Answering This Question?

    Internet: Elements with Electron Configurations Other Than Predicted

    Internet: Electron Configurations for Monatomic Ions

♦ Chapter Glossary

    Internet: Glossary Quiz

♦ Chapter Objectives

Review Questions

Key Ideas

Chapter Problems

Section Goals and Introductions

Section 11.1 The Mysterious Electron

Goals

- *To explain why it is very difficult to describe the modern view of the electron.*
- *To give you some understanding of the nature of the electron by describing how it is like a guitar string.*
- *To explain what atomic orbitals are.*
- *To describe the atomic orbitals available to the electron of a hydrogen atom.*
- *To explain what energy levels and sublevels are.*

The electron is extremely tiny, and modern physics tells us that strange things happen in the realm of the very, very small. This makes it difficult for us to get a good understanding of the nature of the extremely tiny electron. For us, it's easier to consider what the electron is *like* rather than what it *is*. This section begins by giving you a glimpse of the modern view of the electron by showing how it is like a guitar string and how atomic orbitals that are possible for an electron in a hydrogen atom are like the possible ways that a guitar string can vibrate.

The most important component of this section is the introduction of the idea of atomic orbitals. Be sure you understand what the electron clouds that we call orbitals represent, both in terms of the effect they have on the space around the nucleus (which relates to their negative charge) and in terms of the probability of finding the electron in any position outside the nucleus. It will be useful for you to know the different shapes and sizes of the possible orbitals for the one electron in a hydrogen atom and to know how these orbitals can be arranged into energy levels and sublevels.

Section 11.2 Multi-Electron Atoms

Goals

- *To show how the knowledge of the atomic orbitals of hydrogen can be applied to atoms of the other elements.*
- *To describe how electrons of atoms are arranged with respect to orbitals, sublevels, and energy levels.*

This section shows you how the information about the energy levels, sublevels, and orbitals for the hydrogen electron can be applied to the electrons in atoms of other elements. It's important that you learn how to describe the arrangement of electrons in these energy levels, sublevels, and orbitals with orbital diagrams and electron configurations. You will see in Chapter 12 that these orbital diagrams and electron configurations will help us explain the bonding patterns of the elements.

See the two sections on our Web site that are related to this section: *Elements with Electron Configurations Other Than Predicted* and *Electron Configurations for Monatomic Ions*.

www.chemplace.com/college/

Chapter 11 Map

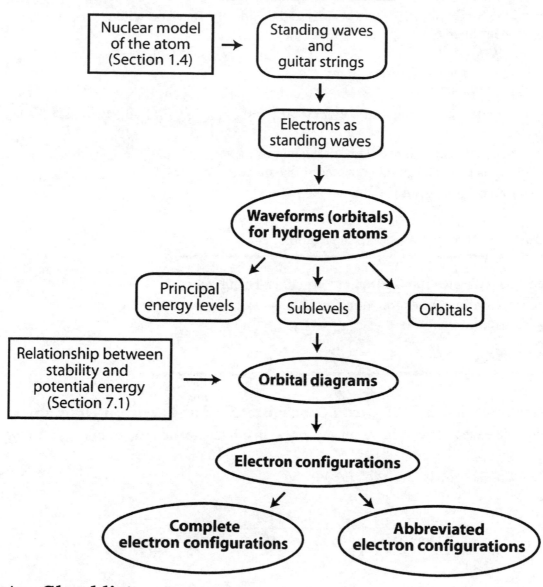

Chapter Checklist

- ☐ Read the Review Skills section. If there is any skill mentioned that you have not yet mastered, review the material on that topic before reading this chapter.
- ☐ Read the chapter quickly before the lecture that describes it.
- ☐ Attend class meetings, take notes, and participate in class discussions.
- ☐ Work the Chapter Exercises, perhaps using the Chapter Examples as guides.
- ☐ Study the Chapter Glossary and test yourself on our Web site:
  - **www.chemplace.com/college/**
- ☐ Study all of the Chapter Objectives. You might want to write a description of how you will meet each objective. (Although it is best to master all of the objectives, the following

objectives are especially important because they pertain to skills that you will need while studying other chapters of this text: 12 and 13.)

☐ Reread the Study Sheets in this chapter and decide whether you will use them or some variation on them to complete the tasks they describe.

    Sample Study Sheet 11.1: Writing Complete Electron Configurations and Orbital Diagrams for Uncharged Atoms

    Sample Study Sheet 11.2: Abbreviated Electron Configurations

☐ To get a review of the most important topics in the chapter, fill in the blanks in the Key Ideas section.

☐ Work all of the selected problems at the end of the chapter, and check your answers with the solutions provided in this chapter of the study guide.

☐ Ask for help if you need it.

Web Resources   www.chemplace.com/college/

Elements with Electron Configurations Other Than Predicted
Electron Configurations for Monatomic Ions
Glossary Quiz

Exercises Key

✍ Exercise 11.1 - Electron Configurations and Orbital Diagrams:

Write the complete electron configuration and draw an orbital diagram for antimony, Sb. *(Objs #12 & 13)*

$$1s^2\, 2s^2\, 2p^6\, 3s^2\, 3p^6\, 4s^2\, 3d^{10}\, 4p^6\, 5s^2\, 4d^{10}\, 5p^3$$

5p ↑ ↑ ↑   4d ↑↓ ↑↓ ↑↓ ↑↓ ↑↓

5s ↑↓

4p ↑↓ ↑↓ ↑↓   3d ↑↓ ↑↓ ↑↓ ↑↓ ↑↓

4s ↑↓

3s ↑↓   3p ↑↓ ↑↓ ↑↓

2s ↑↓   2p ↑↓ ↑↓ ↑↓

1s ↑↓

✍ Exercise 11.2 - Abbreviated Electron Configurations:

Write abbreviated electron configurations for (a) rubidium, Rb, (b) nickel, Ni, and (c) bismuth, Bi. *(Obj #14)*

    a. rubidium, Rb   [Kr] $5s^1$

    b. nickel, Ni   [Ar] $4s^2\, 3d^8$

    b. bismuth, Bi   [Xe] $6s^2\, 4f^{14}\, 5d^{10}\, 6p^3$

Review Questions Key

1. Describe the nuclear model of the atom.

    Protons and neutrons are in a tiny core of the atom called the nucleus, which has a diameter of about 1/100,000 the diameter of the atom. The position and motion of the electrons are uncertain, but they generate a negative charge that is felt in the space that surrounds the nucleus.

2. Describe the relationship between stability and potential energy.

    Increased stability of the components of a system leads to decreased potential energy, and decreased stability of the components of a system leads to increased potential energy.

Key Ideas Answers

3. The electron is extremely tiny, and modern physics tells us that **strange** things happen in the realm of the very, very small.

5. Modern physics tells us that it is **impossible** to know **exactly** where an electron is and what it is doing.

7. In order to accommodate the uncertainty of the electron's position and motion, scientists talk about where the electron **probably** is within the atom, instead of where it **definitely** is.

9. In the wave view, an electron has an effect on the space around it that can be described as a wave of **negative charge** varying in its intensity.

11. Just as the **intensity of movement** of a guitar string can vary, so can the **intensity of the negative charge** of the electron vary at different positions outside the nucleus.

13. As in the case of the guitar string, only certain waveforms are **possible** for the electron in an atom.

15. The information calculated for the hydrogen electron is used to describe the **other elements** as well.

17. The allowed waveforms for the electron are also called orbitals. Another definition of orbital as the volume that contains a given **high percentage** of the electron charge. An orbital can also be defined as the **volume** within which an electron has a high probability of being found.

19. In the particle view, the electron **cloud** can be compared to a multiple-exposure photograph of the electron.

21. Because the **strength** of the attraction between positive and negative charges decreases with increasing distance between the charges, an electron is more strongly attracted to the nucleus and therefore is more stable when it has the smaller 1s waveform than when it has the larger 2s waveform. Increased stability is associated with **decreased** potential energy, so a 1s electron has lower potential energy than a 2s electron.

23. After the electron is excited from the 1s orbital to the 2s orbital, it **spontaneously returns** to its lower-energy 1s form.

25. Orbitals that have the same potential energy, the same size, and the same shape are in the same **sublevel**.

27. Note that the first principal energy level has one sublevel, the second has two, the third has three, and the fourth has four. If *n* is the number associated with the principal energy level, each principal energy level has *n* sublevels.

29. None of the known elements in its ground state has any electrons in a principal energy level higher than the **seventh**.

31. We can visualize the two electrons in a helium atom as **spinning** in opposite directions.

33. An atomic orbital may contain **2** electrons at most, and the electrons must have different **spins**.

35. The highest-energy electrons for all of the elements in groups 1 (1A) and 2 (2A) in the periodic table are in *s* **orbitals**.

37. The last electrons to be added to an orbital diagram for the atoms of the transition metal elements go into *d* orbitals.

Problems Key

Section 11.1: The Mysterious Electron

39. Explain why, in theory, a guitar string can vibrate with an infinite number of possible waveforms but not all waveforms are possible. *(Obj #2)*

    The possible waveforms are limited by the fact that the string is tied down and cannot move at the ends. In theory, there are an infinite number of possible waveforms that allow the string to remain stationary at the ends.

41. Describe the 1*s* orbital in a hydrogen atom in terms of negative charge and in terms of the electron as a particle. *(Obj #4)*

    The negative-charge distribution of an electron in a 1*s* orbital of a hydrogen atom looks like the image in Figure 11.3 on page 457 of the text. The cloud is pictured as surrounding the nucleus and represents the variation in the intensity of the negative charge at different positions outside the nucleus. The negative charge is most intense at the nucleus and diminishes with increasing distance from the nucleus. The variation in charge intensity for this waveform is the same in all directions, so the waveform is a sphere. Theoretically, the charge intensity decreases toward zero as the distance from the nucleus approaches infinity. The 1*s* orbital can be described as a sphere that contains a high percentage (for example 90% or 99%) of the charge of the 1*s* electron.

    According to the particle interpretation of the wave character of the electron, a 1*s* orbital is a surface within which we have a high probability of finding the electron. In the particle view, the electron cloud can be compared to a multiple-exposure photograph of the electron (once again, we must resort to an analogy to describe electron behavior). If we were able to take a series of sharply focused photos of an electron over a period of time without advancing the film, our final picture would look like the image in Figure 11.5 on page 459 of the text. We would find a high density of dots near the nucleus (because most of the times when the shutter snaps, the electron would be near the nucleus) and a decrease in density with increasing distance from the nucleus (because some of the times the shutter snaps, the electron would be farther away from the nucleus). This arrangement of dots would bear out the wave equation's prediction of the probability of finding the electron at any given distance from the nucleus.

43. Describe a 2s orbital for a hydrogen atom. *(Obj #6)*

The 2s orbital for an electron in a hydrogen atom is spherical like the 1s orbital, but it is a larger sphere. For an electron in the 2s orbital, the charge is most intense at the nucleus, it diminishes in intensity to a minimum with increasing distance from the nucleus, it increases again to a maximum, and finally it diminishes again. The section of the 2s orbital where the charge intensity goes to zero is called a node. Figure 11.7, on page 460 of the text, shows cutaway, quarter section views of the 1s and 2s orbitals.

45. Which is larger, a 2p orbital or a 3p orbital? Would the one electron in a hydrogen atom be more strongly attracted to the nucleus in a 2p orbital or in a 3p orbital? Would the electron be more stable in a 2p orbital or in a 3p orbital? Would the electron have higher potential energy when it is in a 2p orbital or a 3p orbital?

The 3p orbital is larger than the 2p orbital. Because the average distance between the positively charged nucleus and the negative charge of an electron in a 2p orbital would be less than for an electron in a 3p orbital, the attraction between a 2p electron and the nucleus would be stronger. This makes an electron in a 2p orbital more stable and gives it lower potential energy than an electron in a 3p orbital.

47. Describe the three 2p orbitals for a hydrogen atom. *(Obj #8)*

The three 2p orbitals are identical in shape and size, but each is 90° from the other two. Because they can be viewed as being on the x, y and z axes of a three-dimensional coordinate system, they are often called the $2p_x$, $2p_y$, and $2p_x$ orbitals. One electron with a 2p waveform has its negative charge distributed in two lobes on opposite sides of the nucleus. We will call this a dumbbell shape. Figures 11.8 and 11.9, on pages 461 and 462 of the text, show two ways to visualize these orbitals, and Figure 11.10, on page 462 of the text, shows the three 2p orbitals together.

50. How many orbitals are there in the 3p sublevel for the hydrogen atom?   **3**

52. How many orbitals are there in the third principal energy level for the hydrogen atom?

There are **9** orbitals in the third principal energy level: 1 in the 3s sublevel, 3 in the 3p sublevel, and 5 in the 3d sublevel.

54. Which of the following sublevels do not exist?

   a. 5p   **exists**   c. 3f   **not exist**
   b. 2s   **exists**   d. 6d   **exists**

Section 11.2: Multi-Electron Atoms

57. What is the maximum number of electrons that can be placed in a $3p$ orbital?   in a $3d$ orbital?

    2   The maximum number of electrons in *any* orbital is 2.

59. What is the maximum number of electrons that can be placed in a $3p$ sublevel?   in a $3d$ sublevel?

    The maximum number of electrons in any p sublevel is **6**. The maximum number of electrons in any d sublevel is **10**.

61. What is the maximum number of electrons that can be placed in the third principal energy level?

    The third principal energy level can hold up to **18** electrons: 2 in the $3s$, 6 in the $3p$, and 10 in the $3d$.

63. For each of the following pairs, identify the sublevel that is filled first.
    a. $2s$ or $3s$ **2s**   c. $3d$ or $4s$ **4s**
    b. $3p$ or $3s$ **3s**   d. $4f$ or $6s$ **6s**

65. Write the complete electron configuration and orbital diagram for each of the following. *(Objs #11 & 12)*

    a. carbon, C

       $1s^2\ 2s^2\ 2p^2$

       2s $\uparrow\downarrow$ 2p $\underline{\uparrow}\ \underline{\uparrow}\ \underline{\quad}$

       1s $\underline{\uparrow\downarrow}$

    b. phosphorus, P

       $1s^2\ 2s^2\ 2p^6\ 3s^2\ 3p^3$

       3s $\underline{\uparrow\downarrow}$ 3p $\underline{\uparrow\downarrow}\ \underline{\uparrow}\ \underline{\uparrow}$

       2s $\underline{\uparrow\downarrow}$ 2p $\underline{\uparrow\downarrow}\ \underline{\uparrow\downarrow}\ \underline{\uparrow\downarrow}$

       1s $\underline{\uparrow\downarrow}$

    c. vanadium, V

       $1s^2\ 2s^2\ 2p^6\ 3s^2\ 3p^6\ 4s^2\ 3d^3$

       4s $\underline{\uparrow\downarrow}$ 3d $\underline{\uparrow}\ \underline{\uparrow}\ \underline{\uparrow}\ \underline{\uparrow}\ \underline{\uparrow}$

       3s $\underline{\uparrow\downarrow}$ 3p $\underline{\uparrow\downarrow}\ \underline{\uparrow\downarrow}\ \underline{\uparrow\downarrow}$

       2s $\underline{\uparrow\downarrow}$ 2p $\underline{\uparrow\downarrow}\ \underline{\uparrow\downarrow}\ \underline{\uparrow\downarrow}$

       1s $\underline{\uparrow\downarrow}$

   d. iodine, I

$$1s^2\,2s^2\,2p^6\,3s^2\,3p^6\,4s^2\,3d^{10}\,4p^6\,5s^2\,4d^{10}\,5p^5$$

5p ↑↓ ↑↓ ↑   4d ↑↓ ↑↓ ↑↓ ↑↓ ↑↓
5s ↑↓
4p ↑↓ ↑↓ ↑↓   3d ↑↓ ↑↓ ↑↓ ↑↓ ↑↓
4s ↑↓
3s ↑↓ 3p ↑↓ ↑↓ ↑↓
2s ↑↓ 2p ↑↓ ↑↓ ↑↓
1s ↑↓

   e. mercury, Hg

$$1s^2\,2s^2\,2p^6\,3s^2\,3p^6\,4s^2\,3d^{10}\,4p^6\,5s^2\,4d^{10}\,5p^6\,6s^2\,4f^{14}\,5d^{10}$$

5d ↑↓ ↑↓ ↑↓ ↑↓ ↑↓ 4f ↑↓ ↑↓ ↑↓ ↑↓ ↑↓ ↑↓ ↑↓
6s ↑↓
5p ↑↓ ↑↓ ↑↓ 4d ↑↓ ↑↓ ↑↓ ↑↓ ↑↓
5s ↑↓
4p ↑↓ ↑↓ ↑↓ 3d ↑↓ ↑↓ ↑↓ ↑↓ ↑↓
4s ↑↓
3s ↑↓ 3p ↑↓ ↑↓ ↑↓
2s ↑↓ 2p ↑↓ ↑↓ ↑↓
1s ↑↓

67. Which element is associated with each of the ground state electron configurations listed below?

   a. $1s^2\,2s^2$ **Be**

   b. $1s^2\,2s^2\,2p^6\,3s^1$ **Na**

   c. $1s^2\,2s^2\,2p^6\,3s^2\,3p^6\,4s^2\,3d^{10}\,4p^5$ **Br**

   d. $1s^2\,2s^2\,2p^6\,3s^2\,3p^6\,4s^2\,3d^{10}\,4p^6\,5s^2\,4d^{10}\,5p^6\,6s^2\,4f^{14}\,5d^{10}\,6p^2$ **Pb**

69. Would the following electron configurations represent ground states or excited states?

   a. $1s^2\,2s^1\,2p^5$ **excited**   c. $1s^2\,2s^2\,2p^4\,3s^1$ **excited**

   b. $1s^2\,2s^2\,2p^4$ **ground**   d. $1s^2\,2s^2\,2p^5$ **ground**

71. Write the abbreviated electron configurations for each of the following. *(Obj #13)*

   a. fluorine, F   **[He] $2s^2\,2p^5$**

   b. silicon, Si   **[Ne] $3s^2\,3p^2$**

   c. cobalt, Co   **[Ar] $4s^2\,3d^7$**

   d. indium, In   **[Kr] $5s^2\,4d^{10}\,5p^1$**

   e. polonium, Po   **[Xe] $6s^2\,4f^{14}\,5d^{10}\,6p^4$**

Additional Problems

73. Which sublevel contains:

    a. the highest-energy electron for francium, Fr?   **7s**

    b. the 25th electron added to an orbital diagram for elements larger than chromium, Cr? **3d**

    c. the 93rd electron added to an orbital diagram for elements larger than uranium, U? **5f**

    d. the 82nd electron added to an orbital diagram for elements larger than lead, Pb? **6p**

75. What is the first element on the periodic table to have

    b. an electron in the $3p$ sublevel.   **Al**

    c. a filled $4s$ sublevel.   **Ca**

    d. a half-filled $3d$ sublevel.   **Mn**

77. Which pair of the following ground-state, abbreviated electron configurations corresponds to elements in the same group on the periodic table? What elements are they? What is the name of the group to which they belong?

    a. [Ne] $3s^2$ c. [Kr] $5s^2$

    b. [Ar] $4s^2\ 3d^{10}$ d. [Xe] $6s^2\ 4f^{14}\ 5d^{10}\ 6p^1$

    The pair "a" and "c" represent the alkaline earth metals magnesium and strontium.

79. What is the maximum number of electrons in each of the following?

    a. the $8j$ sublevel   **30**

    b. a $6h$ orbital   **2**

    c. the n = 8 principle energy level   **128**

82. Write the expected abbreviated electron configuration for the as-yet-undiscovered element with the atomic number of 121. Use Uuo for the symbol of the noble gas below xenon, Xe. (*Hint*: See Figure 11.17, on page 472.)

    [Uuo] $8s^2\ 5g^1$ or [Uuo] $5g^1\ 8s^2$

Chapter 12
Molecular Structure

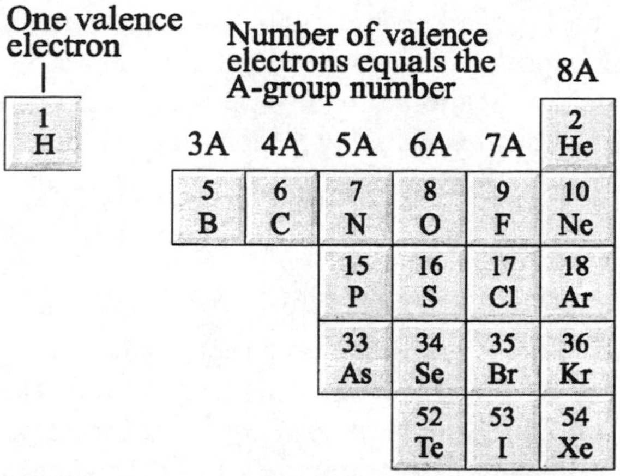

One valence electron		Number of valence electrons equals the A-group number					8A
1 H		3A	4A	5A	6A	7A	2 He
		5 B	6 C	7 N	8 O	9 F	10 Ne
				15 P	16 S	17 Cl	18 Ar
				33 As	34 Se	35 Br	36 Kr
					52 Te	53 I	54 Xe

♦ Review Skills

12.1 A New Look at Molecules and the Formation of Covalent Bonds
- The Strengths and Weaknesses of Models
- The Valence-Bond Model

12.2 Drawing Lewis Structures
- General Procedure
- More Than One Possible Structure

12.3 Resonance
    Internet: Resonance

12.4 Molecular Geometry from Lewis Structures

♦ Chapter Glossary
    Internet: Glossary Quiz

♦ Chapter Objectives

Review Questions

Key Ideas

Chapter Problems

Section Goals and Introductions

Section 12.1 A New Look at Molecules and the Formation of Covalent Bonds
Goals
- *To describe the strengths and weaknesses of scientific models.*
- *To introduce a model, called the valence-bond model, which is very useful for describing the formation of covalent bonds.*

- *To explain the most common covalent bonding patterns for the nonmetallic atoms in terms of the valence-bond model.*

This section shows how the information learned in Chapter 11 can be combined with a model for covalent bonding called the valence-bond model to explain the common bonding patterns of the nonmetallic atoms. The most common of these bonding patterns were listed in Chapter 3, but now you have the background necessary for understanding why atoms have the bonding patterns that they do. It's important to recognize that although the valence bond model is only a model (and therefore a simplification of reality), it is extremely useful. You will find the information in *Table 12.1: Covalent Bonding Patterns* very helpful when drawing Lewis structures (the task described in Section 12.2).

Section 12.2 Drawing Lewis Structures

Goal: To show how Lewis structures can be drawn from chemical formulas.

In Chapter 3, you learned to draw simple Lewis structures by arranging the atoms to yield the most common bonding pattern for each atom. This technique works very well for many molecules, but it is limited. For example, it does not work for polyatomic ions. This section describes a procedure for drawing Lewis structures from chemical formulas that works for a broader range of molecules and polyatomic ions.

Section 12.3 Resonance

Goal: To introduce a concept called resonance and show how it can be used to explain the characteristics of certain molecules and polyatomic ions.

The Lewis structures derived for some molecules and polyatomic ions by the technique described in Section 12.2 do not explain their characteristics adequately. One way that the valence-bond model has been expanded to better explain some of these molecules and polyatomic ions is by introducing the concept of resonance described in this section. See the related section on our Web site called *Resonance*.

www.chemplace.com/college/

Section 12.4 Molecular Geometry from Lewis Structures

Goals

- *To show how you can predict the arrangement of atoms in molecules and polyatomic ions (called molecular geometry).*

- *To show how to make sketches of the molecular geometry of atoms in molecules and polyatomic ions.*

The arrangement of atoms in a molecule or polyatomic ion (that is, its molecular geometry) plays a significant role in determining its properties. This section explains why molecules and polyatomic ions have the geometry that they do, and *Sample Study Sheet 12.2: Predicting Molecular Geometry* and *Table 12.3: Electron Group and Molecular Geometry* show you how to predict and sketch these geometries.

Chapter 12 Map

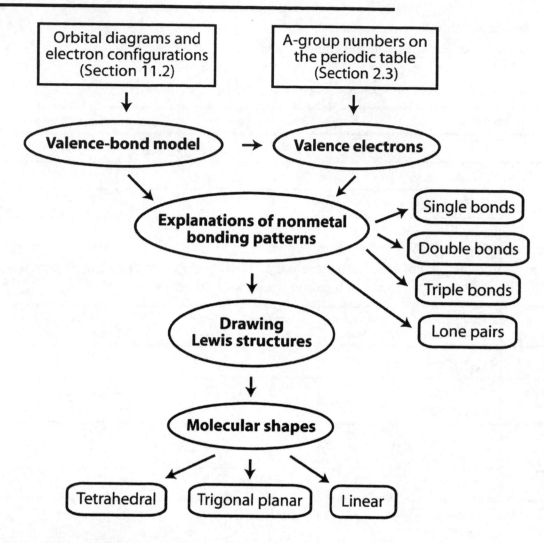

Chapter Checklist

- ☐ Read the Review Skills section. If there is any skill mentioned that you have not yet mastered, review the material on that topic before reading this chapter.
- ☐ Read the chapter quickly before the lecture that describes it.
- ☐ Attend class meetings, take notes, and participate in class discussions.
- ☐ Work the Chapter Exercises, perhaps using the Chapter Examples as guides.
- ☐ Study the Chapter Glossary and test yourself on our Web site:
  www.chemplace.com/college/
- ☐ Study all of the Chapter Objectives. You might want to write a description of how you will meet each objective. (Although it is best to master all of the objectives, the following objectives are especially important because they pertain to skills that you will need while studying other chapters of this text: 7 and 9.)
- ☐ Reread the Study Sheets in this chapter and decide whether you will use them or some variation on them to complete the tasks they describe.

Sample Study Sheet 12.1: Drawing Lewis Structures from Formulas

Sample Study Sheet 12.2: Predicting Molecular Geometry

☐ Memorize the following.

Be sure to check with your instructor to determine how much you are expected to know of the following.

- The most common bonding patterns for the nonmetallic elements.

Elements	Number of covalent bonds	Number of lone pairs
C	4	0
N, P, & As	3	1
O, S, Se	2	2
F, Cl, Br, & I	1	3

- Although it's not absolutely necessary, it will help you to draw Lewis structures to know the expanded list of bonding patterns listed below.

Element	Frequency of pattern	Number of bonds	Number of lone pairs	Example
H	always	1	0	H—
B	most common	3	0	—B—
C	most common	4	0	—C— or —C= or —C≡
	rare	3	1	≡C:
N, P, & As	most common	3	1	—N̈—
	common	4	0	—N—
O, S, & Se	most common	2	2	—Ö— or Ö
	common	1	3	—Ö:
	rare	3	1	≡O:
F, Cl, Br, & I	most common	1	3	—Ẍ:

- The information found in the table below.

e⁻ groups	e⁻ group geometry	General Geometric Sketch	Bond angles	Bond groups	Lone pairs	molecular geometry
2	linear	180°	180°	2	0	linear
3	trigonal planar	120°	120°	3	0	trigonal planar
				2	1	bent
4	tetrahedral	109.5°	109.5°	4	0	tetrahedral
				3	1	trigonal pyramid
				2	2	bent

☐ To get a review of the most important topics in the chapter, fill in the blanks in the Key Ideas section.

☐ Work all of the selected problems at the end of the chapter, and check your answers with the solutions provided in this chapter of the study guide.

☐ Ask for help if you need it.

Web Resources   www.chemplace.com/college/

Resonance
Glossary Quiz

Exercises Key

Exercise 12.1 - Lewis Structures: Draw a reasonable Lewis structure for each of the following formulas.   *(Obj #7)*

a. CCl₄

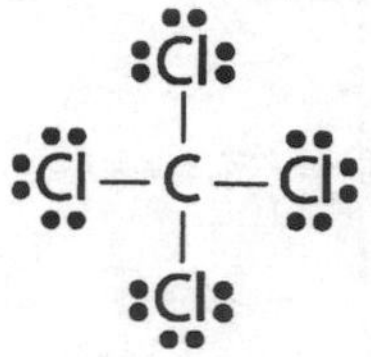

b. Cl₂O

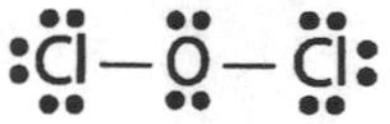

c. COF₂

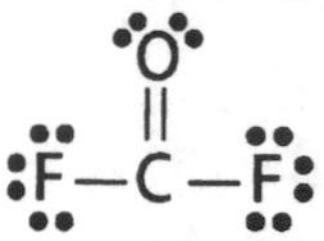

d. C₂Cl₆

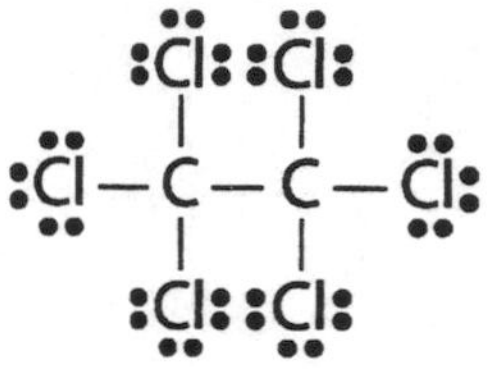

e. BCl₃

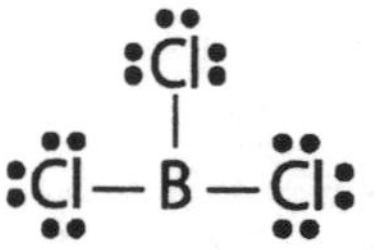

f. N₂H₄

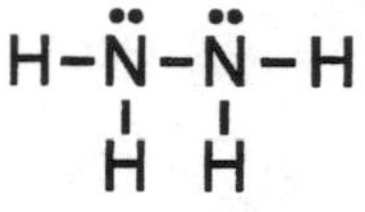

g. H₂O₂

h. NH₂OH

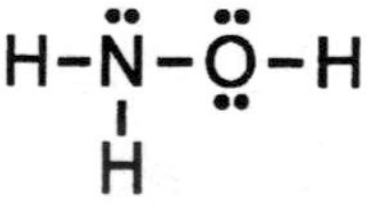

i. NCl₃

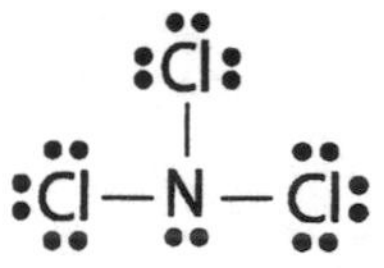

Exercise 12.2 - Resonance: Draw all of the reasonable resonance structures and the resonance hybrid for the carbonate ion, CO_3^{2-}. A reasonable Lewis structure for the carbonate ion is   *(Obj #8)*

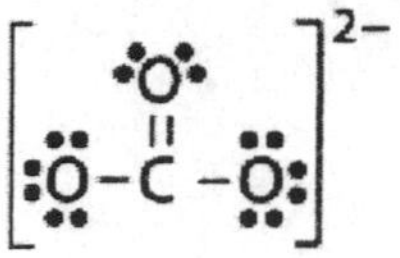

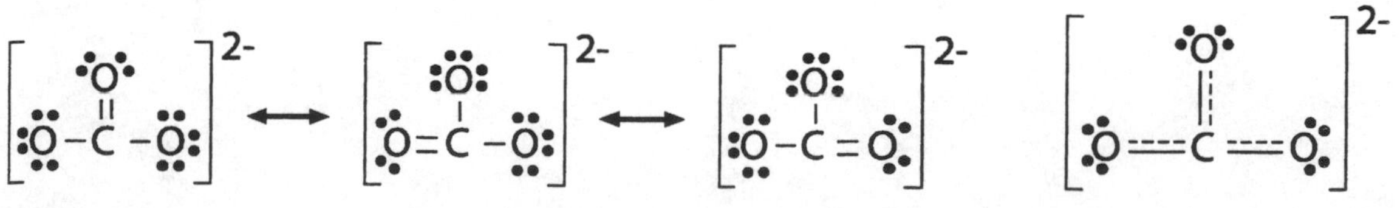

✍ Exercise 12.3 - Molecular Geometry: For each of the Lewis structures that follow, (a) write the name of the electron group geometry around each atom that has two or more atoms attached to it, (b) draw the geometric sketch of the molecule, including bond angles, and (c) write the name of the molecular geometry around each atom that has two or more atoms attached to it. *(Obj #9)*

a.	Electron Group Geometry - **tetrahedral** ≈109.5° Molecular Geometry - **tetrahedral**
b.	Electron Group Geometry - **tetrahedral** ≈109.5° Molecular Geometry – **trigonal pyramid**
c.	Electron Group Geometry - **linear** 180° Molecular Geometry - **linear**
d.	Electron Group Geometry - **tetrahedral** ≈109.5° Molecular Geometry - **bent**
e.	Electron Group Geometry – **trigonal planar** ≈120° Molecular Geometry – **trigonal planar**
f.	Electron Group Geometry – **trigonal planar** ≈120° Molecular Geometry – **trigonal planar**

g. H–C–C≡C–H (with H atoms above, below left C)	Electron Group Geometry for left carbon - **tetrahedral** Electron Group Geometry for right carbon - **linear** 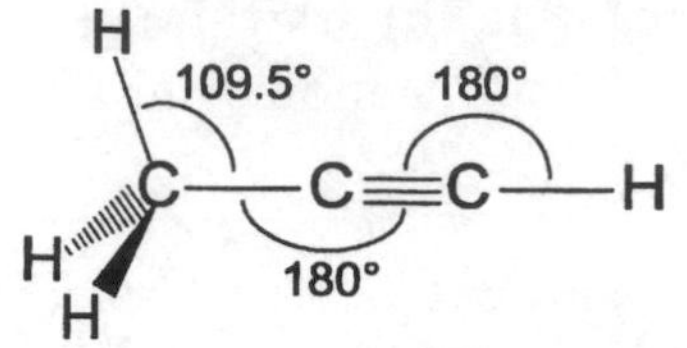 H 109.5° 180° C—C≡C—H 180° Molecular Geometry for left carbon - **tetrahedral** Molecular Geometry for right carbon - **linear**
h. 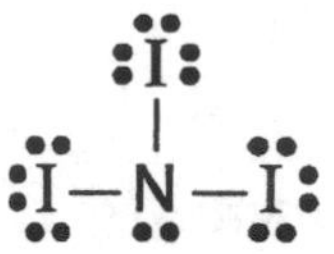 :I̤—N—I̤: (with :Ï: above N)	Electron Group Geometry - **tetrahedral** 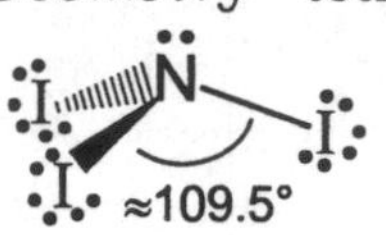 N with I atoms ≈109.5° Molecular Geometry – **trigonal pyramid**

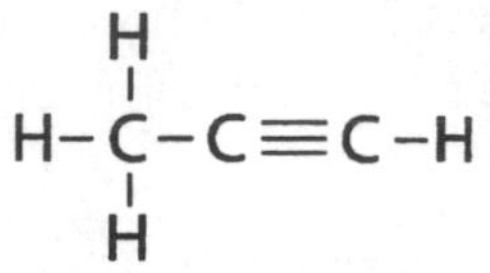

Review Questions Key

1. Using the A-group convention, what is the group number of the column in the periodic table that includes the element chlorine, Cl? (See Section 2.3.)   **7A**

2. Draw Lewis structures for CH_4, NH_3, and H_2O. (See Section 3.3.)

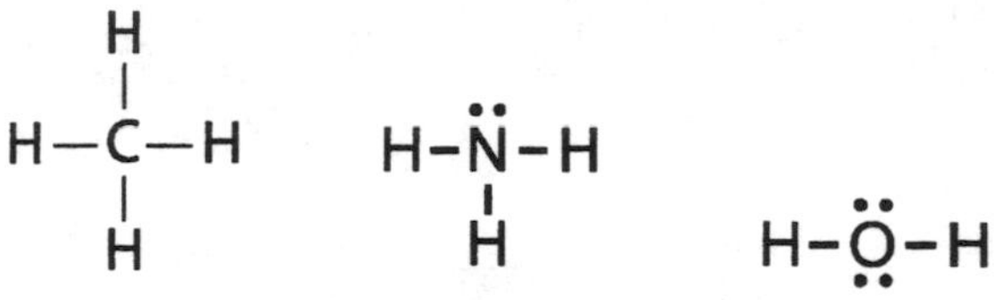

    H–C–H (with H above and below)   H–N̈–H (with H below)   H–Ö–H

3. Define the term *orbital*. (See Section 11.1.)

    Orbital can be defined as the volume that contains a high percentage of the electron charge generated by an electron in an atom. It can also be defined as the volume within which an electron has a high probability of being found.

4. Write a complete electron configuration and an orbital diagram for each of the following. (See Section 11.2.)

    a. oxygen, O

    $1s^2\ 2s^2\ 2p^4$

    2s ↑↓   2p ↑↓ ↑ ↑

    1s ↑↓

b. phosphorus, P

$$1s^2\ 2s^2\ 2p^6\ 3s^2\ 3p^3$$

3s $\uparrow\downarrow$ 3p $\uparrow$ $\uparrow$ $\uparrow$

2s $\uparrow\downarrow$ 2p $\uparrow\downarrow$ $\uparrow\downarrow$ $\uparrow\downarrow$

1s $\uparrow\downarrow$

Key Ideas Answers

5. When developing a model of physical reality, scientists take what they think is **true** and simplify it enough to make it **useful**.

7. One characteristic of models is that they **change** with time.

9. Valence electrons are the highest-energy *s* **and** *p* electrons in an atom.

11. Paired valence electrons are called **lone pairs**.

13. Carbon atoms frequently form double bonds, in which they share **four** electrons with another atom, often another carbon atom.

15. The shortcut for drawing Lewis structures for which we try to give each atom its **most common** bonding pattern works well for many simple uncharged molecules, but it does not work reliably for molecules that are more complex or for **polyatomic ions**.

17. For polyatomic **cations**, the total number of valence electrons is the sum of the valence electrons for each atom minus the charge.

19. Hydrogen and fluorine atoms are **never** in the center of a Lewis structure.

21. The element with the **fewest** atoms in the formula is often in the center of a Lewis structure.

23. Oxygen atoms rarely bond to other **oxygen** atoms.

25. In a reasonable Lewis structure, hydrogen will always have a total of **two** electrons from its one bond.

27. Substances that have the same molecular formula but different **structural** formulas are called isomers.

29. The most stable arrangement for electron groups is one in which they are **as far apart** as possible.

31. In this book, we call the geometry that describes all of the electron groups, including the lone pairs, the **electron group** geometry. The shape that describes the arrangement of the atoms only—treating the lone pairs as invisible—will be called the **molecular** geometry.

Problems Key

Section 12.1 A New Look at Molecules and the Formation of Covalent Bonds

32. Describe the advantages and disadvantages of using models to describe physical reality. *(Obj #2)*

Our models come with advantages and disadvantages. They help us to visualize, explain, and predict chemical changes, but we need to remind ourselves now and then that they are only models, and as models, they have their limitations. For example, because a model is a simplified version of what we think is true, the processes it depicts are sometimes described using the phrase "*as if.*" When you read, "It is *as if* an electron were promoted from one orbital to another," the phrase is a reminder that we do not necessarily think this is what really happens. We find it *useful* to talk about the process *as if* this is the way it happens.

One characteristic of models is that they change with time. Because our models are a simplification of what we think is real, we are not surprised when they sometimes fail to explain experimental observation. When this happens, the model is altered to fit the new observations.

34. How many valence electrons do the atoms of each of the following elements have? Write the electron configuration for these electrons. (For example, fluorine has 7 valence electrons, which can be described as $2s^2\, 2p^5$.) *(Obj #4)*
    a. nitrogen, N   **5 valence electrons – $2s^2\, 2p^3$**
    b. sulfur, S   **6 valence electrons – $3s^2\, 3p^4$**
    c. iodine, I   **7 valence electrons – $5s^2\, 5p^5$**
    a. argon, Ar   **8 valence electrons – $3s^2\, 3p^6$**

36. Draw electron-dot symbols for each of the following elements.   *(Objs #4 & 6)*

    a. nitrogen, N   5 $\cdot \overset{\displaystyle ..}{\underset{\displaystyle .}{N}} \cdot$

    b. sulfur, S   6 $\cdot \overset{\displaystyle ..}{\underset{\displaystyle ..}{S}} \cdot$

    c. iodine, I   7 $\cdot \overset{\displaystyle ..}{\underset{\displaystyle ..}{I}} :$

    d. argon, Ar   8 $: \overset{\displaystyle ..}{\underset{\displaystyle ..}{Ar}} :$

38. To which group on the periodic table would atoms with the following electron-dot symbols belong? List the group numbers using the 1-18 convention and using the A-group convention.

    a. $\cdot \overset{\displaystyle ..}{\underset{\displaystyle .}{X}} \cdot$ **Group 15 or 5A**

    b. $: \overset{\displaystyle ..}{\underset{\displaystyle ..}{X}} :$ **Group 18 or 8A**

    c. $\cdot \underset{\displaystyle .}{X} \cdot$ **Group 13 or 3A**

40. For each of the following elements, sketch *all* of the ways mentioned in Section 12.1 that their atoms could look in a Lewis structure. For example, fluorine has only one bonding pattern, and it looks like $-\ddot{\ddot{F}}:$.

   a. nitrogen, N $-\overset{|}{\underset{|}{N}}-$ or $-\overset{..}{\underset{|}{N}}-$

   b. boron, B $-\overset{|}{\underset{|}{B}}-$

   c. carbon, C $-\overset{|}{\underset{|}{C}}-$ or $-\overset{|}{C}=$ or $-C\equiv$ or $\equiv C:$

42. Use the valence-bond model to explain the following observations. *(Obj #5)*

   The answers to each of these problems is based on the following assumptions of the valence bond model.

   • Only the highest energy electrons participate in bonding.

   • Covalent bonds usually form to pair unpaired electrons.

   a. Fluorine atoms have 1 bond and 3 lone pairs in F_2.

   Fluorine is in group 7A, so it has seven valence electrons per atom. The orbital diagram for the valence electrons of fluorine is below.

   $2s \underline{\uparrow\downarrow}$ $2p \underline{\uparrow\downarrow}\,\underline{\uparrow\downarrow}\,\underline{\uparrow}$ $\cdot\ddot{\ddot{F}}:$

   The 1 unpaired electron leads to one bond, and the 3 pairs of electrons give fluorine atoms three lone pairs.

   $:\!\ddot{F}\!-\!\ddot{F}\!:$

   b. Carbon atoms have 4 bonds and no lone pairs in CH_4.

   Carbon is in group 4A, so it has 4 valence electrons per atom. It is *as if* 1 electron is promoted from the 2s orbital to the 2p orbital.

   $2s \underline{\uparrow\downarrow}$ $2p \underline{\uparrow}\,\underline{\uparrow}\,\underline{\quad}$ $\longrightarrow$ $2s \underline{\uparrow}$ $2p \underline{\uparrow}\,\underline{\uparrow}\,\underline{\uparrow}$

   $:\!C\cdot\cdot\dot{C}\cdot$

   The 4 unpaired electrons lead to 4 covalent bonds. Because there are no pairs of electrons, carbon atoms have no lone pairs when they form 4 bonds.

   $4H\cdot\ +\ \cdot\dot{C}\cdot\ \longrightarrow\ H\!:\!\overset{H}{\underset{H}{\ddot{C}}}\!:\!H\quad$ or $\quad H\!-\!\overset{\overset{H}{|}}{\underset{\underset{H}{|}}{C}}\!-\!H$

  c.   Nitrogen atoms have 3 bonds and 1 lone pair in NH_3.

Nitrogen is in group 5A, so it has 5 valence electrons per atom. The orbital diagram for the valence electrons of nitrogen is below.

$$2s \; \uparrow\downarrow \quad 2p \; \uparrow \;\; \uparrow \;\; \uparrow \qquad \cdot \ddot{N} \cdot$$

The 3 unpaired electrons lead to 3 bonds, and the 1 pair of electrons gives nitrogen atoms 1 lone pair.

$$\text{H}-\overset{\cdot\cdot}{\text{N}}-\text{H}$$
$$|$$
$$\text{H}$$

  d.   Sulfur atoms have 2 bonds and 2 lone pairs in H_2S.

Sulfur is in group 6A, so it has 6 valence electrons per atom. The orbital diagram for the valence electrons of sulfur is below.

$$3s \; \uparrow\downarrow \quad 3p \; \uparrow\downarrow \;\; \uparrow \;\; \uparrow \qquad \colon\!\ddot{S}\cdot$$

The 2 unpaired electrons lead to 2 bonds, and the 2 pairs of electrons give sulfur atoms 2 lone pairs.

$$\text{H}-\overset{\cdot\cdot}{\underset{\cdot\cdot}{\text{S}}}-\text{H}$$

  e.   Oxygen atoms have 1 bond and 3 lone pairs in OH^-.

Oxygen is in group 6A, so it has 6 valence electrons per atom. If it gains 1 electron, it will have a total of 7.

$$2s \; \uparrow\downarrow \quad 2p \; \uparrow\downarrow \;\; \uparrow \;\; \uparrow \qquad \colon\!\ddot{O}\cdot \qquad \xrightarrow{+1e^-} \qquad 2s \; \uparrow\downarrow \quad 2p \; \uparrow\downarrow \;\; \uparrow\downarrow \;\; \uparrow \qquad \colon\!\ddot{O}\cdot$$

The 1 unpaired electron leads to one bond, and the 3 pairs of electrons give oxygen atoms with an extra electron 3 lone pairs.

$$\text{H}\cdot \; + \; \colon\!\ddot{O}\cdot \; \longrightarrow \; \colon\!\ddot{O}\!:\!\text{H} \quad \text{or} \quad \left[\colon\!\ddot{\underset{\cdot\cdot}{O}}\!-\!\text{H}\right]^-$$

44. Based on your knowledge of the most common bonding patterns for the nonmetallic elements, predict the formulas with the lowest subscripts for the compounds that would form from the following pairs of elements. (For example, hydrogen and oxygen can combine to form H_2O and H_2O_2, but H_2O has lower subscripts.)

  a.   C and H   **CH_4**

  b.   S and H   **H_2S**

  c.   B and F   **BF_3**

Section 12.2 Drawing Lewis Structures

46. Copy the following Lewis structure and identify the single bonds, the double bond, and the lone pairs.

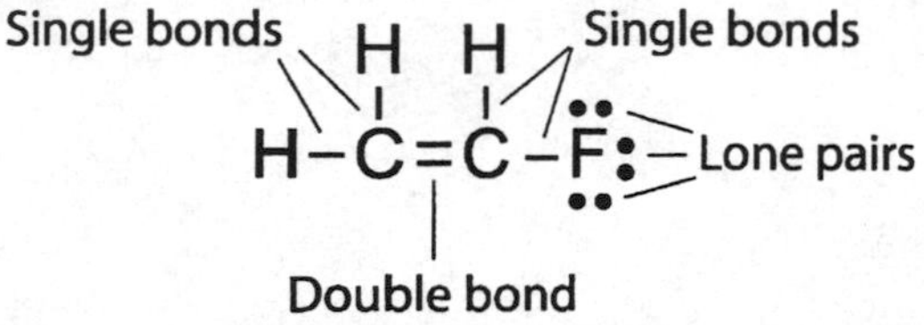

Single bonds   H   H   Single bonds

$$H-C=C-\ddot{F}:—Lone\ pairs$$

Double bond

48. For each of the following molecular compounds, identify the atom that is most likely to be found in the center of its Lewis structure. Explain why.

a. CBr_4 **Carbon** –The element with the fewest atoms in the formula is often in the center. The atom that is capable of making the most bonds is often in the center. Carbon atoms usually form 4 bonds, and bromine atoms usually form 1 bond.

b. SO_2 **Sulfur** –The element with the fewest atoms in the formula is often in the center. Oxygen atoms are rarely in the center. Oxygen atoms rarely bond to other oxygen atoms.

c. H_2S **Sulfur** –The element with the fewest atoms in the formula is often in the center. The atom that is capable of making the most bonds is often in the center. Sulfur atoms usually form 2 bonds, and hydrogen atoms form 1 bond. Hydrogen atoms are never in the center.

d. NOF   **Nitrogen** –The atom that is capable of making the most bonds is often in the center. Nitrogen atoms usually form 3 bonds, oxygen atoms usually form 2 bonds, and fluorine atoms form 1 bond. Fluorine atoms are never in the center.

50. Calculate the total number of valence electrons for each of the following formulas.

a. HNO_3 **1 + 5 + 3(6) = 24 valence electrons**
b. CH_2CHF **2(4) + 3(1) + 7 = 18 valence electrons**

52. Draw a reasonable Lewis structure for each of the following formulas.   *(Obj #7)*

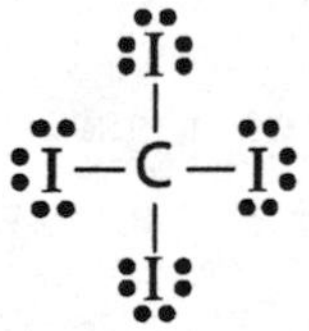

a. CI_4

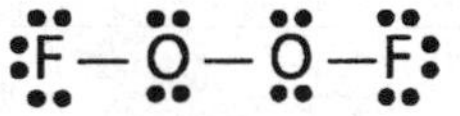

b. O_2F_2

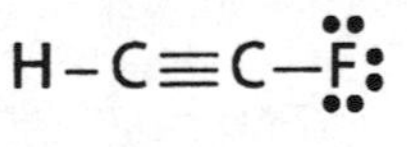

c. HC_2F

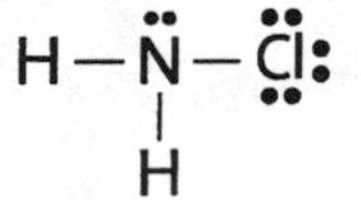

d. NH_2Cl

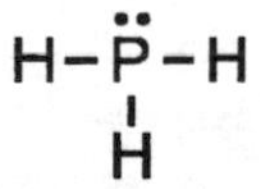

e. PH_3

f. S_2F_2

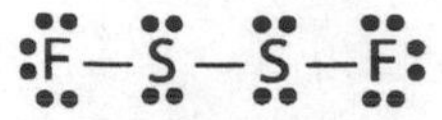

g. HNO_2

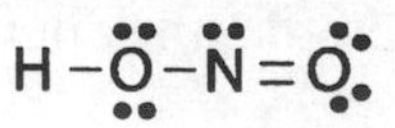

h. N_2F_4

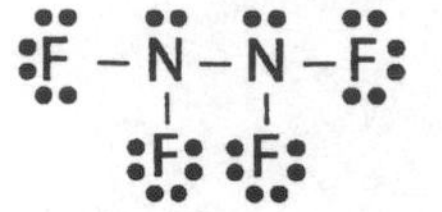

i. CH_2CHCH_3

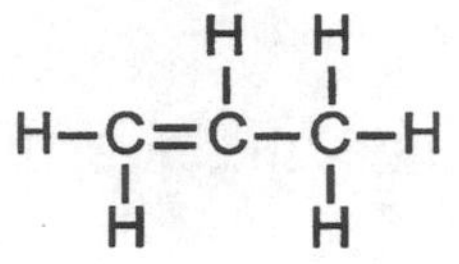

Section 12.3 Resonance

54. Draw a reasonable Lewis structure for the ozone molecule, O_3, using the skeleton that follows. The structure is best described in terms of resonance, so draw all of its reasonable resonance structures and the resonance hybrid that summarizes these structures. *(Obj #8)*

O–O–O

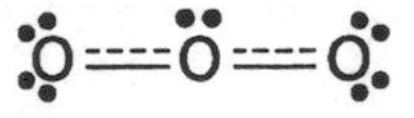

Section 12.4 Molecular Geometry from Lewis Structures

56. Although both CO_2 molecules and H_2O molecules have 3 atoms, CO_2 molecules are linear, and H_2O molecules are bent. Why?

Atoms are arranged in molecules to keep the electron groups around the central atom as far apart as possible. An electron group is either (1) a single bond, (2) a multiple bond (double or triple), or (3) a lone pair. The Lewis structure for CO_2 shows that the carbon atom has 2 electron groups around it. The best way to get two things as far apart as possible is in a linear arrangement.

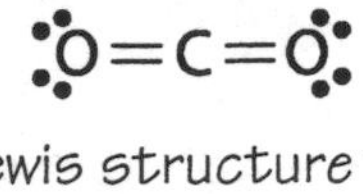

The Lewis structure for H_2O shows that the oxygen atom has 4 electron groups around it. The best way to get four things as far apart as possible is in a tetrahedral arrangement.

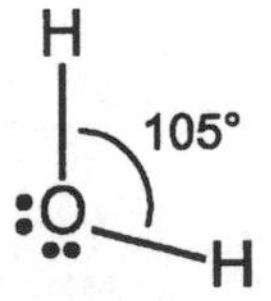

58. Using the symbol X for the central atom and Y for the outer atoms, draw the general geometric sketch for a 3-atom molecule with linear geometry.

61. For each of the Lewis structures that follows,   *(Obj #9)*

- Write the name of the electron group geometry around each atom that has two or more atoms attached to it.

- Draw the geometric sketch of the molecule, including bond angles (or approximate bond angles).

- Write the name of the molecular geometry around each atom that has two or more atoms attached to it.

a. (Lewis structure of CF_4)   tetrahedral   (geometric sketch, ≈109.5°)   tetrahedral

b. $O=N-Cl$ trigonal planar   (geometric sketch, ≈120°)   Bent

c. $Br-B-Br$ with I   trigonal planar   (geometric sketch, ≈120°)   triangular planar

d. $Br-As-Br$ with Br   tetrahedral   (geometric sketch, ≈109.5°)   trigonal pyramid

e. $Br-C-Br$ with O double bond   trigonal planar   (geometric sketch, ≈120°)   trigonal planar

f. $O=C=S$ linear $O=C=S$, 180°   linear

Chapter 13 - Gases

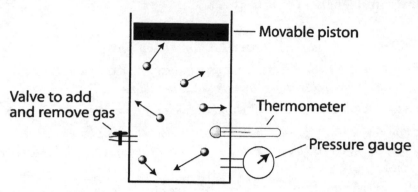

- ◆ Review Skills
- 13.1 Gases and Their Properties
  - • Ideal Gases
    - *Internet: Real Gases*
  - • Properties of Gases
  - • Discovering the Relationships Between Properties
  - • The Relationship Between Volume and Pressure
  - • The Relationship Between Pressure and Temperature
  - • The Relationship Between Volume and Temperature
  - • The Relationship Between Moles of Gas and Pressure
  - • The Relationship Between Moles of Gas and Volume
  - • Gases and the Internal Combustion Engine
  - • Explanations for Other Real-World Situations
- 13.2 Ideal Gas Calculations
  - • Calculations Using the Ideal Gas Equation
  - • When Properties Change
- 13.3 Equation Stoichiometry and Ideal Gases
  - *Internet: Gas Stoichiometry Shortcut*
- 13.4 Dalton's Law of Partial Pressures
  - ***Special Topic 13.1: A Greener Way to Spray Paint***
  - ***Special Topic 13.2: Green Decaf Coffee***
- ◆ Chapter Glossary
  - *Internet: Glossary Quiz*
- ◆ Chapter Objectives
- Review Questions
- Key Ideas
- Chapter Problems

Section Goals and Introductions

Section 13.1 Gases and Their Properties

Goals

- *To describe the particle nature of both real and ideal gases.*
- *To describe the properties of gases that can be used to explain their characteristics: volume, number of particles, temperature, and pressure.*
- *To describe and explain the relationships between the properties of gases.*
- *To use the understanding of the relationships between gas properties to explain real-world things, such as the mechanics of a gasoline engine and the process of breathing.*

This section will increase your understanding of gases by adding more detail to the description of gases first presented in Chapter 2. As you know, it is often useful for scientists and science students to use simplified versions of reality (models) to explain scientific phenomena. This section introduces the ideal gas model that helps us to explain the characteristics of most gases.

The next portion of this section describes the properties of gases (number of gas particles, volume, temperature, and gas pressure) with an emphasis on gas pressure. You will discover what gas pressure is and what causes it. The rest of this section shows how the ideal gas model can be used to explain the relationships between the properties. For example, you will learn why increased temperature leads to increased pressure for a constant amount of gas in a constant volume. Spend the time it takes to develop a mental image of the particle nature of gases, and be sure that you can use that image to *see* how changing one property of a gas leads to changes in others.

See the related section on our Web site called *Real Gases*.

www.chemplace.com/college/

Section 13.2 Ideal Gas Calculations

Goal: To show how the properties of gases can be calculated.

This section derives two equations that relate to ideal gases (called the ideal gas equation and the combined gas law equation) and shows how you can use these equations to calculate values for gas properties. Pay special attention to the two sample study sheets that will help you to develop logical procedures for these calculations.

Section 13.3 Equation Stoichiometry and Ideal Gases

Goal: To show how gas-related calculations can be applied to equation stoichiometry problems.

This section shows how we can combine calculations such as those found in Chapter 10 with the gas calculations described in Section 13.2 to do equation stoichiometry problems that include gaseous reactants and products. It is a good idea to review *Study Sheet 10.3: Equation Stoichiometry Problems* and perhaps other parts of Chapter 10 before reading this section. *Sample Study Sheet 13.3: Equation Stoichiometry Problems* summarizes the procedures for equation stoichiometry problems that allow you to convert between amount of one reactant or product and amount of another reactant or product whether these substances are pure solids, pure liquids, pure gases, or in water solutions (aqueous).

See the related section on our Web site called *Gas Stoichiometry Shortcut.*
www.chemplace.com/college/

Section 13.4 Dalton's Law of Partial Pressures
Goals

- *To describe the properties of mixtures of gases.*
- *To describe calculations that deal with mixtures of gases.*

In the real world, gases are usually mixtures. This section describes how mixing gases affects the properties of the resulting mixture. Be sure that you can visualize mixtures of gases and that you can use this image to help you understand the effect that mixing gases has on the overall pressure created by the mixture. The section also derives two equations that allow you to calculate the pressures of gaseous mixtures.

Chapter 13 Map

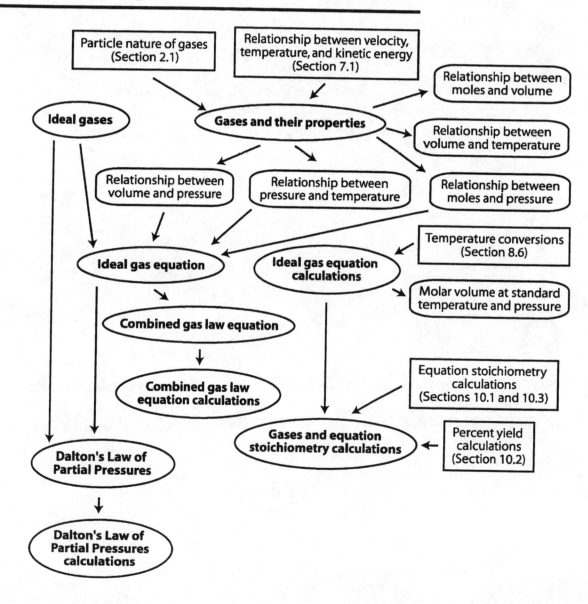

Chapter Checklist

☐ Read the Review Skills section. If there is any skill mentioned that you have not yet mastered, review the material on that topic before reading this chapter.

☐ Read the chapter quickly before the lecture that describes it.

☐ Attend class meetings, take notes, and participate in class discussions.

☐ Work the Chapter Exercises, perhaps using the Chapter Examples as guides.

☐ Study the Chapter Glossary and test yourself on our Web site:

www.chemplace.com/college/

☐ Study all of the Chapter Objectives. You might want to write a description of how you will meet each objective.

☐ Reread the Study Sheets in this chapter and decide whether you will use them or some variation on them to complete the tasks they describe.

Study Sheet 13.1: Using the Ideal Gas Equation

Sample Study Sheet 13.2: Using the Combined Gas Law Equation

Sample Study Sheet 13.3: Equation Stoichiometry Problems

Sample Study Sheet 13.4: Using Dalton's Law of Partial Pressures

☐ This chapter has logic sequences in Figures 13.3, 13.4, 13.5, 13.6, and 13.7. Convince yourself that each of the statements in these sequences logically lead to the next statement.

☐ Memorize the following equations.

$$PV = nRT$$

$$PV = \frac{g}{M} RT$$

$$\frac{P_1 V_1}{n_1 T_1} = \frac{P_2 V_2}{n_2 T_2}$$

$$P_{total} = \sum P_{partial}$$

$$P_{total} = \left(\sum n_{each\ gas} \right) \frac{RT}{V}$$

☐ To get a review of the most important topics in the chapter, fill in the blanks in the Key Ideas section.

☐ Work all of the selected problems at the end of the chapter, and check your answers with the solutions provided in this chapter of the study guide.

☐ Ask for help if you need it.

Web Resources   www.chemplace.com/college/

Real Gases
Gas Stoichiometry Shortcut
Glossary Quiz

Exercise Key

Exercise 13.1 – Using the Ideal Gas Equation: Krypton gas does a better job than argon of slowing the evaporation of the tungsten filament in an incandescent light bulb. Because of its higher cost, however, krypton is used only when longer life is considered to be worth the extra expense.   *(Objs #15, 16, & 17)*

a. How many moles of krypton gas must be added to a 175 mL incandescent light bulb to yield a gas pressure of 117 kPa at 21.6 °C?

$$PV = nRT \qquad n = \frac{PV}{RT} = \frac{117 \text{ kPa } (175 \text{ mL})}{\dfrac{8.3145 \text{ L} \cdot \text{kPa}}{\text{K} \cdot \text{mol}} \; 294.8 \text{ K}} \left(\frac{1 \text{ L}}{10^3 \text{ mL}}\right)$$

$$= \mathbf{8.35 \times 10^{-3} \text{ mol Kr}}$$

b. What is the volume of an incandescent light bulb that contains 1.196 g of Kr at a pressure of 1.70 atm and a temperature of 97 °C?

$$V = ? \qquad g = 1.196 \text{ g} \qquad T = 97 \,°C + 273.15 = 370 \text{ K} \qquad P = 1.70 \text{ atm}$$

$$PV = \frac{g}{M}RT \qquad V = \frac{gRT}{PM} = \frac{1.196 \text{ g Kr} \left(\dfrac{0.082058 \text{ L} \cdot \text{atm}}{\text{K} \cdot \text{mole}}\right) 370 \text{ K}}{1.70 \text{ atm} \left(83.80 \dfrac{g}{\text{mole}}\right)}$$

$$= \mathbf{0.255 \text{ L Kr}}$$

c. What is the density of krypton gas at 18.2 °C and 762 mmHg?

$$\frac{g}{V} = ? \qquad P = 762 \text{ mmHg} \qquad T = 18.2 \,°C + 273.15 = 291.4 \text{ K}$$

$$PV = \frac{g}{M}RT \qquad \frac{g}{V} = \frac{PM}{RT} = \frac{762 \text{ mmHg} \left(83.80 \dfrac{g}{\text{mole}}\right)}{0.082058 \dfrac{\text{L atm}}{\text{K mole}} (291.4 \text{ K})} \left(\frac{1 \text{ atm}}{760 \text{ mmHg}}\right)$$

$$= \mathbf{3.51 \text{ g/L}}$$

Exercise 13.2 – Using the Combined Gas Law Equation: A helium weather balloon is filled in Monterey, California, on a day when the atmospheric pressure is 102 kPa and the temperature is 18 °C. Its volume under these conditions is 1.6×10^4 L. Upon being released, it rises to an altitude where the temperature is –8.6 °C, and its volume increases to 4.7×10^4 L. Assuming that the internal pressure of the balloon equals the atmospheric pressure, what is the pressure at this altitude? *(Obj #18)*

$$P_1 = 102 \text{ kPa} \qquad T_1 = 18 \text{ °C} + 273.15 = 291 \text{ K} \qquad V_1 = 1.6 \times 10^4 \text{ L}$$

$$P_2 = ? \qquad T_2 = -8.6 \text{ °C} + 273.15 = 264.6 \text{ K} \qquad V_2 = 4.7 \times 10^4 \text{ L}$$

$$\frac{P_1 V_1}{n_1 T_1} = \frac{P_2 V_2}{n_2 T_2} \quad \text{to} \quad \frac{P_1 V_1}{T_1} = \frac{P_2 V_2}{T_2}$$

$$P_2 = P_1 \left(\frac{T_2}{T_1}\right)\left(\frac{V_1}{V_2}\right) = 102 \text{ kPa} \left(\frac{264.6 \text{ K}}{291 \text{ K}}\right)\left(\frac{1.6 \times 10^4 \text{ L}}{4.7 \times 10^4 \text{ L}}\right) = \textbf{32 kPa}$$

Exercise 13.3 – Equation Stoichiometry: Iron is combined with carbon in a series of reactions to form pig iron, which is about 4.3% carbon.

$$2C + O_2 \rightarrow 2CO$$

$$Fe_2O_3 + 3CO \rightarrow 2Fe + 3CO_2$$

$$2CO \rightarrow C \text{ (in iron)} + CO_2$$

Pig iron is easier to shape than pure iron, and the presence of carbon lowers its melting point from the 1539 °C required to melt pure iron to 1130 °C. *(Obj #19)*

a. In the first reaction, what minimum volume of oxygen at STP is necessary to convert 125 Mg of carbon to carbon monoxide?

$$? \text{ L } O_2 = 125 \text{ Mg } C \left(\frac{10^6 \text{ g}}{1 \text{ Mg}}\right)\left(\frac{1 \text{ mol } C}{12.011 \text{ g } C}\right)\left(\frac{1 \text{ mol } O_2}{2 \text{ mol } C}\right)\left(\frac{22.414 \text{ L } O_2}{1 \text{ mol } O_2}\right)_{STP}$$

$$= \textbf{1.17} \times \textbf{10}^8 \textbf{ L } \textbf{O}_2 \quad \text{or} \quad \textbf{1.17} \times \textbf{10}^5 \textbf{ m}^3 \textbf{ O}_2$$

b. In the first reaction, what is the maximum volume of carbon monoxide at 1.05 atm and 35 °C that could form from the conversion of 8.74×10^5 L of oxygen at 0.994 atm and 27 °C?

$$? \text{ L CO} = 8.74 \times 10^5 \text{ L } O_2 \left(\frac{K \cdot mol}{0.082058 \text{ L} \cdot atm}\right)\left(\frac{0.994 \text{ atm}}{300 \text{ K}}\right)\left(\frac{2 \text{ mol CO}}{1 \text{ mol } O_2}\right)\left(\frac{0.082058 \text{ L} \cdot atm}{K \cdot mol}\right)\left(\frac{308 \text{ K}}{1.05 \text{ atm}}\right)$$

$$= \textbf{1.70} \times \textbf{10}^6 \textbf{ L CO}$$

Exercise 13.4 – Equation Stoichiometry: Sodium hypochlorite, NaOCl, which is found in household bleaches, can be made from a reaction using chlorine gas and aqueous sodium hydroxide:

$$Cl_2(g) + 2NaOH(aq) \rightarrow NaOCl(aq) + NaCl(aq) + H_2O(l)$$

What minimum volume of chlorine gas at 101.4 kPa and 18.0 °C must be used to react with all the sodium hydroxide in 3525 L of 12.5 M NaOH? *(Obj #19)*

$$? \text{ L Cl}_2 = 3525 \text{ L NaOH soln} \left(\frac{12.5 \text{ mol NaOH}}{1 \text{ L NaOH soln}}\right)\left(\frac{1 \text{ mol Cl}_2}{2 \text{ mol NaOH}}\right)\left(\frac{8.3145 \text{ L}\bullet\text{kPa}}{K\bullet\text{mol}}\right)\left(\frac{291.0 \text{ K}}{101.4 \text{ kPa}}\right)$$

$$= 5.26 \times 10^5 \text{ L Cl}_2$$

Exercise 13.5 – Dalton's Law of Partial Pressures: A typical "neon light" contains neon gas mixed with argon gas. *(Objs 21 & 22)*

a. If the total pressure of the mixture of gases is 1.30 kPa and the partial pressure of neon gas is 0.27 kPa, what is the partial pressure of the argon gas?

$$P_{total} = P_{Ne} + P_{Ar} \qquad P_{Ar} = P_{total} - P_{Ne} = 1.30 \text{ kPa} - 0.27 \text{ kPa} = \textbf{1.03 kPa}$$

b. If 6.3 mg of Ar and 1.2 mg Ne are added to the 375-mL tube at 291 K, what is the total pressure of the gases in millimeters of mercury?

$$? \text{ mol Ar} = 6.3 \text{ mg Ar}\left(\frac{1 g}{10^3 \text{ mg}}\right)\left(\frac{1 \text{ mol Ar}}{39.948 \text{ g Ar}}\right) = 1.6 \times 10^{-4} \text{ mol Ar}$$

$$? \text{ mol Ne} = 1.2 \text{ mg Ne}\left(\frac{1 g}{10^3 \text{ mg}}\right)\left(\frac{1 \text{ mol Ne}}{20.1797 \text{ g Ne}}\right) = 5.9 \times 10^{-5} \text{ mol Ne}$$

$$P_{total} = \left(\sum n\right)\frac{RT}{V}$$

$$= \left(1.6 \times 10^{-4} \text{ mol} + 5.9 \times 10^{-5} \text{ mol}\right)\frac{\frac{0.082058 \text{ L}\bullet\text{atm}}{K\bullet\text{mol}}(291 \text{ K})}{375 \text{ mL}}\left(\frac{10^3 \text{ mL}}{1 \text{ L}}\right)\left(\frac{760 \text{ mmHg}}{1 \text{ atm}}\right)$$

$$= \textbf{11 mmHg}$$

Review Questions Key

1. Describe the particle nature of gases. (See Section 2.1.)

   The Kinetic Molecular Theory provides a simple model of the nature of matter. It has the following components:

   - All matter is composed of tiny particles.

   - These particles are in constant motion. The amount of motion is proportional to temperature. Increased temperature means increased motion.

   - Solids, gases, and liquids differ in the degree of motion of their particles and the extent to which the particles interact.

   Because the particles of a gas are much farther apart than those of the solid or liquid, the particles do not have significant attractions between them. The particles in a gas move freely in straight-line paths until they collide with another particle or the walls of the container. If you were riding on a particle in the gas state, your ride would be generally boring with regular interruptions caused by violent collisions. Between these collisions, you would not even know there were other particles in the container. Because the particles are moving with different velocities and in different directions, the collisions lead to constant changes in the direction and velocity of the motion of each particle. The rapid, random movement of the gas particles allows gases to adjust to the shape and volume of their container.

2. What is 265.2 °C on the kelvin scale? Convert 565.7 K into °C.

   $? K = 265.2\ °C + 273.15\ K$ = **538.4 K** $?\ °C = 565.7\ K - 273.15$ = **292.6 °C**

3. About 55% of industrially produced sodium sulfate is used to make detergents. It is made from the reaction

   $$4NaCl + 2SO_2 + 2H_2O + O_2 \rightarrow 2Na_2SO_4 + 4HCl$$

   a. What is the maximum mass of sodium sulfate that can be produced in the reaction of 745 Mg of sodium chloride with excess SO_2, H_2O, and O_2?

   $$? Mg\ Na_2SO_4 = 745\ Mg\ NaCl \left(\frac{10^6\ g}{1\ Mg}\right)\left(\frac{1\ mol\ NaCl}{58.4425\ g\ NaCl}\right)\left(\frac{2\ mol\ Na_2SO_4}{4\ mol\ NaCl}\right)\left(\frac{142.043\ g\ Na_2SO_4}{1\ mol\ Na_2SO_4}\right)\left(\frac{1\ Mg}{10^6\ g}\right)$$

   or using a shortcut $? Mg\ Na_2SO_4 = 745\ Mg\ NaCl \left(\dfrac{2 \times 142.043\ Mg\ Na_2SO_4}{4 \times 58.4425\ Mg\ NaCl}\right)$ = **905 Mg**

b. What is the maximum mass of sodium sulfate that can be produced in the reaction of 745 Mg of sodium chloride with 150 Mg H_2O and excess SO_2 and O_2?

$$? \text{ Mg Na}_2\text{SO}_4 = 745 \text{ Mg NaCl} \left(\frac{2 \times 142.043 \text{ Mg Na}_2\text{SO}_4}{4 \times 58.4425 \text{ Mg NaCl}} \right) = \textbf{905 Mg}$$

$$? \text{ Mg Na}_2\text{SO}_4 = 150 \text{ Mg H}_2\text{O} \left(\frac{2 \times 142.043 \text{ Mg Na}_2\text{SO}_4}{2 \times 18.0153 \text{ Mg H}_2\text{O}} \right) = 1.18 \times 10^3 \text{ Mg}$$

c. If 868 Mg Na_2SO_4 is formed in the reaction of 745 Mg of sodium chloride with 150 Mg of H_2O and excess SO_2 and O_2, what is the percent yield?

$$\% \text{ yield} = \frac{\text{actual yield}}{\text{theoretical yield}} \times 100 = \frac{868 \text{ Mg}}{905 \text{ Mg}} \times 100 = \textbf{95.9\% yield}$$

d. What volume of 2.0 M Na_2SO_4 can be formed from the reaction of 745 Mg of sodium chloride with excess SO_2, H_2O, and O_2?

$$? \text{ Mg Na}_2\text{SO}_4 = 745 \text{ Mg NaCl} \left(\frac{10^6 \text{ g}}{1 \text{ Mg}} \right) \left(\frac{1 \text{ mol NaCl}}{58.4425 \text{ g NaCl}} \right) \left(\frac{2 \text{ mol Na}_2\text{SO}_4}{4 \text{ mol NaCl}} \right) \left(\frac{1 \text{ L Na}_2\text{SO}_4 \text{ soln}}{2.0 \text{ mol Na}_2\text{SO}_4} \right)$$

$$= \textbf{3.2} \times \textbf{10}^6 \textbf{ L Na}_2\textbf{SO}_4 \textbf{ solution}$$

Key Ideas Answers

4. Under typical conditions, the average distance between gas particles is about **ten** times their diameter.

6. For a gas at room temperature and pressure, the gas particles themselves occupy about **0.1%** of the total volume. The other **99.9%** of the total volume is empty space (whereas in liquids, about **70%** of the volume is occupied by particles).

8. The particles in a gas are in **rapid** and **continuous** motion.

10. The particles in a gas are constantly colliding with the walls of the container and with each other. Because of these collisions, the gas particles are constantly changing their **direction of motion** and their **velocity**.

12. The particles of an ideal gas are assumed to be **point-masses**, that is, particles that have a mass but occupy no volume.

14. The ideal gas model is used to predict changes in four related gas properties: **pressure, volume, number of particles**, and **temperature**.

16. Although gas temperatures are often measured with thermometers that report temperatures in **degrees Celsius, °C**, scientists generally use **Kelvin** temperatures for calculations.

18. The accepted SI unit for gas pressure is the **pascal, Pa.**

20. The observation that the pressure of an ideal gas is inversely proportional to the volume it occupies if the **moles of gas** and the temperature are constant is a statement of Boyle's Law.

22. The pressure of an ideal gas is directly proportional to the **Kelvin temperature** of the gas if the volume and moles of gas are constant. This relationship is sometimes called Gay-Lussac's Law.

24. For an ideal gas, volume and temperature described in kelvins are **directly** proportional if moles of gas and pressure are constant. This is a statement of Charles' Law.

26. If the **temperature and volume** of an ideal gas are held constant, the moles of gas in a container and the gas pressure are directly proportional.

28. It is always a good idea to include the units in a solved equation as well as the numbers. If the units cancel to yield a reasonable unit for the unknown property, you can feel confident that you picked the **correct equation**, that you did the **algebra** correctly to solve for your unknown, and that you have made the **necessary unit conversions**.

30. There are three different ways to convert between a measurable property and moles in equation stoichiometry problems. For pure liquids and solids, we can convert between mass and moles, using the **molar mass** as a conversion factor. For gases, we can convert between volume of gas and moles using the methods described above. For solutions, **molarity** provides a conversion factor that enables us to convert between moles of solute and volume of solution.

32. Assuming ideal gas character, the partial pressure of any gas in a mixture is the pressure that the gas would exert if it were **alone** in the container.

Problems Key

Section 13.1 Gases and Their Properties

34. For a gas under typical conditions, what approximate percentage of the volume is occupied by the gas particles themselves? *(Obj #2)*

    0.1%

36. Why is it harder to walk through water than to walk through air?

    When we walk through air, we push the air particles out of the way as we move. Because the particles in a liquid occupy about 70% of the space that contains the liquid (as opposed to 0.1% of the space occupied by gas particles), there are a lot more particles to push out of the way as you move through water.

38. What are the key assumptions that distinguish an ideal gas from a real gas? *(Obj #4)*

    - The particles are assumed to be point-masses, that is, particles that have a mass but occupy no volume.

    - There are no attractive or repulsive forces between the particles in an ideal gas.

40. A TV weather person predicts a storm for the next day and refers to the dropping barometric pressure to support the prediction. What causes the pressure in the air?

    The particles in the air (N_2, O_2, Xe, CO_2, and others) are constantly moving and constantly colliding with everything surrounded by the air. Each of these collisions exerts a tiny force against the object with which they collide. The total force of these collisions per unit area is the atmospheric pressure.

42. The pressure at the center of the earth is 4×10^{11} pascals, Pa. What is this pressure in kilopascals, atmospheres, millimeters of mercury, and torr? *(Obj #8)*

$$? \, kPa = 4 \times 10^{11} \, Pa \left(\frac{1 \, kPa}{10^3 \, Pa} \right) = 4 \times 10^8 \, \textbf{kPa}$$

$$? \, atm = 4 \times 10^{11} \, Pa \left(\frac{1 \, atm}{101325 \, Pa} \right) = \textbf{4} \times \textbf{10}^\textbf{6} \, \textbf{atm}$$

$$? \, mmHg = 4 \times 10^{11} \, Pa \left(\frac{1 \, atm}{101325 \, Pa} \right) \left(\frac{760 \, mmHg}{1 \, atm} \right) = 3 \times 10^9 \, \textbf{mmHg} = 3 \times 10^9 \, \textbf{torr}$$

44. What does *inversely proportional* mean?

X and Y are inversely proportional if a decrease in X leads to a proportional increase in Y or an increase in X leads to a proportional decrease in Y. For example, volume of gas and its pressure are inversely proportional if the temperature and moles of gas are constant. If the volume is decreased to one-half its original value, the pressure of the gas will double. If the volume is doubled, the pressure decreases to one-half its original value. The following expression summarizes this inverse relationship:

$$P \, \alpha \, \frac{1}{V} \quad \text{if n and T are constant}$$

47. Ammonia, NH_3, is a gas that can be dissolved in water and used as a cleaner. When an ammonia solution is used to clean the wax off a floor, ammonia gas escapes from the solution, mixes easily with the gas particles in the air, and spreads throughout the room. Very quickly, everyone in the room can smell the ammonia gas. Explain why gaseous ammonia mixes so easily with air.

In gases, there is plenty of empty space between the particles and essentially no attractions between them, so there is nothing to stop gases, like ammonia and the gases in air, from mixing readily and thoroughly.

49. Explain why air moves in and out of our lungs as we breathe. *(Obj #12)*

When the muscles of your diaphragm contract and your chest expands, the volume of your lungs increases. This leads to a decrease in the number of particles per unit volume inside the lungs, which leaves fewer particles near any given area of the inner surface of the lungs. There are then fewer collisions per second per unit area of lungs and a decrease in force per unit area or gas pressure. During quiet, normal breathing, this increase in volume decreases the pressure in the lungs to about 0.4 kilopascals lower than the atmospheric pressure. The larger volume causes air to move into the lungs faster than it moves out, bringing in fresh oxygen. When the muscles relax, the lungs return to their original volume, and the decrease in volume causes the pressure in the lungs to increase to about 0.4 kilopascals above atmospheric pressure. Air now goes out of the lungs faster than it comes in. See Figure 13.9 on page 542 of the textbook.

51. With reference to the relationships between properties of gases, answer the following questions.
    a. In a common classroom demonstration, water is heated in a 1-gallon can, covered with a tight lid, and allowed to cool. As the can cools, it collapses. Why?

    As the can cools, the water vapor in the can condenses to liquid, leaving fewer moles of gas. Decreased moles of gas and decreased temperature both lead to decreased gas pressure in the can. Because the external pressure pushing on the outside of the can is then greater than the internal pressure pushing outward, the can collapses.

    b. Dents in ping-pong balls can often be removed by placing the balls in hot water. Why?

    The increased temperature causes the internal pressure of the ping-pong ball to increase. This leads to the pressure pushing out on the shell of the ball to be greater than the external pressure pushing in on the shell. If the difference in pressure is enough, the dents are pushed out.

Section 13.3 Calculations Involving Ideal Gases

53. Neon gas produced for use in luminous tubes comes packaged in 1.0-L containers at a pressure of 1.00 atm and a temperature of 19 °C. How many moles of Ne do these cylinders hold? *(Obj #15)*

$$PV = nRT \qquad n = \frac{PV}{RT} = \frac{1.00 \text{ atm } (1.0 \text{ L})}{\left(\frac{0.082058 \text{ L} \cdot \text{atm}}{\text{K} \cdot \text{mol}}\right) 292 \text{ K}} = \textbf{0.042 moles Ne}$$

55. A typical aerosol can is able to withstand 10-12 atm without exploding.
    a. If a 375-mL aerosol can contains 0.062 mole of gas, at what temperature would the gas pressure reach 12 atm of pressure? *(Obj #15)*

$$PV = nRT \qquad T = \frac{PV}{nR} = \frac{12 \text{ atm } (375 \text{ mL})}{0.062 \text{ mol} \left(\frac{0.082058 \text{ L} \cdot \text{atm}}{\text{K} \cdot \text{mol}}\right)} \left(\frac{1 \text{ L}}{10^3 \text{ mL}}\right)$$

$$= \textbf{8.8} \times \textbf{10}^2 \textbf{ K or 6.1} \times \textbf{10}^2 \textbf{ °C}$$

    b. Aerosol cans usually contain liquids as well as gas. If the 375-mL can described in Part (a) contained liquid along with the 0.062 moles of gas, why would it explode at a lower temperature than if it contained only the gas? (*Hint:* What happens to a liquid when it is heated?) *(Obj #15)*

    As the temperature is increased, the liquid would evaporate more rapidly, increasing the amount of gas in the container and increasing the pressure.

57. Bromomethane, CH_3Br, commonly called methyl bromide, is used as a soil fumigant, but because it is a threat to the ozone layer, its use is being phased out. This colorless gas is toxic by ingestion, inhalation, or skin absorption, so the American Conference of Government Industrial Hygienists has assigned it a threshold limit value, or TLV, a maximum concentration to which a worker may be repeatedly exposed day after day without experiencing adverse effects. (TLV standards are meant to serve as guides in control of health hazards, not to provide definitive dividing lines between safe and dangerous concentrations.) Bromomethane reaches its TLV when 0.028 mole escapes into a room that has a volume of 180 m^3 and a temperature of 18 °C. What is the pressure in atmospheres of the gas under those conditions? (There are 10^3 L per cubic meter.) *(Obj #15)*

$$PV = nRT \qquad P = \frac{nRT}{V} = \frac{0.028 \text{ mol} \left(\dfrac{0.082058 \text{ L} \cdot \text{atm}}{K \cdot \text{mol}} \right) 291 \text{ K}}{180 \text{ m}^3} \left(\frac{1 \text{ m}^3}{10^3 \text{ L}} \right)$$

$$= \mathbf{3.7 \times 10^{-6} \text{ atm}}$$

59. Gaseous chlorine, Cl_2, which is used in water purification, is a dense, greenish-yellow substance with an irritating odor. It is formed from the electrolysis of molten sodium chloride. Chlorine reaches its TLV (defined in Problem 57) in a laboratory when 0.017 mole of Cl_2 is released, yielding a pressure of 7.5×10^{-4} mmHg at a temperature of 19 °C. What is the volume, in liters, of the room? *(Obj #15)*

$$PV = nRT \qquad V = \frac{nRT}{P} = \frac{0.017 \text{ mol} \left(\dfrac{0.082058 \text{ L} \cdot \text{atm}}{K \cdot \text{mol}} \right) 292 \text{ K}}{7.5 \times 10^{-4} \text{ mmHg}} \left(\frac{760 \text{ mmHg}}{1 \text{ atm}} \right)$$

$$= \mathbf{4.1 \times 10^5 \text{ L}}$$

61. Hydrogen sulfide, H_2S, which is used to precipitate sulfides of metals, is a colorless gas that smells like rotten eggs. It can be detected by the nose at about 0.6 mg/m^3. To reach this level in a 295-m^3 room, 0.177 g of H_2S must escape into the room. What is the pressure in pascals of 0.177 g of H_2S that escapes into a 295-m^3 laboratory room at 291 K? *(Obj #16)*

$$PV = \frac{g}{M}RT \qquad P = \frac{gRT}{MV} = \frac{0.177 \text{ g} \left(\dfrac{8.3145 \text{ L} \cdot \text{kPa}}{K \cdot \text{mol}} \right) 291 \text{ K}}{\dfrac{34.082 \text{ g}}{\text{mol}} \, 295 \text{ m}^3} \left(\frac{1 \text{ m}^3}{10^3 \text{ L}} \right) \left(\frac{10^3 \text{ Pa}}{1 \text{ kPa}} \right)$$

$$= \mathbf{0.0426 \text{ Pa}}$$

63. Hydrogen chloride, HCl, is used to make vinyl chloride, which is then used to make polyvinyl chloride (PVC) plastic. Hydrogen chloride can be purchased in liquefied form in cylinders that contain 8 lb of HCl. What volume, in liters, will 8.0 lb of HCl occupy if the compound becomes a gas at 1.0 atm and 85 °C? *(Obj #16)*

$$PV = \frac{g}{M}RT \qquad V = \frac{gRT}{MP} = \frac{8.0 \text{ lb} \left(\dfrac{0.082058 \text{ L} \cdot \text{atm}}{K \cdot \text{mol}} \right) 358 \text{ K}}{\dfrac{36.4606 \text{ g}}{1 \text{ mol}} \, 1.0 \text{ atm}} \left(\frac{453.6 \text{ g}}{1 \text{ lb}} \right)$$

$$= \mathbf{2.9 \times 10^3 \text{ L}}$$

65. The bulbs used for fluorescent lights have a mercury gas pressure of 1.07 Pa at 40 °C. How many milligrams of liquid mercury must evaporate at 40 °C to yield this pressure in a 1.39-L fluorescent bulb? *(Obj #16)*

$$PV = \frac{g}{M}RT \qquad g = \frac{PVM}{RT} = \frac{1.07\ Pa\ (1.39\ L)\ \dfrac{200.59\ g}{1\ mol}}{\left(\dfrac{8.3145\ L \cdot kPa}{K \cdot mol}\right)313\ K}\left(\frac{1\ kPa}{10^3\ Pa}\right)\left(\frac{10^3\ mg}{1\ g}\right)$$

$$= \mathbf{0.115\ mg}$$

67. Flashtubes are light bulbs that produce a high-intensity flash of light that is very short in duration. They are used in photography and for airport approach lighting. The xenon gas, Xe, in flashtubes has large atoms that can quickly carry the heat away from the filament to stop the radiation of light. At what temperature does 1.80 g of Xe in a 0.625-L flashtube have a pressure of 75.0 kPa? *(Obj #16)*

$$PV = \frac{g}{M}RT \qquad T = \frac{PVM}{gR} = \frac{75.0\ kPa\ (0.625\ L)\ \dfrac{131.29\ g}{1\ mol}}{1.80\ g\left(\dfrac{8.3145\ L \cdot kPa}{K \cdot mol}\right)} = \mathbf{411\ K\ or\ 138\ °C}$$

69. When 0.690 g of an unknown gas is held in an otherwise-empty 285-mL container, the pressure is 756.8 mmHg at 19 °C. What is the molecular mass of the gas? *(Obj #16)*

$$PV = \frac{g}{M}RT$$

$$M = \frac{gRT}{PV} = \frac{0.690\ g\left(\dfrac{0.082058\ L \cdot atm}{K \cdot mol}\right)292\ K}{756.8\ mmHg\ (285\ mL)}\left(\frac{10^3\ mL}{1\ L}\right)\left(\frac{760\ mmHg}{1\ atm}\right)$$

$$= \mathbf{58.3\ g/mol}$$

71. Consider two helium tanks used to fill children's balloons. The tanks have equal volume and are at the same temperature, but tank A is new and therefore has a higher pressure of helium than tank B, which has been used to fill many balloons. Which tank holds gas with a greater density, tank A or tank B? Support your answer.

Tank A – The following shows that density (g/V) is proportional to pressure when temperature is constant. Higher pressure means higher density at a constant temperature.

$$PV = \frac{g}{M}RT \qquad \frac{g}{V} = P\left(\frac{M}{RT}\right)$$

73. Butadiene, $CH_2CHCHCH_2$, a suspected carcinogen, is used to make plastics such as styrene-butadiene rubber (SBR). Because it is highly flammable and forms explosive peroxides when exposed to air, it is often shipped in insulated containers at about 2 °C. What is the density of butadiene when its pressure is 102 kPa and its temperature is 2 °C? *(Obj #17)*

$$PV = \frac{g}{M}RT \qquad \frac{g}{V} = \frac{PM}{RT} = \frac{102\ kPa\left(\dfrac{54.092\ g}{1\ mol}\right)}{\left(\dfrac{8.3145\ L \cdot kPa}{K \cdot mol}\right)275\ K} = \textbf{2.41 g/L}$$

75. In order to draw air into your lungs, your diaphragm and other muscles contract, increasing the lung volume. This lowers the air pressure of the lungs to below atmospheric pressure, and air flows in. When your diaphragm and other muscles relax, the volume of the lungs decreases, and air is forced out. *(Obj #28)*

a. If the total volume of your lungs at rest is 6.00 L and the initial pressure is 759 mmHg, what is the new pressure if the lung volume is increased to 6.02 L?

$$\frac{P_1V_1}{n_1T_1} = \frac{P_2V_2}{n_2T_2} \quad \text{Assuming constant moles of gas and temperature, } P_1V_1 = P_2V_2$$

$$P_2 = P_1\left(\frac{V_1}{V_2}\right) = 759\ mmHg\left(\frac{6.00\ L}{6.02\ L}\right) = \textbf{756 mmHg}$$

b. If the total volume of your lungs at rest is 6.00 L and the initial pressure is 759 mmHg, at what volume will the pressure be 763 mmHg?

$$\frac{P_1V_1}{n_1T_1} = \frac{P_2V_2}{n_2T_2} \quad \text{Assuming constant moles of gas and temperature, } P_1V_1 = P_2V_2$$

$$V_2 = V_1\left(\frac{P_1}{P_2}\right) = 6.00\ L\left(\frac{759\ mmHg}{763\ mmHg}\right) = \textbf{5.97 L}$$

77. A scuba diver has 4.5 L of air in his lungs 66 ft below the ocean surface, where the pressure is 3.0 atm. What would the volume of this gas be at the surface, where the pressure is 1.0 atm? If the diver's lungs can hold no more than 7.0 L without rupturing, is he in trouble if he doesn't exhale as he rises to the surface? *(Obj #18)*

$$\frac{P_1V_1}{n_1T_1} = \frac{P_2V_2}{n_2T_2} \qquad \text{Assuming constant moles of gas and temperature, } P_1V_1 = P_2V_2$$

$$V_2 = V_1\left(\frac{P_1}{P_2}\right) = 4.5\ L\left(\frac{3.0\ atm}{1.0\ atm}\right) = \textbf{13.5 L}$$

The diver had better exhale as he goes up.

78. Picture a gas in the apparatus shown in Figure 13.2 on page 534 of the textbook. If the volume and temperature remain constant, how would you change the number of gas particles to increase the pressure of the gas?

Increased number of gas particles leads to increased pressure when temperature and volume are constant.

80. A balloon containing 0.62 mole of gas has a volume of 15 L. Assuming constant temperature and pressure, how many moles of gas does the balloon contain when enough gas leaks out to decrease the volume to 11 L? *(Obj #18)*

$$\frac{P_1 V_1}{n_1 T_1} = \frac{P_2 V_2}{n_2 T_2} \qquad \text{Assuming constant pressure and temperature,} \qquad \frac{V_1}{n_1} = \frac{V_2}{n_2}$$

$$n_2 = n_1 \left(\frac{V_2}{V_1}\right) = 0.62 \text{ mol} \left(\frac{11 \text{ L}}{15 \text{ L}}\right) = \mathbf{0.45 \ mol}$$

82. Consider a weather balloon with a volume on the ground of 113 m^3 at 101 kPa and 14 °C. *(Obj #18)*

   a. If the balloon rises to a height where the temperature is –50 °C and the pressure is 2.35 kPa, what will its the volume be? (In 1958, a balloon reached 101,500 ft, or 30.85 km, the height at which the temperature is about –50 °C.)

   $$\frac{P_1 V_1}{n_1 T_1} = \frac{P_2 V_2}{n_2 T_2} \qquad \text{Assuming constant moles of gas,} \qquad \frac{P_1 V_1}{T_1} = \frac{P_2 V_2}{T_2}$$

   $$V_2 = V_1 \left(\frac{T_2}{T_1}\right)\left(\frac{P_1}{P_2}\right) = 113 \text{ m}^3 \left(\frac{223 \text{ K}}{287 \text{ K}}\right)\left(\frac{101 \text{ kPa}}{2.35 \text{ kPa}}\right) = \mathbf{3.77 \times 10^3 \ m^3}$$

   b. When the balloon rises to 10 km, where the temperature is –40 °C, its volume increases to 129 m^3. If the pressures inside and outside the balloon are equal, what is the atmospheric pressure at this height?

   $$\frac{P_1 V_1}{n_1 T_1} = \frac{P_2 V_2}{n_2 T_2} \qquad \text{Assuming constant moles of gas,} \qquad \frac{P_1 V_1}{T_1} = \frac{P_2 V_2}{T_2}$$

   $$P_2 = P_1 \left(\frac{T_2}{T_1}\right)\left(\frac{V_1}{V_2}\right) = 101 \text{ kPa} \left(\frac{233 \text{ K}}{287 \text{ K}}\right)\left(\frac{113 \text{ m}^3}{129 \text{ m}^3}\right) = \mathbf{71.8 \ kPa}$$

   c. If the balloon has a volume of 206 m^3 at 25 km, where the pressure is 45 kPa, what is the temperature at this height?

   $$\frac{P_1 V_1}{n_1 T_1} = \frac{P_2 V_2}{n_2 T_2} \qquad \text{Assuming constant moles of gas,} \qquad \frac{P_1 V_1}{T_1} = \frac{P_2 V_2}{T_2}$$

   $$T_2 = T_1 \left(\frac{P_2}{P_1}\right)\left(\frac{V_2}{V_1}\right) = 287 \text{ K} \left(\frac{45 \text{ kPa}}{101 \text{ kPa}}\right)\left(\frac{206 \text{ m}^3}{113 \text{ m}^3}\right) = \mathbf{233 \ K}$$

Section 13.3 Equation Stoichiometry and Ideal Gases

85. Chlorine gas is used in the production of many other chemicals, including carbon tetrachloride, polyvinyl chloride plastic, and hydrochloric acid. It is produced from the electrolysis of molten sodium chloride. What minimum mass of sodium chloride, in megagrams, is necessary to make 2.7×10^5 L of chlorine gas at standard temperature and pressure, STP? *(Obj #18)*

$$2NaCl(l) \quad \xrightarrow{\text{Electrolysis}} \quad 2Na(l) + Cl_2(g)$$

$$? \text{ Mg NaCl} = 2.7 \times 10^5 \text{ L Cl}_2 \left(\frac{1 \text{ mol Cl}_2}{22.414 \text{ L Cl}_2} \right)_{STP} \left(\frac{2 \text{ mol NaCl}}{1 \text{ mol Cl}_2} \right) \left(\frac{58.4425 \text{ g NaCl}}{1 \text{ mol NaCl}} \right) \left(\frac{1 \text{ Mg}}{10^6 \text{ g}} \right)$$

$$= \textbf{1.4 Mg NaCl}$$

87. Air bags in cars inflate when sodium azide, NaN_3, decomposes to generate nitrogen gas. How many grams of NaN_3 must react to generate 112 L N_2 at 121 kPa and 305 K? *(Obj #18)*

$$2NaN_3(s) \rightarrow 2Na(s) + 3N_2(g)$$

$$? \text{ g NaN}_3 = 112 \text{ L N}_2 \left(\frac{\text{K} \cdot \text{mol}}{8.3145 \text{ L} \cdot \text{kPa}} \right) \left(\frac{121 \text{ kPa}}{305 \text{ K}} \right) \left(\frac{2 \text{ mol NaN}_3}{3 \text{ mol N}_2} \right) \left(\frac{65.0099 \text{ g NaN}_3}{1 \text{ mol NaN}_3} \right)$$

$$= \textbf{232 g NaN}_3$$

89. The hydrogen chloride gas used to make polyvinyl chloride (PVC) plastic is made by reacting hydrogen gas with chlorine gas. What volume of $HCl(g)$ at 105 kPa and 296 K can be formed from 1150 m^3 of $H_2(g)$ at STP? *(Obj #18)*

$$H_2(g) + Cl_2(g) \rightarrow 2 \text{ HCl}(g)$$

$$? \text{ m}^3 \text{ HCl} = 1150 \text{ m}^3 \text{ H}_2 \left(\frac{10^3 \text{ L}}{1 \text{ m}^3} \right) \left(\frac{1 \text{ mol H}_2}{22.414 \text{ L H}_2} \right)_{STP} \left(\frac{2 \text{ mol HCl}}{1 \text{ mol H}_2} \right) \left(\frac{8.3145 \text{ L} \cdot \text{kPa}}{\text{K} \cdot \text{mol}} \right) \left(\frac{296 \text{ K}}{105 \text{ kPa}} \right) \left(\frac{1 \text{ m}^3}{10^3 \text{ L}} \right)$$

$$= \textbf{2.41} \times \textbf{10}^3 \text{ m}^3 \text{ HCl}$$

91. Ammonia and carbon dioxide combine to produce urea, NH_2CONH_2, and water through the following reaction. About 90% of the urea is used to make fertilizers, but some is also used to make animal feed, plastics, and pharmaceuticals. To get the optimum yield of urea, the ratio of NH_3 to CO_2 is carefully regulated at a 3:1 molar ratio of NH_3 to CO_2.

$$2NH_3(g) + CO_2(g) \rightarrow NH_2CONH_2(s) + H_2O(l)$$

a. What volume of NH_3 at STP must be combined with 1.4×10^4 L of CO_2 at STP to yield the 3:1 molar ratio? *(Obj #18)*

$$? \text{ L NH}_3 = 1.4 \times 10^4 \text{ CO}_2 \left(\frac{1 \text{ mol CO}_2}{22.414 \text{ L CO}_2} \right)_{STP} \left(\frac{3 \text{ mol NH}_3}{1 \text{ mol CO}_2} \right) \left(\frac{22.414 \text{ L NH}_3}{1 \text{ mol NH}_3} \right)_{STP}$$

$$= \textbf{4.2} \times \textbf{10}^4 \text{ L NH}_3$$

b. The concentration of urea in a typical liquid fertilizer is about 1.0 M NH_2CONH_2. What minimum volume of CO_2 at 190 °C and 34.0 atm is needed to make 3.5×10^4 L of 1.0 M NH_2CONH_2? *(Obj #18)*

$$? \text{ L CO}_2 = 3.5 \times 10^4 \text{ L NH}_2CONH_2 \text{ soln} \left(\frac{1.0 \text{ mol NH}_2CONH_2}{1 \text{ L NH}_2CONH_2 \text{ soln}} \right) \left(\frac{1 \text{ mol CO}_2}{1 \text{ mol NH}_2CONH_2} \right) \left(\frac{0.082058 \text{ L} \cdot \text{atm}}{\text{K} \cdot \text{mol}} \right) \left(\frac{463 \text{ K}}{34.0 \text{ atm}} \right)$$

$$= \textbf{3.9} \times \textbf{10}^4 \text{ L CO}_2$$

93. Some chlorofluorocarbons, CFCs, are formed from carbon tetrachloride, CCl_4, which can be made from the methane in natural gas, as shown below. What minimum volume in cubic meters of methane gas at STP must be added to react completely 1.9×10^6 L of $Cl_2(g)$ at STP? *(Obj #18)*

$$CH_4(g) + 4Cl_2(g) \rightarrow CCl_4(l) + 4HCl(g)$$

$$? \, m^3 \, CH_4 = 1.9 \times 10^6 \, L \, Cl_2 \left(\frac{1 \, mol \, Cl_2}{22.414 \, L \, Cl_2} \right)_{STP} \left(\frac{1 \, mol \, CH_4}{4 \, mol \, Cl_2} \right) \left(\frac{22.414 \, L \, CH_4}{1 \, mol \, CH_4} \right)_{STP} \left(\frac{1 \, m^3}{10^3 \, L} \right)$$

$$= \textbf{475 m}^3 \textbf{ CH}_4$$

95. Ethylene oxide, C_2H_4O, is a cyclic compound; its two carbon atoms and its oxygen atom form a three-member ring. The small ring is somewhat unstable, so ethylene oxide is a very reactive substance, useful as a rocket propellant, to sterilize medical plastic tubing, and to make polyesters and antifreeze. It is formed from the reaction of ethylene, CH_2CH_2, with oxygen at 270-290 °C and 8-20 atm in the presence of a silver catalyst.

$$2CH_2CH_2(g) + O_2(g) \rightarrow 2 \overset{\ddot{O}}{\overset{\diagup\diagdown}{CH_2-CH_2}} (g)$$

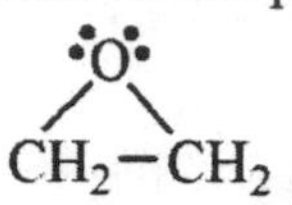

a. What minimum mass, in megagrams, of liquefied ethylene, CH_2CH_2, is necessary to form 2.0×10^2 m^3 of ethylene oxide, C_2H_4O, at 107 kPa and 298 K? *(Obj #18)*

$$? \, Mg \, CH_2CH_2 = 2.0 \times 10^2 \, m^3 \, C_2H_4O \left(\frac{10^3 \, L}{1 \, m^3} \right) \left(\frac{K \cdot mol}{8.3145 \, L \cdot kPa} \right) \left(\frac{107 \, kPa}{298 \, K} \right) \left(\frac{2 \, mol \, CH_2CH_2}{2 \, mol \, C_2H_4O} \right) \left(\frac{28.054 \, g \, CH_2CH_2}{1 \, mol \, CH_2CH_2} \right) \left(\frac{1 \, Mg}{10^6 \, g} \right)$$

$$= \textbf{0.24 Mg CH}_2\textbf{CH}_2$$

b. Ethylene can be purchased in cylinders containing 30 lb of liquefied ethylene. Assuming complete conversion of ethylene into ethylene oxide, determine how many of these cylinders would be necessary to make 2.0×10^2 m^3 of ethylene oxide at 107 kPa and 298 K? *(Obj #18)*

$$? \, cylinders = 0.24 \, Mg \, CH_2CH_2 \left(\frac{10^6 \, g}{1 \, Mg} \right) \left(\frac{1 \, lb}{453.6 \, g} \right) \left(\frac{1 \, cylinder}{30 \, lb \, CH_2CH_2} \right)$$

$$= \textbf{18 cylinders}$$

c. How many liters of ethylene oxide gas at 2.04 atm and 67 °C can be made from 1.4×10^4 L of CH_2CH_2 gas at 21 °C and 1.05 atm? *(Obj #18)*

$$? \, L \, C_2H_4O = 1.4 \times 10^4 \, L \, CH_2CH_2 \left(\frac{K \cdot mol}{0.082058 \, L \cdot atm} \right) \left(\frac{1.05 \, atm}{294 \, K} \right) \left(\frac{2 \, mol \, C_2H_4O}{2 \, mol \, CH_2CH_2} \right) \left(\frac{0.082058 \, L \cdot atm}{K \cdot mol} \right) \left(\frac{340 \, K}{2.04 \, atm} \right)$$

$$= \textbf{8.3} \times \textbf{10}^3 \textbf{ L C}_2\textbf{H}_4\textbf{O}$$

d. How many kilograms of liquefied ethylene oxide can be made from the reaction of 1.4×10^4 L of CH_2CH_2 gas at 1.05 atm and 21 °C and 8.9×10^3 L of oxygen gas at 0.998 atm and 19 °C? *(Obj #18)*

$$? \text{ kg } C_2H_4O = 1.4 \times 10^4 \text{ L } CH_2CH_2 \left(\frac{K \cdot mol}{0.082058 \text{ L} \cdot atm} \right) \left(\frac{1.05 \text{ atm}}{294 \text{ K}} \right) \left(\frac{2 \text{ mol } C_2H_4O}{2 \text{ mol } CH_2CH_2} \right) \left(\frac{44.053 \text{ g } C_2H_4O}{1 \text{ mol } C_2H_4O} \right) \left(\frac{1 \text{ kg}}{10^3 \text{ g}} \right)$$

$$= \textbf{27 kg } C_2H_4O$$

$$? \text{ kg } C_2H_4O = 8.9 \times 10^3 \text{ L } O_2 \left(\frac{K \cdot mol}{0.082058 \text{ L} \cdot atm} \right) \left(\frac{0.998 \text{ atm}}{292 \text{ K}} \right) \left(\frac{2 \text{ mol } C_2H_4O}{1 \text{ mol } O_2} \right) \left(\frac{44.053 \text{ g } C_2H_4O}{1 \text{ mol } C_2H_4O} \right) \left(\frac{1 \text{ kg}}{10^3 \text{ g}} \right)$$

$$= 33 \text{ kg } C_2H_4O$$

97. Rutile ore is about 95% titanium(IV) oxide, TiO_2. The TiO_2 is purified in the two reactions that follow. In the first reaction, what minimum volume, in cubic meters, of Cl_2 at STP is necessary to convert all of the TiO_2 in 1.7×10^4 metric tons of rutile ore into $TiCl_4$? *(Obj #18)*

$$3TiO_2 + 4C + 6Cl_2 \xrightarrow{900 \text{ °C}} 3TiCl_4 + 2CO + 2CO_2$$

$$TiCl_4 + O_2 \xrightarrow{1200\text{-}1400 \text{ °C}} TiO_2 + 2Cl_2$$

$$? \text{ m}^3 \text{ } Cl_2 = 1.7 \times 10^4 \text{ met. ton ore} \left(\frac{10^3 \text{ kg}}{1 \text{ met. ton}} \right) \left(\frac{10^3 \text{ g}}{1 \text{ kg}} \right) \left(\frac{95 \text{ g } TiO_2}{100 \text{ g ore}} \right) \left(\frac{1 \text{ mol } TiO_2}{79.866 \text{ g } TiO_2} \right)$$

$$\left(\frac{6 \text{ mol } Cl_2}{3 \text{ mol} TiO_2} \right) \left(\frac{22.414 \text{ L } Cl_2}{1 \text{ mol } Cl_2} \right)_{STP} \left(\frac{1 \text{ m}^3}{10^3 \text{ L}} \right)$$

$$= \textbf{9.1} \times \textbf{10}^6 \text{ m}^3 \text{ } Cl_2$$

Section 13.4 Dalton's Law of Partial Pressures

100. A typical 100-watt incandescent light bulb contains argon gas and nitrogen gas. *(Objs #21 & 22)*

a. If enough argon and nitrogen are added to this light bulb to yield a partial pressure of Ar of 98.5 kPa and a partial pressure of N_2 of 10.9 kPa, what is the total gas pressure in the bulb?

$$P_{total} = P_{Ar} + P_{N_2} = 98.5 \text{ kPa} + 10.9 \text{ kPa} = \textbf{109.4 kPa}$$

b. If a 125-mL light bulb contains 5.99×10^{-4} mole of N_2 and 5.39×10^{-3} mole of Ar, what is the pressure in kPa in the bulb at a typical operating temperature of 119 °C?

$$P_{total} = \left(\sum n \right) \frac{RT}{V}$$

$$= \left(5.99 \times 10^{-4} \text{ mol} + 5.39 \times 10^{-3} \text{ mol} \right) \frac{\frac{8.3145 \text{ L} \cdot \text{kPa}}{K \cdot mol} (392 \text{ K})}{125 \text{ mL}} \left(\frac{10^3 \text{ mL}}{1 \text{ L}} \right) = \textbf{156 kPa}$$

102. If a 135-mL luminous tube contains 1.2 mg of Ne and 1.2 mg of Ar, at what temperature will the total gas pressure be 13 mmHg?

$$? \text{ mol Ne} = 1.2 \text{ mg Ne} \left(\frac{1 \text{ g}}{10^3 \text{ mg}} \right) \left(\frac{1 \text{ mol Ne}}{20.1797 \text{ g Ne}} \right) = 5.9 \times 10^{-5} \text{ mol Ne}$$

$$? \text{ mol Ar} = 1.2 \text{ mg Ar} \left(\frac{1 \text{ g}}{10^3 \text{ mg}} \right) \left(\frac{1 \text{ mol Ar}}{39.948 \text{ g Ar}} \right) = 3.0 \times 10^{-5} \text{ mol Ar}$$

$$P_{total} = \left(\sum n \right) \frac{RT}{V}$$

$$T = \frac{P_{total} V}{\left(\sum n \right) R} = \frac{13 \text{ mmHg } (135 \text{ mL})}{\left(5.9 \times 10^{-5} \text{ mol} + 3.0 \times 10^{-5} \text{ mol} \right) \dfrac{0.082058 \text{ L} \cdot \text{atm}}{K \text{ mol}}} \left(\frac{1 \text{ atm}}{760 \text{ mmHg}} \right) \left(\frac{1 \text{ L}}{10^3 \text{ mL}} \right)$$

= 316 K or 43 °C

104. Natural gas is a mixture of gaseous hydrocarbons and other gases in small quantities. The percentage of each component varies depending on the source. Consider a 21.2-m^3 storage container that holds 11.8 kg of methane gas, CH_4, 2.3 kg of ethane gas, C_2H_6, 1.1 kg of propane gas, $C_3H_8(g)$, and an unknown amount of other gases.

a. If the total pressure in the container is 1.00 atm at 21 °C, how many moles of gases other than CH_4, C_2H_6, and C_3H_6 are in the container?

$$P_{total} = \left(\sum n \right) \frac{RT}{V} \qquad \left(\sum n \right) = \frac{P_{total} V}{RT} = \frac{1.00 \text{ atm} \left(21.2 \text{ m}^3 \right)}{\left(\dfrac{0.082058 \text{ L} \cdot \text{atm}}{K \cdot \text{mol}} \right) 294 \text{ K}} \left(\frac{10^3 \text{ L}}{1 \text{ m}^3} \right) = 879 \text{ mol}$$

$$? \text{ mol } CH_4 = 11.8 \text{ kg } CH_4 \left(\frac{10^3 \text{ g}}{1 \text{ kg}} \right) \left(\frac{1 \text{ mol } CH_4}{16.043 \text{ g } CH_4} \right) = 736 \text{ mol } CH_4$$

$$? \text{ mol } C_2H_6 = 2.3 \text{ kg } C_2H_6 \left(\frac{10^3 \text{ g}}{1 \text{ kg}} \right) \left(\frac{1 \text{ mol } C_2H_6}{30.070 \text{ g } C_2H_6} \right) = 76 \text{ mol } C_2H_6$$

$$? \text{ mol } C_3H_8 = 1.1 \text{ kg } C_3H_8 \left(\frac{10^3 \text{ g}}{1 \text{ kg}} \right) \left(\frac{1 \text{ mol } C_3H_8}{44.097 \text{ g } C_3H_8} \right) = 25 \text{ mol } C_3H_8$$

$$? \text{ mol other} = n_{total} - n_{CH_4} - n_{C_2H_6} - n_{C_3H_8} = 879 \text{ mol} - 736 \text{ mol} - 76 \text{ mol} - 25 \text{ mol}$$

= 42 mol other

b. What percentage of the gas particles in the cylinder are methane molecules?

$$? \% \, CO_2 = \frac{736 \text{ mol } CH_4}{879 \text{ mol total}} \times 100 = \mathbf{83.7\%}$$

Additional Problems

106. Although the temperature decreases with increasing altitude in the troposphere (the lowest portion of the earth's atmosphere), it increases again in the upper portion of the stratosphere. Thus, at about 50 km, the temperature is back to about 20 °C. Compare the following properties of the air along the California coastline at 20 °C and air at the same temperature but an altitude of 50 km. Assume that the percent composition of the air in both places is essentially the same.

   a. Do the particles in the air along the California coastline and the air at 50 km have the same average kinetic energy? If not, which has particles with the higher average kinetic energy? Explain your answer.

   > Yes. The average kinetic energy is dependent on the temperature of the particles, so the average kinetic energy is the same when the temperature is the same.

   b. Do the particles in the air along the California coastline and the air at 50 km have the same velocity? If not, which has particles with the higher average velocity? Explain your answer.

   > Yes. If the composition of the air is the same, the average mass of the particles is the same. Because the average kinetic energy of the particles is the same for the same temperature, the average velocity must also be the same.
   >
   > $$KE_{average} = \tfrac{1}{2}\, m\mu^2_{average}$$

   c. Does the air along the California coastline and the air at 50 km have the same average distance between the particles? If not, which has the greater average distance between the particles? Explain your answer.

   > No. Because there are fewer particles per unit volume at 50 km, the average distance between the particles is greater than that at sea level.

   d. Do the particles in the air along the California coastline and the air at 50 km have the same average frequency of particle collisions? If not, which has the greater frequency of between particle collisions? Explain your answer.

   > No. Because the average velocity of the particles is the same and the average distance between the particles at 50 km is longer, the average distance between the collisions at 50 km is greater than that at sea level. Thus the frequency of collisions is greater at sea level.

   e. Does the air along the California coastline and the air at 50 km have the same density? If not, which has the higher density? Explain your answer.

   > No, because there are fewer particles per unit volume, the density of the air at 50 km is much less than at sea level. Thus the sea level air has the higher density.

   f. Does the air along the California coastline and the air at 50 km have the same gas pressure? If not, which has the higher gas pressure? Explain your answer.

   > No. Because the particles in the air in the two locations would have the same average mass and the same average velocity, they would collide with the walls of

a container with the same force per collision. Because there are more gas particles per unit volume at sea level than at 50 km, there would be more collisions with the walls of a container that holds a sample of the air along the California coastline. This would lead to a greater force pushing on the walls of the container and a greater gas pressure for the gas at sea level.

108. With reference to the relationships between properties of gases, answer the following questions.

   a. The pressure in the tires of a car sitting in a garage is higher at noon than at midnight. Why?

   The higher temperature at noon causes the pressure to increase.

   b. The pressure in a car's tires increases as it is driven from home to school. Why?

   The interaction between the moving tires and the stationary road causes the particles in the tires to increase their velocity, so the temperature of the tires and ultimately of the gas in the tires goes up. The increase in the temperature of the gas in the tire increases the pressure of the gas.

110. When people inhale helium gas and talk, their voices sound strange, as if it had been recorded and played back at a faster rate. Why do you think that happens?

   Helium is less dense than air, so it is easier for our vocal cords to vibrate. Because of this, they vibrate faster, making our voices sound strange.

Chapter 14
Liquids: Condensation, Evaporation, and Dynamic Equilibrium

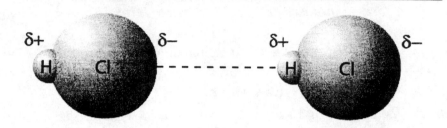

♦ Review Skills

14.1 Change from Gas to Liquid and from Liquid to Gas – An Introduction to Dynamic Equilibrium
- The Process of Condensation
- The Process of Evaporation
- How Evaporation Causes Cooling
- Rate of Evaporation
- Dynamic Equilibrium Between Liquid and Vapor
- Equilibrium Vapor Pressure

Special Topic 14.1: Chemistry Gets the Bad Guys

14.2 Boiling Liquids
- How Do Bubbles Form in Liquids?
- Response of Boiling-Point Temperature to External Pressure Changes
- Relative Boiling-Point Temperatures and Strengths of Attractions

14.3 Particle-Particle Attractions
- Dipole-Dipole Attractions
- Predicting Bond Type
- Predicting Molecular Polarity
  Internet: Molecular Polarity
- Hydrogen Bonds
- London Forces
  Internet: London Forces and Polar Molecules
- Particle Interaction in Pure Elements
- Summary of the Types of Particles and the Attractions Between Them
  Internet: Relative Strengths of Attractions

♦ Chapter Glossary
  Internet: Glossary Quiz
♦ Chapter Objectives
Review Questions
Key Ideas
Chapter Problems

Section Goals and Introductions

Section 14.1 Change from Gas to Liquid and from Liquid to Gas – An Introduction to Dynamic Equilibrium

Goals

- *To describe the process that takes place at the particle level when a gas condenses to a liquid.*
- *To describe the process that takes place at the particle level when a liquid evaporates to become a gas.*
- *To explain why liquids cool as they evaporate.*
- *To describe the factors that affect the rate at which liquids evaporate.*
- *To explain what a dynamic equilibrium is.*
- *To explain why the system in which a liquid in a closed container comes to a dynamic equilibrium between the rate of evaporation and the rate of condensation.*
- *To explain what equilibrium vapor pressure is and why it changes with changing temperature.*

This section continues the effort to help you visualize the changes that take place on the particle level for many everyday processes. The ability to picture these changes for the conversions for liquid to gas and gas to liquid will help explain why you feel cool when you step out of the shower even on a warm day; why certain substances evaporate faster than others; why the liquid in a soft drink doesn't disappear even though it is constantly evaporating into the space above the liquid; and why pressures build up in aerosol cans when they are heated. Promise yourself that you will make a special effort to "see" in your mind's eye the particle changes that accompany each of the situations described in this section.

Section 14.2 Boiling Liquids

Goals

- *To describe the changes that must take place on the particle level for a liquid to boil.*
- *To explain why a liquid must reach a certain minimum temperature before it can boil.*
- *To explain why boiling point temperature for a liquid changes with changing external pressure acting on the liquid.*
- *To explain why different substances boil at different temperatures.*

In this section, you will use your ability to visualize liquid and gaseous particles to understand the mechanics of boiling. This will help you explain why different substances have different temperatures at which they boil and why the boiling point temperature of a specific liquid changes with changes in the external pressure acting on the liquid.

Section 14.3 Particle-Particle Attractions

Goals

- *To describe the different types of attractions that hold particles together in the liquid or solid form.*
- *To describe how to predict the type of attraction holding the particles of a particular substance together in the solid and liquid form.*

- *To explain what it means when we say a molecule is polar and show how you can predict whether a molecule is polar or nonpolar.*
- *To explain why, in general, larger molecules have stronger attractions between them.*

In this section, you get more information about the particles that form the fundamental structure of substances and about the attractions between them. This information will help you understand why certain substances are solids at room temperature, some are liquids, and others are gases; and it will further develop your ability to visualize the changes that take place when substances melt or boil.

See the three related sections on our Web site: *Molecular Polarity, London Forces and Polar Molecules,* and *Relative Strengths of Attractions.*

www.chemplace.com/college/

Chapter 14 Maps

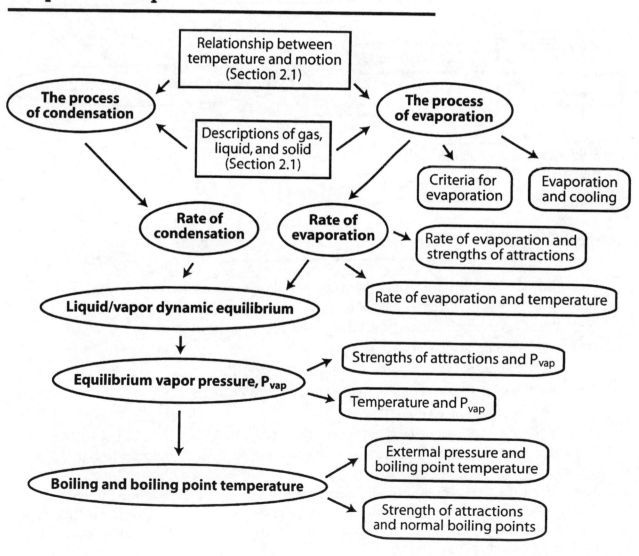

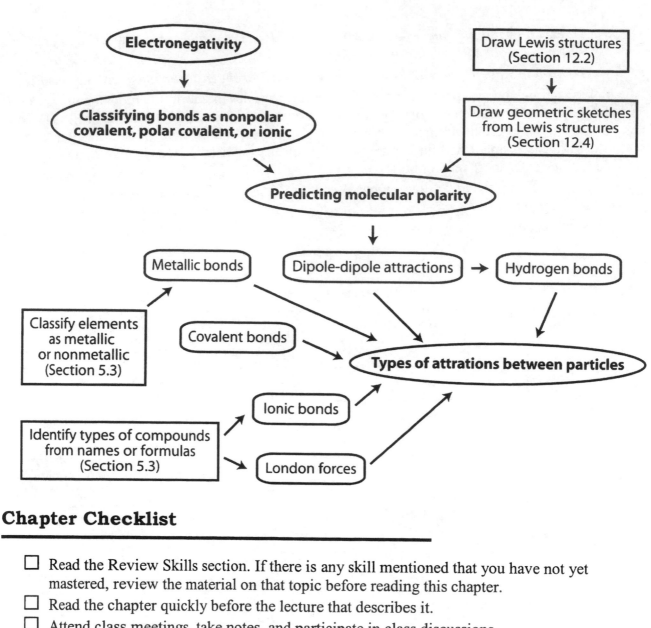

Chapter Checklist

- ☐ Read the Review Skills section. If there is any skill mentioned that you have not yet mastered, review the material on that topic before reading this chapter.
- ☐ Read the chapter quickly before the lecture that describes it.
- ☐ Attend class meetings, take notes, and participate in class discussions.
- ☐ Work the Chapter Exercises, perhaps using the Chapter Examples as guides.
- ☐ Study the Chapter Glossary and test yourself on our Web site:

  www.chemplace.com/college/

- ☐ This chapter has logic sequences in Figures 14.3, 14.4, 14.8, 14.9, 14.14, and 14.16. Convince yourself that each of the statements in these sequences logically lead to the next statement.
- ☐ Study all of the Chapter Objectives. You might want to write a description of how you will meet each objective. (Although it is best to master all of the objectives, the following objectives are especially important because they pertain to skills that you will need while studying other chapters of this text: 11, 22 and 30.)
- ☐ Reread the Study Sheet in this chapter and decide whether you will use it or some variation on it to complete the tasks it describes.

Sample Study Sheet 14.1: Electronegativity, Types of Chemical Bonds, and Bond Polarity

☐ Memorize the following.

Be sure to check with your instructor to determine how much you are expected to know of the following.

- The information on Table 14.1

Type of Substance	Particles	Examples	Attraction in Solid or Liquid
Elements			
Metal	Cations in a sea of electrons	Au	Metallic Bond
Noble Gases	Atoms	Xe	London Forces
Carbon (diamond)	Carbon Atoms	C(dia)	Covalent Bonds
Other Nonmetal Elements	Molecules	H_2, N_2, O_2, F_2, Cl_2, Br_2, I_2, S_8, Se_8, P_4	London Forces
Ionic Compounds	Cations and Anions	NaCl	Ionic Bond
Molecular Compounds			
Nonpolar Molecular	Molecules	CO_2 and Hydrocarbons	London Forces
Polar molecules without H–F, O–H, or N–H bond	Molecules	HF, HCl, HBr, and HI	Dipole-Dipole Forces
Molecules with H–F, O–H or N–H bond	Molecules	HF, H_2O, alcohols, NH_3	Hydrogen Bonds

☐ To get a review of the most important topics in the chapter, fill in the blanks in the Key Ideas section.

☐ Work all of the selected problems at the end of the chapter, and check your answers with the solutions provided in this chapter of the study guide.

☐ Ask for help if you need it.

Web Resources   www.chemplace.com/college/

Molecular Polarity
London Forces and Polar Molecules
Relative Strengths of Attractions
Glossary Quiz

Exercises Key

Exercise 14.1 - Using Electronegativities: Classify the following bonds as nonpolar covalent, polar covalent, or ionic. If a bond is polar covalent, identify which atom has the partial negative charge and which has the partial positive charge. If a bond is ionic, identify which atom is negative and which is positive. *(Obj #22)*

    a. N bonded to H **polar covalent**

        N is partial negative and H is partial positive.

    b. N bonded to Cl **nonpolar covalent**

    c. Ca bonded to O **ionic   O is negative, and Ca is positive.**

    d. P bonded to F **polar covalent**

        F is partial negative and P is partial positive.

Exercise 14.2 - Comparing Bond Polarities: Which bond would you expect to be more polar, P–H or P–F? Why? *(Obj #22)*

    P–F bond; greater difference in electronegativity

Exercise 14.3 - Types of Particles and Types of Attractions:
For (a) iron, (b) iodine, (c) CH_3OH, (d) NH_3, (e) hydrogen chloride, (f) KF, and (g) carbon in the diamond form, specify (1) the type of particle that forms the substance's basic structure and (2) the name of the type of attraction that holds these particles in the solid and liquid form. *(Obj #30)*

Substance	Particles to visualize	Type of attraction
iron	**Fe cations in a sea of electrons**	**Metallic bonds**
iodine	I_2 **molecules**	**London forces**
CH_3OH	CH_3OH **molecules**	**Hydrogen bonds**
NH_3	NH_3 **molecules**	**Hydrogen bonds**
hydrogen chloride	**HCl molecules**	**Dipole-dipole attractions**
KF	**Cations and anions**	**Ionic bonds**
C (diamond)	**Atoms**	**Covalent bonds**

Review Questions Key

1. For each of the following pairs of elements, decide whether a bond between them would be covalent or ionic.

    a. Ni and F   Metal-nonmetal combinations usually lead to **ionic bonds**.

    b. b. S and O Nonmetal-nonmetal combinations lead to **covalent bonds**.

2. Classify each of the following as either a molecular compound or an ionic compound.
    a. oxygen difluoride OF_2
        all nonmetallic elements (no ammonium), so molecular
    b. Na_2O **metal-nonmetal, so ionic**
    c. calcium carbonate $CaCO_3$ **metal-polyatomic ion, so ionic**
    d. C_3H_8 **all nonmetallic elements (no ammonium), so molecular**
3. Classify each of the following compounds as (1) a binary ionic compound, (2) an ionic compound with polyatomic ion(s), (3) a binary covalent compound, (4) a binary acid, (5) an alcohol, or (6) an oxyacid. Write the chemical formula that corresponds to each name.
    a. magnesium chloride   **ionic $MgCl_2$**
    b. hydrogen chloride   **binary covalent   HCl**
    c. sodium nitrate   **ionic with polyatomic ion $NaNO_3$**
    d. methane   **binary covalent (hydrocarbon) CH_4**
    e. ammonia   **binary covalent NH_3**
    f. hydrochloric acid   **binary acid   HCl(*aq*)**
    g. nitric acid   **oxyacid HNO_3**
    h. ethanol   **alcohol C_2H_5OH**
4. Classify each of the following compounds as (1) a binary ionic compound, (2) an ionic compound with polyatomic ion(s), (3) a binary covalent compound, (4) a binary acid, (5) an alcohol, or (6) an oxyacid. Write the name that corresponds to each chemical formula.
    a. HF   **binary covalent   hydrogen fluoride**
    b. CH_3OH **alcohol   methanol**
    c. LiBr   **ionic   lithium bromide**
    d. NH_4Cl **ionic with polyatomic ion   ammonium chloride**
    e. C_2H_6 **binary covalent (hydrocarbon)   ethane**
    f. BF_3 **binary covalent   boron trifluoride**
    g. H_2SO_4 **oxyacid   sulfuric acid**
5. For each of the formulas listed below,
    *   Draw a reasonable Lewis structure.
    *   Write the name of the electron group-geometry around the central atom.
    *   Draw the geometric sketch of the molecule, including bond angles.
    *   Write the name of the molecular geometry around the central atom.
        a. CCl_4

tetrahedral   tetrahedral

b. SF_2

tetrahedral   ≈109.5°   bent

c. PF_3

tetrahedral   ≈109.5°   trigonal pyramid

d. BCl_3

trigonal planar   ≈120°   trigonal planar

Key Ideas Answers

6. Gases can be converted into liquids by a decrease in temperature. At a high temperature, there are no **significant** attractions between particles of a gas. As the temperature is lowered, attractions between particles lead to the formation of very small clusters that **remain** in the gas phase. As the temperature is lowered further, the particles move slowly enough to form clusters so **large** that they drop to the bottom of the container and combine to form a liquid.

8. During the evaporation of a liquid, the **more rapidly** moving particles escape, leaving the particles left in the liquid with a lower average velocity, and lower average velocity means **lower** temperature.

10. Greater surface area means **more** particles at the surface of a liquid, which leads to a greater rate of evaporation.

12. Increased temperature increases the average velocity and momentum of the particles. As a result, a greater **percentage** of particles will have the minimum momentum necessary to escape, so the liquid will evaporate more quickly.

14. For a dynamic equilibrium to exist, the rates of the two **opposing** changes must be equal, so that there are constant changes between state A and state B but no **net** change in the components of the system.

16. The **weaker** the attractions between particles of a substance, the higher the equilibrium vapor pressure for that substance at a given temperature.

18. As a liquid is heated, the vapor pressure of the liquid increases until the temperature gets high enough to make the pressure within the bubbles that form equal to the **external pressure**. At this temperature, the bubbles can maintain their volume, and the liquid boils.

20. If the external pressure acting on a liquid changes, then the **internal vapor pressure** needed to preserve a bubble changes, and therefore the boiling-point temperature changes.

22. We know that increased strength of attractions leads to decreased rate of evaporation, decreased rate of condensation at equilibrium, decreased concentration of vapor, and decreased vapor pressure at a given temperature. This leads to a(n) **increased temperature** necessary to reach a vapor pressure of 1 atmosphere.

24. A dipole-dipole attraction is an attraction between the **partial positive** end of one molecule and the **partial negative** end of another molecule.

26. The higher an element's electronegativity, the greater its ability to **attract** electrons from other elements.

28. If the difference in electronegativity (ΔEN) between two atoms is between 0.4 and 1.7, we expect the bond between them to be a(n) **polar** covalent bond.

30. The atom with the higher electronegativity has the partial **negative** charge. The atom with the lower electronegativity has the partial **positive** charge.

32. When comparing two covalent bonds, the bond with the greater **difference** in electronegativity (ΔEN) is likely to be the more polar bond.

34. When there are no polar bonds in a molecule, there is no permanent charge difference between one part of the molecule and another, and the molecule is **nonpolar**.

36. **Hydrogen** bonds are attractions that occur between a nitrogen, oxygen, or fluorine atom of one molecule and a hydrogen atom bonded to a nitrogen, oxygen, or fluorine atom in another molecule.

38. The larger the molecules of a substance, the **stronger** the London forces between them. A larger molecule has more electrons and a greater **chance** of having its electron cloud distorted from its nonpolar shape. Thus instantaneous dipoles are more likely to form in larger molecules. The electron clouds in larger molecules are also larger, so the average distance between the nuclei and the electrons is greater; as a result, the electrons are held **less tightly** and shift more easily to create a dipole.

Problems Key

Section 14.1 Change from Gas to Liquid and from Liquid to Gas – An Introduction to Dynamic Equilibrium

40. Why is dew more likely to form on a lawn at night than in the day? Describe the changes that take place as dew forms.

> At the lower temperature during the night, the average velocity of the water molecules in the air is lower, making it more likely that they will stay together when they collide. They stay together long enough for other water molecules to collide with them forming clusters large enough for gravity to pull them down to the grass where they combine with other clusters to form the dew.

42. Consider two test tubes, each containing the same amount of liquid acetone. A student leaves one of the test tubes open overnight and covers the other one with a balloon so that gas cannot escape. When the student returns to the lab the next day, all of the acetone is gone from the open test tube, but most of it remains in the covered tube.

    a. Explain why the acetone is gone from one test tube and not from the other.

        In the closed test tube, the vapor particles that have escaped from the liquid are trapped in the space above the liquid. The concentration of acetone vapor rises quickly to the concentration that makes the rate of condensation equal to the rate of evaporation, so there is no net change in the amount of liquid or vapor in the test tube.

        In the open test tube, the acetone vapor escapes into the room. The concentration of vapor never gets high enough to balance the rate of evaporation, so all of the liquid finally disappears.

    b. Was the initial rate at which liquid changed to gas (the rate of evaporation) greater in one test tube than in the other? Explain your answer.

        No, the rate of evaporation is dependent on the strengths of attractions between particles in the liquid, the liquid's surface area, and temperature. All of these factors are the same for the two systems, so the initial rate of evaporation is the same for each.

    c. Consider the system after 30 minutes, with liquid remaining in both test tubes. Is condensation (vapor to liquid) taking place in both test tubes? Is the rate of condensation the same in both test tubes? Explain your answer.

        There will be some vapor above both liquids, so some vapor molecules will collide with the surface of the liquid and return to the liquid state. Thus, there will be condensation in both test tubes. The concentration of acetone vapor above the liquid in the closed container will be higher, so the rate of collision between the vapor particles and the liquid surface will be higher. Thus, the rate of condensation in the closed container will be higher.

    d. Describe the submicroscopic changes in the covered test tube that lead to a constant amount of liquid and vapor. *(Obj #10)*

        The liquid immediately begins to evaporate with a rate of evaporation that is dependent on the surface area of the liquid, the strengths of attractions between the liquid particles, and temperature. If these three factors remain constant, the rate of evaporation will be constant. If we assume that the container initially holds no vapor particles, there is no condensation of vapor when the liquid is first added. As the liquid evaporates, the number of vapor particles above the liquid increases, and the condensation process begins. As long as the rate of evaporation of the liquid is greater than the rate of condensation of the vapor, the concentration of vapor particles above the liquid will increase. As the concentration of vapor particles increases, the rate of collisions of vapor particles with the liquid increases, increasing the rate of condensation. If there is enough liquid in the container to avoid all of it evaporating, the rising rate of condensation eventually becomes equal to the rate of evaporation. At this point, for every particle that leaves the liquid, a particle somewhere else in the container returns to the liquid. Thus, there is no net change in the amount of substance in the liquid form or the amount of substance in the vapor form.

e.   The balloon expands slightly after it is placed over the test tube, suggesting an initial increase in pressure in the space above the liquid. Why? After this initial expansion, the balloon stays inflated by the same amount. Why doesn't the pressure inside the balloon change after the initial increase? *(Obj #12)*

While the rate of evaporation is greater than the rate of condensation, there is a steady increase in the amount of vapor above the liquid. This increases the total pressure of gas above the liquid, so the balloon expands. When the rates of evaporation and condensation become equal, the amount of vapor and the total gas pressure remain constant, so the balloon maintains the same degree of inflation.

f.   If the covered test tube is heated, the balloon expands. Part of this expansion is due to the increase in gas pressure that results from the rise in temperature of the gas, but the increase is greater than expected from this factor alone. What other factor accounts for the increase in pressure? Describe the submicroscopic changes that take place that lead to this other factor. *(Obj #14)*

The rate of evaporation is dependent on the temperature of the liquid. Increased temperature increases the average velocity and momentum of the particles in the liquid. This increases the percentage of particles that have the minimum velocity necessary to escape and increases the rate of evaporation. More particles escape per second, and the partial pressure due to the vapor above the liquid increases.

43. The attractions between ethanol molecules, C_2H_5OH, are stronger than the attractions between diethyl ether molecules, $CH_3CH_2OCH_2CH_3$.

a.   Which of these substances would you expect to have the higher rate of evaporation at room temperature? Why?

The weaker attractions between diethyl ether molecules are easier to break, allowing a higher percentage of particles to escape from the surface of liquid diethyl ether than from liquid ethanol. If the surface area and temperature is the same for both liquids, more particles will escape per second from the diethyl ether than from the ethanol.

b. Which of these substances would you expect to have the higher equilibrium vapor pressure? Why? *(Obj #13)*

> Because the attractions between diethyl ether molecules are weaker than those between ethanol molecules, it is easier for a diethyl ether molecule to break them and move into the vapor phase. Therefore, the rate of evaporation from liquid diethyl ether is greater than for ethanol at the same temperature. When the dynamic equilibrium between evaporation and condensation for the liquids is reached and the two rates become equal, the rate of condensation for the diethyl ether is higher than for ethanol. Because the rate of condensation is determined by the concentration of vapor above the liquid, the concentration of diethyl ether vapor at equilibrium is higher than for the ethanol. The higher concentration of diethyl ether particles leads to a higher equilibrium vapor pressure.

45. Picture a half-empty milk bottle in the refrigerator. The water in the milk will be constantly evaporating into the gas-filled space above the liquid, and the water molecules in this space will be constantly colliding with the liquid and returning to the liquid state. If the milk is tightly closed, a dynamic equilibrium forms between the rate of evaporation and the rate of condensation. If the bottle is removed from the refrigerator and left out in the room with its cap still on tightly, what happens to the rates of evaporation and condensation? An hour later when the milk has reached room temperature, will a dynamic equilibrium exist between evaporation and condensation?

> As the temperature of the liquid milk increases, its rate of evaporation increases. This will disrupt the equilibrium, making the rate of evaporation greater than the rate of condensation. This leads to an increase in the concentration of water molecules in the gas space above the liquid, which increases the rate of condensation until it increases enough to once again become equal to the rate of evaporation. At this new dynamic equilibrium, the rates of evaporation and condensation will both be higher.

Section 14.2 Boiling Liquids

47. The normal boiling point of ethanol, C_2H_5OH, is 78.3 °C.

a. Describe the submicroscopic events that occur when a bubble forms in liquid ethanol. *(Obj #15)*

> Ethanol molecules are moving constantly, sometimes at very high velocity. When they collide with other particles, they push them out of their positions, leaving small spaces in the liquid. Other particles move across the spaces and collide with other particles, and the spaces grow in volume. These spaces can be viewed as tiny bubbles.
>
> The surface of each of these tiny bubbles is composed of a spherical shell of liquid particles. Except for shape, this surface is the same as the surface at the top of the liquid. Particles can escape from the surface (evaporate) into the

vapor phase in the bubble, and when particles in the vapor phase collide again with the surface of the bubble, they return to the liquid state (condense). A dynamic equilibrium between the rate of evaporation and the rate of condensation is set up in the bubble just like the liquid-vapor equilibrium above the liquid in the closed container.

b. Consider heating liquid ethanol in a system where the external pressure acting on the liquid is 1 atm. Explain why bubbles cannot form and escape from the liquid until the temperature reaches 78.3 °C. *(Obj #16)*

Each time a particle moves across a bubble and collides with the surface of the bubble, it exerts a tiny force pushing the wall of the bubble out. All of the collisions with the shell of the bubble combine to yield a gas pressure inside the bubble. This pressure is the same as the equilibrium vapor pressure for the vapor above the liquid in a closed container.

If the 1 atm of external pressure pushing on the bubble is greater than the vapor pressure of the bubble, the liquid particles are pushed closer together, and the bubble collapses. If the vapor pressure of the bubble is greater than the external pressure, the bubble will grow. If the two pressures are equal, the bubble maintains its volume. The vapor pressure of the bubbles in ethanol does not reach 1 atm until the temperature rises to 78.3 °C.

c. If the external pressure on the surface of the ethanol is increased to 2 atm, will its boiling-point temperature increase, decrease, or stay the same? Why? *(Obj #17)*

If the external pressure acting on the bubbles in ethanol rises to 2 atm, the vapor pressure inside the bubbles must rise to 2 atm also to allow boiling. This requires an increase in the temperature, so the boiling point increases.

49. At 86 m below sea level, Death Valley is the lowest point in the western hemisphere. The boiling point of water in Death Valley is slightly higher than water's normal boiling point. Explain why.

The pressure of the earth's atmosphere decreases with increasing distance from the center of the earth. Therefore, the average atmospheric pressure in Death Valley is greater than at sea level where it is 1 atmosphere. This greater external pressure acting on liquid water increases the vapor pressure necessary for the water to maintain bubbles and boil. This leads to a higher temperature necessary to reach the higher vapor pressure. Because the boiling-point temperature of a liquid is the temperature at which the vapor pressure of the liquid reaches the external pressure acting on it, the boiling-point temperature is higher in Death Valley.

51. Explain why liquid substances with stronger interparticle attractions will have higher boiling points than liquid substances whose particles experience weaker interparticle attractions. *(Obj #19)*

> Increased strength of attractions leads to decreased rate of evaporation, decreased rate of condensation at equilibrium, decreased concentration of vapor, and decreased vapor pressure at a given temperature. This leads to an increased temperature necessary to reach a vapor pressure of one atmosphere.

Section 14.3 Particle-Particle Attractions

54. Complete the following table by classifying each bond as nonpolar covalent, polar covalent, or ionic. If a bond is polar covalent, identify the atom that has the partial negative charge and the atom that has the partial positive charge. If a bond is ionic, identify the ion that has the negative charge and the ion that has the positive charge. *(Obj #22)*

Atoms	Is the bond polar covalent, nonpolar covalent, or ionic?	For polar covalent bonds, which atom is partial negative? For ionic bonds, which atom is negative?
C–N	Polar covalent	N
C–H	Nonpolar covalent	-----
H–Br	Polar covalent	Br
Li–F	Ionic	F
C–Se	Nonpolar covalent	----
Se–S	Nonpolar covalent	----
F–S	Polar covalent	F
O–P	Polar covalent	O
O–K	Ionic	O
F–H	Polar covalent	F

56. Identify the bond in each pair that you would expect to be more polar.   *(Obj #22)*

a. C–O or C–H   **C–O**
b. P–H or H–Cl   **H–Cl**

58. Explain why water molecules are polar, why ethane, C_2H_6, molecules are nonpolar, and why carbon dioxide, CO_2, molecules are nonpolar. *(Obj #23)*

> Water molecules have an asymmetrical distribution of polar bonds, so they are polar.

> All of the bonds in ethane molecules are nonpolar, so the molecules are nonpolar.

> Carbon dioxide molecules have a symmetrical distribution of polar bonds, so they are nonpolar.

60. Ammonia has been used as a refrigerant. In the cooling cycle, gaseous ammonia is alternately compressed into a liquid and allowed to expand back to the gaseous state. What are the particles that form the basic structure of ammonia? What type of attraction holds these particles together? Draw a rough sketch of the structure of liquid ammonia. *(Obj #26)*

> Ammonia is composed of NH_3 molecules that are attracted by hydrogen bonds between the partially positive hydrogen atoms and the partially negative nitrogen atoms of other molecules. See Figure 14.25 on page 609 of the textbook.
>
> The liquid would look much like the image for liquid water shown in Figure 3.14 on page 115 of the textbook, except with NH_3 molecules in the place of H_2O molecules.

62. Bromine, Br_2, is used to make ethylene bromide, which is an antiknock additive in gasoline. The Br_2 molecules have nonpolar covalent bonds between the atoms, so we expect isolated Br_2 molecules to be nonpolar. Despite the nonpolar character of isolated Br_2 molecules, attractions form between bromine molecules that are strong enough to hold the particles in the liquid form at room temperature and pressure. What is the nature of these attractions? How do they arise? Describe what you would see if you were small enough to ride on a Br_2 molecule in liquid bromine. *(Obj #27)*

> The attractions are London forces. Because the Br–Br bond is nonpolar, the expected distribution of the electrons in the Br_2 molecule is a symmetrical arrangement around the two bromine nuclei, but this arrangement is far from static. Even though the most probable distribution of charge in an isolated Br_2 molecule is balanced, in a sample of bromine that contains many billions of molecules, there is a chance that a few of these molecules will have their electron clouds shifted more toward one bromine atom than the other. The resulting dipoles are often called *instantaneous dipoles* because they may be short-lived. Remember also that in all states of matter, there are constant collisions

between molecules. When Br_2 molecules collide, the repulsion between their electron clouds will distort the clouds and shift them from their nonpolar state. The dipoles that form are also called instantaneous dipoles.

An instantaneous dipole can create a dipole in the molecule next to it. For example, the negative end of one instantaneous dipole will repel the negative electron cloud of a nonpolar molecule next to it, pushing the cloud to the far side of the neighboring molecule. The new dipole is called an *induced dipole*. The induced dipole can then induce a dipole in the molecule next to it. This continues until there are many polar molecules in the system. The resulting partial charges on these polar molecules lead to attractions between the opposite charges on the molecules. See Figure 14.26 on page 611 of the textbook.

64. Carbon disulfide, CS_2, which is used to make rayon, is composed of nonpolar molecules that are similar to carbon dioxide molecules, CO_2. Unlike carbon dioxide, carbon disulfide is liquid at room temperature. Why?

Both of these substances have nonpolar molecules held together by London forces. Because the CS_2 molecules are larger, they have stronger London forces that raise carbon disulfide's boiling point to above room temperature.

66. Methanol, CH_3OH, is used to make formaldehyde, CH_2O, which is used in embalming fluids. The molecules of these substances have close to the same atoms and about the same molecular mass, so why is methanol a liquid at room temperature and formaldehyde a gas?

Because of the O–H bond in methanol, the attractions between CH_3OH molecules are hydrogen bonds. The hydrogen atoms in CH_2O molecules are bonded to the carbon atom, not the oxygen atom, so there is no hydrogen bonding for formaldehyde. The C–O bond in each formaldehyde molecule is polar, and when there is only one polar bond in a molecule, the molecule is polar. Therefore, CH_2O molecules are held together by dipole-dipole attractions. For molecules of about the same size, hydrogen bonds are stronger than dipole-dipole attractions. The stronger hydrogen bonds between CH_3OH molecules raise its boiling point above room temperature, making it a liquid.

$$\begin{array}{c} \text{H} \\ | \\ \text{H--C--}\ddot{\text{O}}\text{--H} \\ | \\ \text{H} \end{array} \qquad \begin{array}{c} \ddot{\text{O}} \\ \| \\ \text{H--C--H} \end{array}$$

methanol   formaldehyde

68. Complete the following table by specifying (1) the name for the type of particle viewed as forming the structure of a solid, liquid, or gas of each substance and (2) the name of the type of attraction that holds these particles in the solid or liquid form. *(Obj #30)*

Substance	Particles to visualize	Type of attraction
Silver	Ag cations in a sea of electrons	Metallic bonds
HCl	HCl molecules	Dipole-Dipole attractions
C_2H_5OH	C_2H_5OH molecules	Hydrogen bonds
NaBr	Cations and anions	Ionic bonds
Carbon (diamond)	Carbon atoms	Covalent bonds
C_5H_{12}	C_5H_{12} molecules	London forces
water	H_2O molecules	Hydrogen bonds

70. Have you ever broken a mercury thermometer? If you have, you probably noticed that the mercury forms droplets on the surface on which it falls rather than spreading out and wetting it like water. Describe the difference between liquid mercury and liquid water that explains this different behavior. (*Hint*: Consider the attractions between particles.)

As a liquid spreads out on a surface, some of the attractions between liquid particles are broken. Because the metallic bonds between mercury atoms are much stronger than the hydrogen bonds between water molecules, they keep mercury from spreading out like water.

Chapter 15
Solution Dynamics

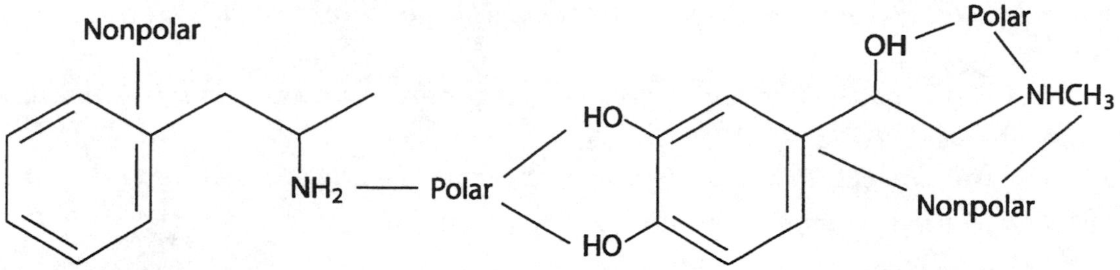

amphetamine   epinephrine

- ◆ Review Skills

15.1 Entropy, Solutions, and Solubility
- • Entropy and the Second Law of Thermodynamics
- • Why Do Solutions Form?
    Internet: Ethanol and Water Mixing
    Internet: Entropy and Solubility
- • Predicting Solubility
- • Hydrophobic and Hydrophilic Substances

15.2 Fats, Oils, Soaps, and Detergents

15.3 Saturated Solutions and Dynamic Equilibrium
- • Two Opposing Rates of Change
- • Net Rate of Solution
- • Saturated Solutions
    Internet: Supersaturated Solutions

15.4 Solutions of Gases in Liquids
- • Gas Solubility
- • Partial Pressure and Gas Solubility
- • Gas Solubility and Breathing
    Special Topic 15.1: Gas Solubility, Scuba Diving, and Soft Drinks
    Internet: Temperature and Solubility
    Special Topic 15.2: Global Warming, the Oceans, and CO_2 Torpedoes

- ◆ Chapter Glossary
    Internet: Glossary Quiz
- ◆ Chapter Objectives
Review Questions
Key Ideas
Chapter Problems

Section Goals and Introductions

Section 15.1 Entropy, Solutions, and Solubility
Goals

- *To introduce the concept of entropy and explain why the entropy of the universe increases.*
- *To explain why changes that lead to greater disorder are more likely to take place than changes that lead to greater order.*
- *To explain why changes that lead to greater dispersal of matter are likely to take place.*
- *To explain why liquids tend to mix and form solutions.*
- *To show how you can predict whether substances are soluble in water.*
- *To show how you can predict whether substances are soluble in hexane.*
- *To show how you can predict which of two substances would be more soluble in water.*
- *To show how you can predict which of two substances would be more soluble in hexane.*

This section introduces some key ideas that relate to entropy and the Second Law of Thermodynamics. We just scratch the surface of this large and important topic, but it is meant to give you a fundamental understanding of how scientists explain why changes take place. This understanding is extremely important, so plan to spend some extra time on it, if necessary.

We use the Second Law to help us understand why some substances mix to form solutions and why others do not. This leads us to some guidelines that allow us to predict the relative solubilities of substances in water and hexane.

See the two related sections on our Web site: *Ethanol and Water Mixing* and *Entropy and Solubility*.

www.chemplace.com/college/

Section 15.2 Fats, Oils, Soaps, and Detergents
Goals

- *To describe fats, oils, soaps, and detergents.*
- *To explain how soap is made.*
- *To explain why soaps and detergents aid the cleaning process.*

This section uses what was described in Section 15.1 to explain why soaps and detergents aid the cleaning process. You will also learn about the structure of fats and oils, how soap is made, and the similarities and differences between soap and detergent.

Section 15.3 Saturated Solutions and Dynamic Equilibrium
Goals

- *To describe the factors that affect the rate at which substances dissolve and to explain why these factors have the effect they do.*
- *To describe the factors that affect the rate at which substances return from solution to the undissolved form.*
- *To describe saturated and unsaturated solutions.*
- *To explain how saturated solutions form.*

This section strengthens your ability to visualize the particle-nature of matter by describing the changes that take place on the submicroscopic level when substances dissolve and when they return to the undissolved state. This mental image will help you understand why certain factors affect the rate at which substances dissolve and also help you understand why some substances have a solubility limit.

See the related section on our Web site called *Supersaturated Solutions*.

www.chemplace.com/college/

Section 15.4 Solutions of Gases in Liquids

Goals

- *To describe the process by which gases dissolve in liquids.*
- *To explain why gases have a solubility limit in liquids.*
- *To explain why increased partial pressure of a gas over a liquid will lead to an increase in the solubility of that gas.*

The ability to visualize the changes described in this section that take place when gases dissolve in liquids will help you understand how carbonated beverages get their bubbles and also help you understand some of the issues that concern scuba divers and athletes.

See the related section on our Web site called *Temperature and Solubility*.

www.chemplace.com/college/

Chapter 15 Maps

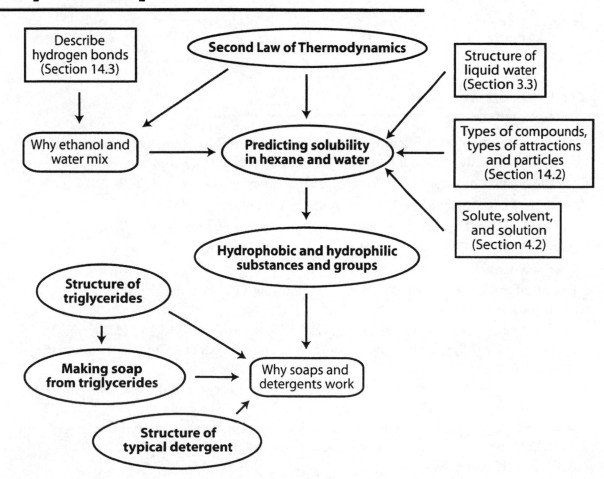

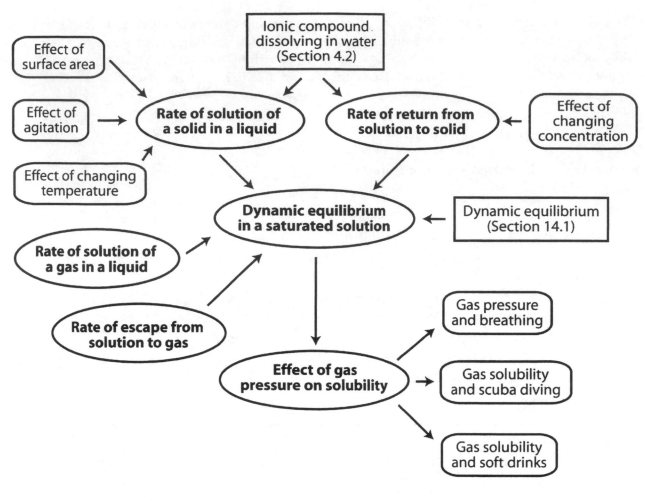

Chapter Checklist

- ☐ Read the Review Skills section. If there is any skill mentioned that you have not yet mastered, review the material on that topic before reading this chapter.
- ☐ Read the chapter quickly before the lecture that describes it.
- ☐ Attend class meetings, take notes, and participate in class discussions.
- ☐ Work the Chapter Exercises, perhaps using the Chapter Examples as guides.
- ☐ Study the Chapter Glossary and test yourself on our Web site:

  www.chemplace.com/college/

- ☐ Study all of the Chapter Objectives. You might want to write a description of how you will meet each objective.
- ☐ This chapter has logic sequences in Figures 15.15, 15.18, 15.20, 15.21, 15.24, and 15.25. Convince yourself that each of the statements in these sequences logically leads to the next statement.
- ☐ To get a review of the most important topics in the chapter, fill in the blanks in the Key Ideas section.
- ☐ Work all of the selected problems at the end of the chapter, and check your answers with the solutions provided in this chapter of the study guide.
- ☐ Ask for help if you need it.

Web Resources   www.chemplace.com/college/

Ethanol and Water Mixing
Entropy and Solubility
Saturated Solutions
Temperature and Solubility
Glossary Quiz

Exercises Key

Exercise 15.1 – Predicting Water Solubility: Predict whether each of the following compounds is soluble in water. (See Section 14.3 if you need to review the list of compounds that you are expected to recognize as composed of either polar or nonpolar molecules.) *(Obj #6)*

   a.   sodium fluoride, NaF (used to fluoridate water) **soluble**

   b.   acetic acid, CH_3CO_2H (added to foods for tartness) **soluble**

   c.   acetylene, C_2H_2 (used in torches designed for welding and cutting) **insoluble**

   d.   methanol, CH_3OH (a common solvent) **soluble**

Exercise 15.2 – Predicting Solubility in Hexane: Predict whether each of the following compounds is soluble in hexane, C_6H_{14}. *(Obj #7)*

   a.   sodium perchlorate, $NaClO_4$ (used to make explosives) *ionic compound, so* **insoluble**

   b.   propylene, CH_3CHCH_2 (used to make polypropylene plastic for children's toys)
     nonpolar molecular compound, so **soluble**

Exercise 15.3 – Predicting Relative Solubility in Water and Hexane: Predict whether each of the following compounds is more soluble in water or in hexane. *(Obj #8)*

   a.   sodium iodate, $NaIO_3$ (used as a disinfectant) *ionic compound, so more soluble in* **water**

   b.   2,2,4-trimethylpentane (sometimes called isooctane), $(CH_3)_3CCH_2CH(CH_3)_2$ (used as a standard in the octane rating of gasoline) *nonpolar molecular compound, so more soluble in* **hexane**

Exercise 15.4 – Predicting Relative Solubility in Water: The compound to the left below, 2-methyl-2-propanol, which is often called t-butyl alcohol, is an octane booster for unleaded gasoline. The other compound, menthol, when added to foods and medicines affects the cold receptors on the tongue in such a way as to produce a "cool" taste. Which of these two compounds would you expect to be more soluble in water? *(Obj #9)*

2-methyl-2-propanol   menthol

The **2-methyl-2-propanol** has a greater percentage of its structure that is polar, so we expect it to be **more soluble** in water.

Review Questions Key

1. In the past, sodium bromide was used medically as a sedative. Describe the process by which this ionic compound dissolves in water, including the nature of the particles in solution and the attractions between the particles in the solution. Identify the solute and the solvent in this solution.

   When solid sodium bromide is added to water, all of the ions at the surface of the solid can be viewed as shifting back and forth between moving out into the water and returning to the solid surface. Sometimes when an ion moves out into the water, a water molecule collides with it, helping to break the ionic bond, and pushing it out into the solution. Water molecules move into the gap between the ion in solution and the solid and shield the ion from the attraction to the solid.

   The ions are kept stable and held in solution by attractions between them and the polar water molecules. The negatively charged oxygen ends of water molecules surround the sodium ions, and the positively charged hydrogen ends of water molecules surround the bromide ions. (See Figures 4.4 on page 161 of the text and Figure 4.5 on page 162 with Br^- in the place of Cl^-.) The sodium bromide is the solute, and the water is the solvent.

2. Draw a reasonable Lewis structure and a geometric sketch for each of the following molecules. Identify each compound as polar or nonpolar.

   a. C_2H_6

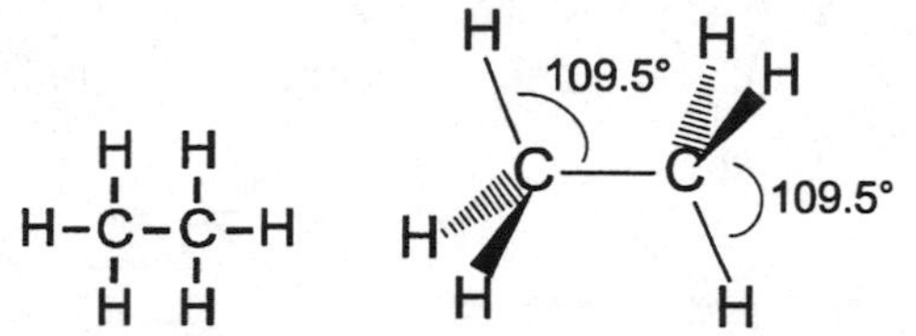

   nonpolar (no polar bonds)

   b. CH_3CH_2OH

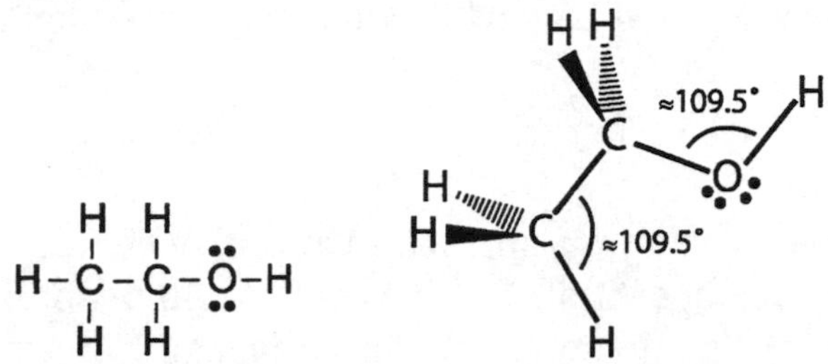

   polar (asymmetrical distribution of polar bonds)

   c. CH_3CO_2H

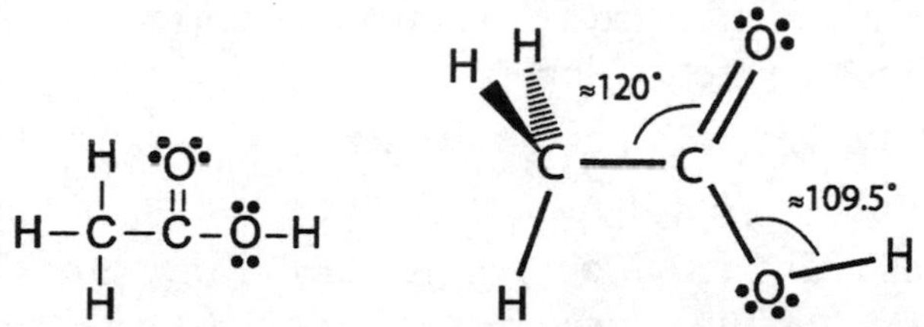

   polar (asymmetrical distribution of polar bonds)

3. For each of the following substances, write the name for the type of particle viewed as forming the structure of its solid, liquid, or gas, and write the name of the type of attraction that holds these particles in the solid or liquid form.

   a. heptane, C_7H_{16} **nonpolar molecules / London forces**
   b. formic acid, HCO_2H **polar molecules / hydrogen bonds**
   c. copper(II) sulfate, $CuSO_4$ **cations and anions / ionic bonds**
   d. methanol, CH_3OH **polar molecules / hydrogen bonds**
   e. iodine, I_2 **nonpolar molecules / London forces**
   f. carbon dioxide, CO_2 **nonpolar molecules / London forces**

4. When liquid propane is pumped into an empty tank, some of the liquid evaporates, and after that, both liquid and gaseous propane are present in the tank. Explain why the system adjusts so that there is a constant amount of liquid and gas in the container. With reference to the constant changes that take place inside the container, explain why this system can be described as a dynamic equilibrium.

   As soon as the liquid propane is added to the tank, the liquid begins to evaporate. In the closed container, the vapor particles that have escaped from the liquid are trapped. The concentration of propane vapor rises quickly to the concentration that makes the rate of condensation equal to the rate of evaporation, so there is no net change in the amount of liquid or vapor in the tank. There are constant changes (from liquid to vapor and vapor to liquid), but because the rates of these two changes are equal (the rate of evaporation equals the rate of condensation), there is no net change in the system (the amount of liquid and vapor remains constant). Thus, the system is a dynamic equilibrium.

Key Ideas Answers

5. The Second Law of Thermodynamics is the key scientific principle that allows us to explain and predict changes. One way to state the Second Law of Thermodynamics is that the entropy of the **universe** increases.

7. A change from state A to state B tends to take place when state **B** has greater disorder than state **A**.

9. Because there are more possible arrangements for gas particles when they are dispersed throughout a container than when they are concentrated in one corner of it, **probability** suggests that they will spread out to fill the total volume available to them.

11. If less than 1 gram of the substance will dissolve in 100 milliliters (or 100 g) of solvent, the substance is considered **insoluble**.

13. If between 1 and 10 grams of a substance will dissolve in 100 milliliters (or 100 g) of solvent, the substance is considered **moderately** soluble.

15. Nonpolar substances are likely to dissolve in **nonpolar solvents**.

17. **Polar substances** are not likely to dissolve to a significant degree in nonpolar solvents.

19. A polar section of a molecule, which is **attracted** to water, is called hydrophilic (literally, "water loving"), and a(n) **nonpolar** part of the molecule, which is not expected to be attracted to water, is called hydrophobic ("water fearing").

21. Soap can be made from animal **fats** and vegetable **oils**, which in turn are composed of triacylglycerols, or triglycerides.

23. A soap is an ionic compound composed of a(n) **cation** (usually sodium or potassium) and an anion with the general form RCO_2^-, in which R represents a(n) **long-chain** hydrocarbon group.

25. **Hard** water is water that contains dissolved calcium ions, Ca^{2+}, magnesium ions, Mg^{2+}, and often iron ions, Fe^{2+} or Fe^{3+}. These ions bind strongly to soap anions, causing the soap to **precipitate** from hard-water solutions.

27. The net rate of solution depends on three factors: **surface area of the solute**, **degree of agitation or stirring**, and **temperature**.

29. If the partial pressure of a gas over a liquid is increased, the **solubility** of that gas is increased.

Problems Key

Section 15.1 Entropy, Solutions, and Solubility

31. The apparatus shown below consists of two containers connected by an opening that is initially blocked. One side contains a gas, and the other side is empty. When the divider between the two containers is removed, the gas moves between the containers until it is evenly distributed. Explain, in terms of the Second Law of Thermodynamics, why this happens. Is the entropy of the system greater before or after? Why? Your explanation should compare the relative (1) disorder in these systems, (2) degree of dispersal of the particles, and (3) number of ways to arrange the particles in the systems. *(Obj #3)*

> The Second Law of Thermodynamics states that the entropy of the universe increases. Entropy is a measure of disorder and degree of dispersal. This leads to the general statement that changes tend to take place when they lead to an increase in the disorder or degree of dispersal of the system. The particles in (b) are more dispersed and more disordered than in (a), so (b) is a higher entropy state than (a). Therefore, we expect that the system will shift from (a) to (b). The reason for this shift is that there are more ways to arrange our system in the more dispersed form—(b)—than in the more concentrated form—(a). Because the particles can move freely in the container, they will shift to the more probable, more dispersed, and more disordered form. See Figure 15.2 on page 632 of the text.

Gas in one chamber	$\rightarrow$	Gas in both chambers
Fewer ways to arrange particles		More ways to arrange particles
Less probable		More probable
More ordered		Less ordered
Particles closer together		Particles more dispersed

33. The primary components of vinegar are acetic acid and water, both of which are composed of polar molecules with hydrogen bonds that link them. These two liquids will mix in any proportion. With reference to the Second Law of Thermodynamics, explain why acetic acid and water are miscible. *(Obj #4)*

> H   :O:
> | ||
> H–C–C–Ö–H
> | ··
> H   Acetic acid

> Picture a layer of acetic acid that is carefully added to water (See Figure 15.3 on page 633 of the text. Picture acetic acid molecules in the place of the ethanol molecules.). Because the particles of a liquid are moving constantly, some of the acetic acid particles at the boundary between the two liquids will immediately move into the water, and some of the water molecules will move into the acetic acid. In this process, water-water and acetic acid-acetic acid attractions are broken and acetic acid-water attractions are

formed. Both acetic acid and water are molecular substances with O–H bonds, so the attractions broken between water molecules and the attractions broken between acetic acid molecules are hydrogen bonds. The attractions that form between the acetic acid and water molecules are also hydrogen bonds. We expect the hydrogen bonds that form between water molecules and acetic acid molecules to be similar in strength to the hydrogen bonds that are broken. Because the attractions between the particles are so similar, the freedom of movement of the acetic acid molecules in the water solution is about the same as their freedom of movement in the pure acetic acid. The same can be said for the water. Because of this freedom of movement, both liquids will spread out to fill the total volume of the combined liquids. In this way, they will shift to the most probable, most disordered, most dispersed, highest entropy state available, the state of being completely mixed. There are many more possible arrangements for this system when the acetic acid and water molecules are dispersed throughout a solution than when they are restricted to separate layers.

36. Would the following combinations be expected to be soluble or insoluble?
    a. polar solute and polar solvent   **soluble**
    b. nonpolar solute and polar solvent   **insoluble**
    c. ionic solute and water   **often soluble**
    d. molecular solute with small molecules and hexane   **often soluble**
    e. hydrocarbon solute and water   **insoluble**

38. Write the chemical formula for the primary solute in each of the following solutions. Explain how the solubility guideline *like dissolves like* leads to the prediction that these substances would be soluble in water.
    a. vinegar   Acetic acid, **$HC_2H_3O_2$**, is like water because both are composed of small, polar molecules with hydrogen bonding between them.
    b. household ammonia   Ammonia, **NH_3**, is like water because both are composed of small, polar molecules with hydrogen bonding between them.

40. Predict whether each of the following compounds is soluble in water. *(Obj #6)*
    a. the polar molecular compound 1-propanol, $CH_3CH_2CH_2OH$ (a solvent for waxes and vegetable oils)
        soluble (Small alcohols and other small polar molecular substances tend to be soluble in water.)
    b. cis-2-pentene, $CH_3CHCHCH_2CH_3$ (a polymerization inhibitor)
        insoluble (Hydrocarbon molecules and other nonpolar molecular substances are insoluble in water.)
    c. the polar molecular compound formic acid, HCO_2H (transmitted in ant bites)
        soluble (Small polar molecular substances tend to be soluble in water.)
    d. strontium chlorate, $Sr(ClO_3)_2$ (used in tracer bullets)
        soluble (Ionic compounds tend to be soluble in water.)

42. Predict whether each of the following compounds is soluble in hexane, C_6H_{14}. *(Obj #7)*

    a. butane, C_4H_{10} (fuel for cigarette lighters)

        soluble (Hydrocarbon molecules and other nonpolar molecular substances tend to be soluble in hexane.)

    b. potassium hydrogen oxalate, KHC_2O_4 (used to remove stains)

        insoluble (Ionic compounds are insoluble in hexane.)

44. Predict whether each of the following compounds would be more soluble in water or hexane. *(Obj #8)*

    a. toluene, $C_6H_5CH_3$ (in aviation fuels)

        hexane (Hydrocarbon molecules and other nonpolar molecular substances are more soluble in hexane than in water.)

    b. lithium perchlorate, $LiClO_4$ (in solid rocket fuels)

        water (Ionic compounds are more soluble in water than in hexane.)

47. Would you expect acetone, CH_3COCH_3 (in some nail polish removers), or 2-hexanone, $CH_3COCH_2CH_2CH_2CH_3$ (a solvent), to be more soluble in water? Why? *(Obj #9)*

    Acetone molecules have a higher percentage of their structure that is polar, so **acetone** is more soluble in water.

49. Would you expect ethane, C_2H_6 (in natural gas), or strontium perchlorate, $Sr(ClO_4)_2$ (in fireworks), to be more soluble in hexane, C_6H_{14}? Why? *(Obj #9)*

    Nonpolar substances, like **ethane**, are **more soluble** in the nonpolar solvent hexane than highly polar ionic compounds like strontium perchlorate.

51. Explain why amphetamine can pass through the blood-brain barrier more easily than epinephrine. *(Obj #10)*

    The three –OH groups and the N–H bond in the epinephrine structure give it a greater percentage of its structure that is polar, so we predict that epinephrine would be more soluble in water than amphetamine. The cell membranes that separate the blood stream from the brain cells have a nonpolar interior that tends to block polar substances from moving into the brain. Epinephrine is too polar to move from the blood stream into the brain, but the stimulant effects of the less polar amphetamine are in part due to its ability to pass through the blood-brain barrier. See Figure 15.5 on page 638 of the text.

Section 15.2 Fats, Oils, Soaps, and Detergents

53. What are the products of the reaction between the following triglyceride and sodium hydroxide? *(Obj #13)*

55. What part of the following soap structure is hydrophilic and what portion is hydrophobic? *(Obj #14)*

hydrophilic   hydrophobic

The very polar, ionic end on the left is hydrophilic (attracted to water), and the nonpolar, hydrocarbon portion of the structure is hydrophobic.

57. If you live in an area that has very hard water, detergents are a better choice for cleaning agents than soap. Describe the difference in structure between a typical anionic detergent and a typical soap. Describe what makes water hard, and explain why detergents work better in hard water than soap does. *(Objs #16 & 17)*

> For soap to work, its anions must stay in solution. Unfortunately, they tend to precipitate from solution when the water is "hard." Hard water is water that contains dissolved calcium ions, Ca^{2+}, magnesium ions, Mg^{2+}, and often iron ions, Fe^{2+} or Fe^{3+}. These ions bind strongly to soap anions, causing the soap to precipitate from hard water solutions.
>
> Detergents have been developed to avoid this problem of soap in hard water. Their structures are similar to (although more varied than) soap but less likely to form insoluble compounds with hard water ions. Some detergents are ionic like soap, and some are molecular. For example, in sodium dodecyl sulfate (SDS), a typical ionic detergent, the $-CO_2^-$ portion of the conventional soap structure is replaced by an $-OSO_3^-$ group, which is less likely to link with hard water cations and precipitate from the solution:

SDS, a typical ionic detergent

Section 15.3 - Saturated Solutions and Dynamic Equilibrium

59. Epsom salts, a common name for magnesium sulfate heptahydrate, $MgSO_4 \cdot 7H_2O$, can be used to help reduce swelling caused by injury. For example, if you want to reduce the swelling of a sprained ankle, you can soak the ankle in a saturated solution of magnesium sulfate prepared by adding an excess of Epsom salts to water and waiting until the greatest possible amount of solid dissolves.

    a. Describe the reversible change that takes place as the $MgSO_4$ dissolves. (You do not need to mention the water of hydration that is attached to the magnesium sulfate.) *(Obj #18)*

> Picture the $MgSO_4$ solid sitting on the bottom of the container. The magnesium and sulfate ions at the surface of the solid are constantly moving out into the water and being pulled back by the attractions to the other ions still on the surface of the solid. Sometimes when an ion moves out into the water, a water molecule collides with it and pushes it farther out into the solution. Other water molecules move into the gap between the ion in solution and the solid and shield the ion from the attraction to the solid. The ion is kept stable and held in solution by attractions between the polar water molecules and the charged ions. The negative oxygen atoms of the water molecules surround the cations, and the positive hydrogen ends of water molecules surround the anions.

Once the ions are in solution, they move throughout the solution like any particle in a liquid. Eventually they collide with the surface of the solid. When this happens, they come back under the influence of the attractions that hold the particles in the solid, and they are likely to return to the solid form.

b. Explain why an increase in the concentration of solute particles in this solution containing undissolved solid leads to an increase in the rate of return of the solute to the solid form. *(Obj #19)*

The more particles of solute there are per liter of solution, the more collisions there will be between solute particles and the solid. More collisions lead to a greater rate of return of solute particles from the solution to the solid (See Figures 15.15 and 15.16 on page 646 of the textbook.)

c. Why does the magnesium sulfate dissolve faster if the solution is stirred? *(Obj #23)*

As particles leave the solid and go into solution, localized high concentrations of dissolved solute form around the surface of the solid. Remember that the higher concentration of solute leads to a higher rate of return to the solid form. A higher rate of return leads to a lower overall *net* rate of solution. If you stir or in some way agitate the solution, the solute particles near the solid will be moved more quickly away from the solid, and the localized high concentrations of solute will be avoided. This will diminish the rate of return and increase the net rate of solution. This is why stirring the mixture of water and Epsom salts dissolves the magnesium sulfate more rapidly. (See Figures 15.19 and 15.20 on pages 648 and 649 of the textbook.)

d. When the maximum amount of magnesium sulfate has dissolved, does the solid stop dissolving? Explain.

The solution becomes saturated when the rate of solution and the rate of return become equal. The solid continues to dissolve, but particles return to the solid from solution at the same rate. Even though the specific particles in solution are constantly changing, there is no net change in the total amount of solid or the amount of ions in solution.

61. If you wanted to make sugar dissolve as quickly as possible, would you use

a. room temperature water or hot water? Why?

Higher temperature increases the rate at which particles escape from a solid, increasing the rate of solution. The higher temperature also helps the escaped particles to move away from the solid more quickly, minimizing the rate of return. Together these two factors make the net rate of solution in the **hot water** higher.

b. powdered sugar or granular sugar? Why?

> Only particles at the surface of the solid sugar have a possibility of escaping into the solution. The particles in the center each sugar crystal have to wait until the particles between them and the surface dissolve to have any chance of escaping. **Powdered sugar** has much smaller crystals, so a much higher percentage of the sugar particles are at the surface. This increases the rate of solution. (See Figures 15.17 and 15.18 on pages 647 and 648 of the textbook.)

63. What does it mean to say a solution is saturated? Describe the changes that take place at the particle level in a saturated solution of sodium chloride that contains an excess of solid NaCl. Is the NaCl still dissolving?

> A saturated solution has enough solute dissolved to reach the solubility limit. In a saturated solution of NaCl, Na^+ and Cl^- ions are constantly escaping from the surface of the solid and moving into solution, but other Na^+ and Cl^- ions in solution are colliding with the solid and returning to the solid at a rate equal to the rate of solution. Because the rate of solution and the rate of return to the solid are equal, there is no net shift to more or less salt dissolved.

65. Can a solution be both concentrated (with a relatively high concentration of solute) and unsaturated? Explain your answer.

> Yes, a solution can be both concentrated and unsaturated. If the solubility of a substance is high, the concentration of the solute in solution can be high even in a solution where the rate of return to the undissolved solute is still below the rate of solution.

Section 15.4 Solutions of Gases in Liquids

67. Consider a soft drink bottle with a screw cap. When the cap is removed, the excess CO_2 in the space above the soft drink escapes into the room, leaving normal air above the liquid. Explain why the soft drink will lose its carbonation more quickly if the cap is left off than if the cap is immediately put back on tightly.

> At the same temperature, the rate at which carbon dioxide molecules escape from the soft drink is the same whether the container is open or closed. When a soft drink is open to the air, the carbon dioxide molecules that escape can move farther from the surface of the liquid and are therefore less likely to collide with the surface of the liquid and return to the solution. This means a lower rate of return in the drink open to the air. Therefore, the difference between the rate of escape of the CO_2 from the soft drink and the rate of return will be greater in the bottle open to the air, so there is a greater net rate of escape of the gas from the soft drink.

Chapter 16 - The Process of Chemical Reactions

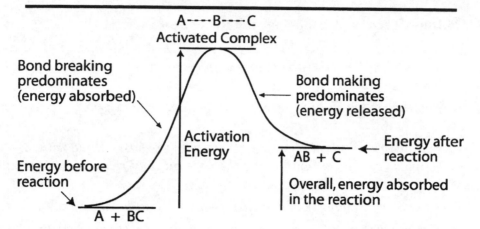

♦ Review Skills

16.1 Collision Theory: A Model for the Reaction Process

- The Basics of Collision Theory
- Endergonic Reactions
- Summary of Collision Theory

16.2 Rates of Chemical Reactions

- Temperature and Rates of Chemical Reactions
- Concentration and Rates of Chemical Reactions
- Catalysts
- Homogeneous and
- Heterogeneous Catalysts

Special Topic 16.1: Green Chemistry - The Development of New and Better Catalysts

16.3 Reversible Reactions and Chemical Equilibrium

- Reversible Reactions and Dynamic Equilibrium
- Equilibrium Constants
- Determination of Equilibrium Constant Values
- Equilibrium Constants and Extent of Reaction
- Heterogeneous Equilibria
- Equilibrium Constants and Temperature

Internet: Calculating Concentrations and Gas Pressures

Internet: pH and pH Calculations

Internet: Weak Acids and Equilibrium Constants

16.4 Disruption of Equilibrium

- The Effect of Changes in Concentrations on Equilibrium Systems

  Internet: Changing Volume and Gas Phase Equilibrium

- Le Châtelier's Principle
- The Effect of Catalysts on Equilibria

  Special Topic 16.2: The Big Question—How Did We Get Here?

♦ Chapter Glossary

Internet: Glossary Quiz

♦ Chapter Objectives

Review Questions

Key Ideas

Chapter Problems

Section Goals and Introductions

Section 16.1 Collision Theory: A Model for the Reaction Process
Goals

- *To describe a model, called collision theory, that helps us to visualize the process of many chemical reactions.*
- *To use collision theory to explain why not all collisions between possible reactants lead to products.*
- *To use collision theory to explain why possible reactants must collide with an energy equal to or above a certain amount to have the possibility of reacting and forming products.*
- *To show how the energy changes in chemical reactions can be described with diagrams.*
- *To use collision theory to explain why possible reactants must collide with a specific orientation to have the possibility of reacting and forming products.*

Once again, this chapter emphasizes that if you develop the ability to visualize changes on the particle level, it will help you understand and explain many different things. This section introduces you to a model for chemical change that is called collision theory, which helps you explain the factors that affect the rates of chemical reactions. These factors are described in Section 16.2.

Section 16.2 Rates of Chemical Reactions
Goals

- *To show how rates of chemical reactions are described.*
- *To explain why increased temperature increases the rates of most chemical reactions.*
- *To explain why increased concentration of reactants increases the rates of chemical reactions.*
- *To describe how catalysts increase the rates of certain chemical reactions.*

This section shows how collision theory helps you explain the factors that affect rates of chemical changes. These factors include amounts of reactants and products, temperature, and catalysts.

Section 16.3 Reversible Reactions and Chemical Equilibrium
Goals

- *To explain why chemical reactions that are reversible come to a dynamic equilibrium with equal forward and reverse rates of reaction.*
- *To show what equilibrium constants are and how they can be determined.*
- *To describe how equilibrium constants can be used to show the relative amounts of reactants and products in the system at equilibrium.*
- *To explain the effect of temperature on equilibrium systems and equilibrium constants.*

This section takes the basic ideas of dynamic equilibrium introduced in Chapter 14 and applies them to reversible chemical changes. This is a very important topic, so plan to spend some extra time on this section, if necessary. You will also learn how equilibrium constants are used to describe the relative amounts of reactants and products for a chemical reaction at equilibrium, and you will learn how these values can be calculated. Finally, you will learn more about the effect of temperature on chemical changes.

See the three related sections on our Web site: *Calculating Concentrations and Gas Pressures, pH and pH Calculations,* and *Weak Acids and Equilibrium Constants.*

www.chemplace.com/college/

Section 16.4 Disruption of Equilibrium

Goal: To describe how equilibrium systems can be disrupted and show you how to predict whether certain changes on an equilibrium will lead to more products, more reactants, or neither.

Although the concept of chemical equilibrium is very important, many reversible reactions in nature never form equilibrium systems. This section's description of the ways that equilibrium systems can be disrupted will help you to understand why this is true. The ability to predict the effects of changes on equilibrium systems will help you understand the ways that research and industrial chemists create conditions for their chemical reactions that maximize the rate at which desirable reactions move to products and minimize that rate at which undesirable reactions take place.

See the section on our Web site that provides information on *Changing Volumes and Gas Phase Equilibrium*

www.chemplace.com/college/

Chapter 16 Map

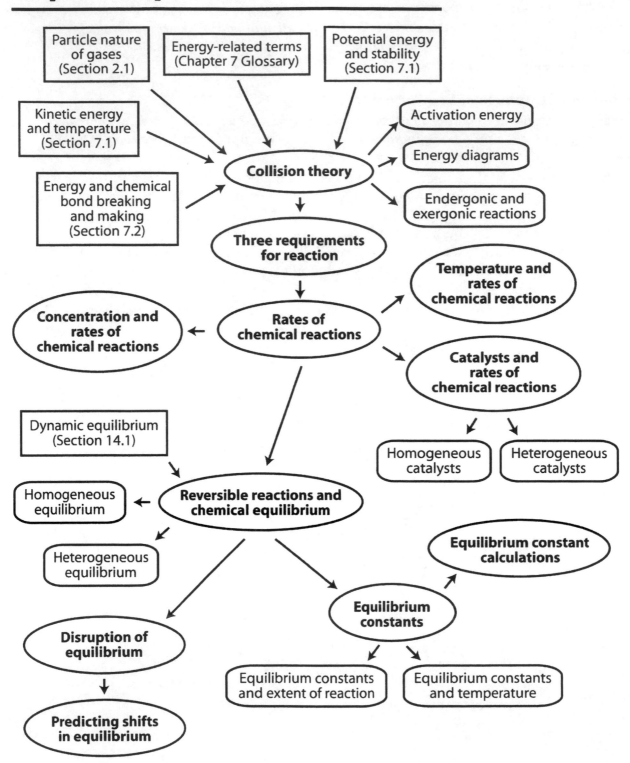

Chapter Checklist

- ☐ Read the Review Skills section. If there is any skill mentioned that you have not yet mastered, review the material on that topic before reading this chapter.
- ☐ Read the chapter quickly before the lecture that describes it.
- ☐ Attend class meetings, take notes, and participate in class discussions.
- ☐ Work the Chapter Exercises, perhaps using the Chapter Examples as guides.
- ☐ Study the Chapter Glossary and test yourself on our Web site:

    www.chemplace.com/college/

- ☐ Study all of the Chapter Objectives. You might want to write a description of how you will meet each objective.
- ☐ This chapter has logic sequences in Figures 16.12, 16.14, 16.16, 16.23, and 16.26. Convince yourself that each of the statements in these sequences logically leads to the next statement.
- ☐ To get a review of the most important topics in the chapter, fill in the blanks in the Key Ideas section.
- ☐ Work all of the selected problems at the end of the chapter, and check your answers with the solutions provided in this chapter of the study guide.
- ☐ Ask for help if you need it.

Web Resources   www.chemplace.com/college/

Calculating Concentrations and Gas Pressures
pH and pH Calculations
Weak Acids and Equilibrium Constants
Changing Volume and Gas Phase Equilibrium
Glossary Quiz

Exercises Key

Exercise 16.1 – Writing Equilibrium Constant Expressions: Sulfur dioxide, SO_2, one of the intermediates in the production of sulfuric acid, can be made from the reaction of hydrogen sulfide gas with oxygen gas. Write the equilibrium constant expressions for K_C and K_P for the following equation for this reaction. *(Objs #24 & 25)*

$$2H_2S(g) + 3O_2(g) \rightleftharpoons 2SO_2(g) + 2H_2O(g)$$

$$K_C = \frac{[SO_2]^2[H_2O]^2}{[H_2S]^2[O_2]^3} \qquad K_P = \frac{P_{SO_2}{}^2 \, P_{H_2O}{}^2}{P_{H_2S}{}^2 \, P_{O_2}{}^3}$$

Exercise 16.2 – Equilibrium Constant Calculation: Ethanol, C_2H_5OH, can be made from the reaction of ethylene gas, C_2H_4, and water vapor. A mixture of $C_2H_4(g)$ and $H_2O(g)$ is allowed to come to equilibrium in a container at 110 °C, and the partial pressures of the gases are found to be 0.35 atm for $C_2H_4(g)$, 0.75 atm for $H_2O(g)$, and 0.11 atm for $C_2H_5OH(g)$. What is K_P for this reaction at 110 °C? *(Obj #26)*

$$C_2H_4(g) + H_2O(g) \rightleftharpoons C_2H_5OH(g)$$

$$K_P = \frac{P_{C_2H_5OH}}{P_{C_2H_4}\, P_{H_2O}} = \frac{0.11 \text{ atm}}{0.35 \text{ atm } (0.75 \text{ atm})} = 0.42 \text{ 1/atm or } 0.42$$

Exercise 16.3 – Predicting the Extent of Reaction: Using the information in Table 16.1, predict whether each of the following reversible reactions favors reactants, products, or neither at 25 °C. *(Obj #27)*

a. This reaction is partially responsible for the release of pollutants from automobiles.

$$2NO(g) + O_2(g) \rightleftharpoons 2NO_2(g)$$

According to Table 16.1, the K_P for this reaction is 2.2×10^{12}, so it **favors products.**

b. The $NO_2(g)$ molecules formed in the reaction in part (a) can combine to form N_2O_4.

$$2NO_2(g) \rightleftharpoons N_2O_4(g)$$

According to Table 16.1, the K_P for this reaction is 6.7. **Neither** reactants nor products are favored.

Exercise 16.4 – Writing Equilibrium Constants for Heterogeneous Equilibria: The following equation describes one of the steps in the purification of titanium dioxide, which is used as a white pigment in paints. Liquid titanium(IV) chloride reacts with oxygen gas to form solid titanium oxide and chlorine gas. Write K_C and K_P expressions for this reaction. *(Objs #24 & 25)*

$$TiCl_4(l) + O_2(g) \rightleftharpoons TiO_2(s) + 2Cl_2(g)$$

$$K_C = \frac{[Cl_2]^2}{[O_2]} \qquad K_P = \frac{P_{Cl_2}^{\,2}}{P_{O_2}}$$

Exercise 16.5 – Predicting the Effect of Disruptions on Equilibrium:
Nitric acid can be made from the exothermic reaction of nitrogen dioxide gas and water vapor in the presence of a rhodium and platinum catalyst at 700-900 °C and 5-8 atm. Predict whether each of the following changes in the equilibrium system will shift the system to more products, to more reactants, or to neither. Explain each answer in two ways: (1) by applying Le Châtelier's principle and (2) by describing the effect of the change on the forward and reverse reaction rates. *(Objs #40- 42 & 44- 46)*

$$3NO_2(g) + H_2O(g) \overset{Rh/Pt}{\underset{750\text{-}920\,°C}{\rightleftharpoons}} 2HNO_3(g) + NO(g) + 37.6\text{ kJ}$$
5-8 atm

a. The concentration of H_2O is increased by the addition of more H_2O.

- (1) Using Le Châtelier's Principle, we predict that the system will **shift to more products** to partially counteract the increase in H_2O.

- (2) The increase in the concentration of water vapor speeds the forward reaction without initially affecting the rate of the reverse reaction. The equilibrium is disrupted, and the system **shifts to more products** because the forward rate is greater than the reverse rate.

b. The concentration of NO_2 is decreased.

- (1) Using Le Châtelier's Principle, we predict that the system will **shift to more reactants** to partially counteract the decrease in NO_2.

- (2) The decrease in the concentration of $NO_2(g)$ slows the forward reaction without initially affecting the rate of the reverse reaction. The equilibrium is disrupted, and the system **shifts toward more reactants** because the reverse rate is greater than the forward rate.

c. The concentration of $HNO_3(g)$ is decreased by removing the nitric acid as it forms.

- (1) Using Le Châtelier's Principle, we predict that the system will **shift to more products** to partially counteract the decrease in HNO_3.

- (2) The decrease in the concentration of $HNO_3(g)$ slows the reverse reaction without initially affecting the rate of the forward reaction. The equilibrium is disrupted, and the system **shifts toward more products** because the forward rate is greater than the reverse rate.

d. The temperature is decreased from 1000 °C to 800 °C.

- (1) Using Le Châtelier's Principle, we predict that the system shifts in the exothermic direction to partially counteract the decrease in temperature. As the system **shifts toward more products**, energy is released, and the temperature increases.

- (2) The decreased temperature decreases the rates of both the forward and reverse reactions, but it has a greater effect on the endothermic reaction. Because the forward reaction is exothermic, the reverse reaction must be endothermic. Therefore, the reverse reaction is slowed more than the forward reaction. The system **shifts toward more products** because the forward rate becomes greater than the reverse rate.

e. The Rh/Pt catalyst is added to the equilibrium system.

- (1) Le Châtelier's Principle does not apply here.

- (2) The catalyst speeds both the forward and the reverse rates equally. Thus, there is no shift in the equilibrium. The purpose of the catalyst is to bring the system to equilibrium faster.

Review Questions Key

1. Describe what you visualize occurring inside a container of oxygen gas, O_2, at room temperature and pressure.

    The gas is composed of O_2 molecules that are moving constantly in the container. For a typical gas, the average distance between particles is about 10 times the diameter of each particle. This leads to the gas particles themselves taking up only about 0.1% of the total volume. The other 99.9% of the total volume is empty space. According to our model, each O_2 molecule moves freely in a straight-line path until it collides with another O_2 molecule or one of the walls of the container. The particles are moving fast enough to break any attraction that might form between them, so after two particles collide, they bounce off each other and continue on alone. Due to collisions, each particle is constantly speeding up and slowing down, but its average velocity stays constant as long as the temperature stays constant.

2. Write in each blank the word that best fits the definition.
    a. **Energy** is the capacity to do work.
    b. **Kinetic energy** is the capacity to do work due to the motion of an object.
    c. A(n) **endergonic** change is a change that absorbs energy.
    d. A(n) **exergonic** change is a change that releases energy.
    e. **Thermal** is the energy associated with the random motion of particles.
    f. **Heat** is thermal energy that is transferred from a region of higher temperature to a region of lower temperature as a result of the collisions of particles.
    g. A(n) **exothermic** change is a change that leads to heat energy being evolved from the system to the surroundings.
    h. A(n) **endothermic** change is a change that leads the system to absorb heat energy from the surroundings.
    i. A(n) **catalyst** is a substance that speeds a chemical reaction without being permanently altered itself.

3. When the temperature of the air changes from 62 °C at 4:00 A.M. to 84 °C at noon on a summer day, does the average kinetic energy of the particles in the air increase, decrease, or stay the same?

    Increased temperature means **increased average kinetic energy**.

4. Explain why it takes energy to break an O–O bond in an O_3 molecule.

    Any time a change leads to decreased forces of attraction, it leads to increased potential energy. The Law of Conservation of Energy states that energy cannot be created or destroyed, so energy must be added to the system. It always takes energy to break attractions between particles.

5. Explain why energy is released when two oxygen atoms come together to form an O_2 molecule.

    Any time a change leads to increased forces of attraction, it leads to decreased potential energy. The Law of Conservation of Energy states that energy cannot be created or destroyed, so energy is released from the system. Energy is always released when new attractions between particles are formed.

6. Explain why some chemical reactions *release heat* to their surroundings.

    If the bonds in the products are stronger and lower potential energy than in the reactants, energy will be released from the system. If the energy released is due to the conversion of potential energy to kinetic energy, the temperature of the products will be higher than the original reactants. The higher temperature products are able to transfer heat to the surroundings, and the temperature of the surroundings increases.

7. Explain why some chemical reactions *absorb heat* from their surroundings.

    If the bonds in the products are weaker and higher potential energy than in the reactants, energy must be absorbed. If the energy absorbed is due to the conversion of kinetic energy to potential energy, the temperature of the products will be lower than the original reactants. The lower temperature products are able to absorb heat from the surroundings, and the temperature of the surroundings decreases.

8. What are the general characteristics of any dynamic equilibrium system?

    The system must have two opposing changes, from state A to state B and from state B to state A. For a dynamic equilibrium to exist, the rates of the two opposing changes must be equal, so there are constant changes between state A and state B but no *net* change in the components of the system.

Key Ideas Answers

9. At a certain stage in the progress of a reaction, bond breaking and bond making are of equal importance. In other words, the energy necessary for bond breaking is **balanced** by the energy supplied by bond making. At this turning point, the particles involved in the reaction are joined in a structure known as the activated complex, or **transition state**.

11. In a chemical reaction, the **minimum** energy necessary for reaching the activated complex and proceeding to products is called the activation energy. Only the collisions that provide a net kinetic energy **equal to** or **greater than** the activation energy can lead to products.

13. The energies associated with endergonic (or endothermic) changes are described with **positive** values.

15. Because the formation of the **new bonds** provides some of the energy necessary to break the **old bonds**, the making and breaking of bonds must occur more or less simultaneously. This is possible only when the particles collide in such a way that the bond-forming atoms are **close to each other**.

17. Increased temperature means an increase in the average kinetic energy of the collisions between the particles in a system. This leads to an increase in the **fraction** of the collisions that have enough energy to reach the activated complex (the activation energy).

19. One of the ways in which catalysts accelerate chemical reactions is by providing a(n) **alternative pathway** between reactants and products that has a(n) **lower activation energy**.

21. If the catalyst is not in the same state as the reactants, the catalyst is called a(n) **heterogeneous** catalyst.

23. The extent to which reversible reactions proceed toward products before reaching equilibrium can be described with a(n) **equilibrium constant**, which is derived from the ratio of the concentrations of products to the concentrations of reactants at equilibrium. For homogeneous equilibria, the concentrations of all reactants and products can be described in **moles per liter**, and the concentration of each is raised to a power equal to its **coefficient** in a balanced equation for the reaction.

25. The **larger** a value for K, the farther the reaction shifts toward products before the rates of the forward and reverse reactions become equal and the concentrations of reactants and products stop changing.

27. Changing temperature always causes a shift in equilibrium systems—sometimes toward more products and sometimes toward more **reactants**.

29. If the forward reaction is endergonic, increased temperature will shift the system toward more **products**.

31. Le Châtelier's principle states that if a system at equilibrium is altered in a way that **disrupts** the equilibrium, the system will shift in such a way as to counter the **change**.

Problems Key

Section 16.1 Collision Theory: A Model for the Reaction Process

33. Assume that the following reaction is a single-step reaction in which one of the O–O bonds in O_3 is broken and a new N–O bond is formed. The heat of reaction is –226 kJ/mol.

$$NO(g) + O_3(g) \rightarrow NO_2(g) + O_2(g) + 226 \text{ kJ}$$

    a. With reference to collision theory, describe the general process that takes place as this reaction moves from reactants to products. *(Obj #2)*

> NO and O_3 molecules are constantly moving in the container, sometimes with a
> high velocity and sometimes more slowly. The particles are constantly colliding,
> changing their direction of motion, and speeding up or slowing down. If the

molecules collide in a way that puts the nitrogen atom in NO near one of the outer oxygen atoms in O_3, one of the O–O bonds in the O_3 molecule begins to break, and a new bond between one of the oxygen atoms in the ozone molecule and the nitrogen atom in NO begins to form. If the collision yields enough energy to reach the activated complex, it proceeds on to products. If the molecules do not have the correct orientation, or if they do not have enough energy, they separate without a reaction taking place.

b. List the three requirements that must be met before a reaction between NO(g) and O_3(g) is likely to take place. *(Obj #11)*

NO and O_3 molecules must collide, they must collide with the correct orientation to form an N–O bond at the same time that an O–O bond is broken, and they must have the minimum energy necessary to reach the activated complex (the activation energy).

c. Explain why NO(g) and O_3(g) must collide before a reaction can take place. *(Obj #3)*

The collision brings the atoms that will form the new bonds close, and the net kinetic energy in the collision provides the energy necessary to reach the activated complex and proceed to products.

d. Explain why it is usually necessary for the new N–O bonds to form at the same time that the O–O bonds are broken. *(Obj # 4)*

It takes a significant amount of energy to break O–O bonds, and collisions between particles are not likely to provide enough. As N–O bonds form, they release energy, so the formation of the new bonds can provide energy to supplement the energy provided by the collisions. The sum of the energy of collision and the energy released in bond formation is more likely to provide enough energy for the reaction.

e. Draw a rough sketch of the activated complex.

N–O------O------O–O
      | |
   Bond Bond
 making breaking

f. Explain why a collision between NO(g) and O_3(g) must have a certain minimum energy (activation energy) in order to proceed to products. *(Obj #5)*

In the initial stage of the reaction, the energy released in bond making is less than the energy absorbed by bond breaking. Therefore, energy must be available from the colliding particles to allow the reaction to proceed. At some point in the change, the energy released in bond formation becomes equal to the energy absorbed in bond breaking. If the colliding particles have enough energy to reach this point (in other words, if they have the activation energy), the reaction proceeds to products.

g. The activation energy for this reaction is 132 kJ/mol. Draw an energy diagram for this reaction, showing the relative energies of the reactants, the activated complex, the products, and (using arrows) the activation energy and heat of reaction. *(Obj #7)*

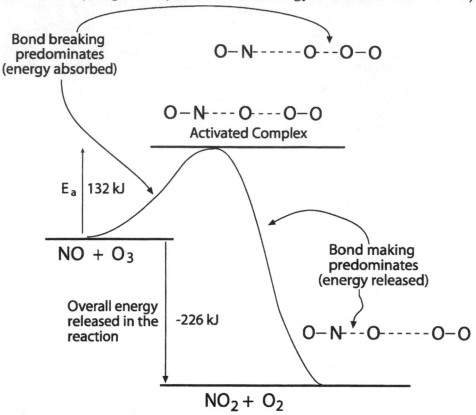

Bond breaking predominates (energy absorbed)

O–N - - - - O - - O–O

O–N - - - O - - - O–O
Activated Complex

E_a | 132 kJ

NO + O$_3$

Overall energy released in the reaction | -226 kJ

Bond making predominates (energy released)

O–N - - O - - - - - O–O

NO$_2$ + O$_2$

h. Is this reaction exothermic or endothermic? *(Objs #6 & 8)*

The negative sign for the heat of reaction shows that energy is released overall, so the reaction is **exothermic**.

i. Explain why NO(*g*) and O$_3$(*g*) molecules must collide with the correct orientation if a reaction between them is likely to take place. *(Obj #10)*

For a reaction to be likely, new bonds must be made at the same time as other bonds are broken. Therefore, the nitrogen atom in NO must collide with one of the outer oxygen atoms in O$_3$.

Section 16.2 Rates of Chemical Reactions

35. Consider the following general reaction for which gases A and B are mixed in a constant volume container.

$$A(g) + B(g) \rightarrow C(g) + D(g)$$

What happens to the rate of this reaction when

   a. more gas A is added to the container?

   > Increased concentration of reactant A leads to increased rate of collision between A and B and therefore to leads to **increased rate of reaction**.

   b. the temperature is decreased?

   > Decreased temperature leads to decreased average kinetic energy of collisions between A and B. This leads to a decrease in the percentage of collisions with the minimum energy necessary for the reaction and therefore leads to **decreased rate of reaction**.

   c. a catalyst is added that lowers the activation energy?

   > With a lower activation energy, there is a greater percentage of collisions with the minimum energy necessary for the reaction and therefore an **increased rate of reaction**.

37. The reactions listed below are run at the same temperature. The activation energy for the first reaction is 132 kJ/mol. The activation energy for the second reaction is 76 kJ/mol. In which of these reactions would a higher fraction of collisions between reactants have the minimum energy necessary to react (the activation energy)? Explain your answer.

   $$NO(g) + O_3(g) \rightarrow NO_2(g) + O_2(g) \qquad \text{Activation energy} = 132 \text{ kJ}$$

   $$I^-(aq) + CH_3Br(aq) \rightarrow CH_3I(aq) + Br^-(aq) \qquad \text{Activation energy} = 76 \text{ kJ}$$

   > At a particular temperature, the lower the activation energy is, the higher the percentage of collisions with at least that energy or more will be. Thus, the **second reaction** would have the higher fraction of collisions with the activation energy.

39. Two reactions can be described by the energy diagrams below. What is the approximate activation energy for each reaction? Which reaction is exothermic and which is endothermic?

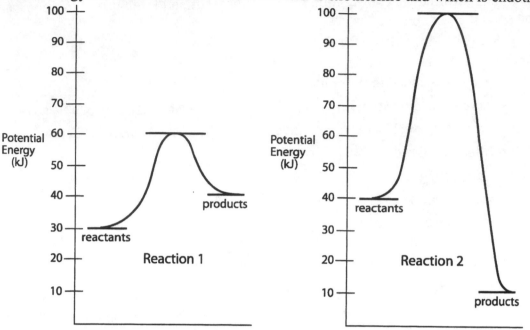

The approximate activation energy for reaction 1 is 30 kJ and for reaction 2 is 60 kJ. Reaction 1 is endothermic, and reaction 2 is exothermic.

41. Explain why chlorine atoms speed the conversion of ozone molecules, O_3, and oxygen atoms, O, into oxygen molecules, O_2. *(Obj #14)*

In part, chlorine atoms are a threat to the ozone layer just because they provide another pathway for the conversion of O_3 and O to O_2, but there is another reason. The reaction between O_3 and Cl that forms ClO and O_2 has an activation energy of 2.1 kJ/mole. At 25 °C, about 3 of every 7 collisions (or 43%) have enough energy to reach the activated complex. The reaction between O and ClO to form Cl and O_2 has an activation energy of only 0.4 kJ/mole. At 25 °C, about 85% of the collisions have at least this energy. The uncatalyzed reaction has an activation energy of about 17 kJ/mole. At 25 °C (298 K), about one of every one thousand collisions (or 0.1%) between O_3 molecules and O atoms has a net kinetic energy large enough to form the activated complex and proceed to products. Thus, a much higher fraction of the collisions have the minimum energy necessary to react for the catalyzed reaction than for the direct reaction between O_3 and O. Thus, a much greater fraction of the collisions has the minimum energy necessary for the reaction to proceed for the catalyzed reaction than for the uncatalyzed reaction. Figures 16.15 and 16.16, on page 678, of the textbook illustrate this.

43. Using the proposed mechanism for the conversion of NO(g) into N$_2$(g) and O$_2$(g) as an example, write a description of the four steps thought to occur in heterogeneous catalysis. *(Obj #16)*

> **Step #1:** The reactants (NO molecules) collide with the surface of the catalyst where they bind to the catalyst. This step is called adsorption. The bonds within the reactant molecules are weakened or even broken as the reactants are adsorbed. (N–O bonds are broken.)
>
> **Step #2:** The adsorbed particles (separate N and O atoms) move over the surface of the catalyst.
>
> **Step #3:** The adsorbed particles combine to form products (N$_2$ and O$_2$).
>
> **Step #4:** The products (N$_2$ and O$_2$) leave the catalyst.
>
> See Figure 16.17, on page 679, of the textbook.

Section 16.3 Reversible Reactions and Chemical Equilibrium

45. Equilibrium systems have two opposing rates of change that are equal. For each of the following equilibrium systems that were mentioned in earlier chapters, describe what is changing in the two opposing rates.

   a. a solution of the weak acid acetic acid, HC$_2$H$_3$O$_2$ (Chapter 5)

   > Acetic acid molecules react with water to form hydronium ions and acetate ions, and at the same time, hydronium ions react with acetate ions to return to acetic acid molecules and water.
   >
   > $$HC_2H_3O_2(aq) + H_2O(l) \rightleftharpoons H_3O^+(aq) + C_2H_3O_2^-(aq)$$

   b. pure liquid in a closed container (Chapter 14)

   > Liquids evaporate to form vapor at a rate that is balanced by the return of vapor to liquid.

   c. a closed bottle of carbonated water with 4 atm of CO$_2$ in the gas space above the liquid (Chapter 15)

   > Carbon dioxide escapes from the solution at a rate that is balanced by the return of CO$_2$ to the solution.

47. Two gases, A and B, are added to an empty container. They react in the following reversible reaction.

   $$A(g) + B(g) \rightleftharpoons C(g) + D(g)$$

   a. When is the forward reaction rate greatest: (1) when A and B are first mixed, (2) when the reaction reaches equilibrium, or (3) sometime between these two events?

   > The forward reaction rate is at its peak when A and B are first mixed. Because A and B concentrations are diminishing as they form C and D, the rate of the forward reaction declines steadily until equilibrium is reached.

b. When is the reverse reaction rate greatest: (1) when A and B are first mixed, (2) when the reaction reaches equilibrium, or (3) sometime between these two events?

> The reverse reaction rate is at its peak when the reaction reaches equilibrium. Because C and D concentrations are increasing as they form from A and B, the rate of the reverse reaction increases steadily until equilibrium is reached.

49. Assume that in the following reversible reaction both the forward and the reverse reactions take place in a single step.

$$I^-(aq) + CH_3Br(aq) \rightleftharpoons CH_3I(aq) + Br^-(aq)$$

a. With reference to the changing forward and reverse reaction rates, explain why this reaction moves toward a dynamic equilibrium with equal forward and reverse reaction rates. *(Obj #20)*

> When I^- ions and CH_3Br molecules are added to a container, they begin to collide and react. As the reaction proceeds, the concentrations of I^- and CH_3Br diminish, so the rate of the forward reaction decreases. Initially, there are no CH_3I molecules or Br^- ions in the container, so the rate of the reverse reaction is initially zero. As the concentrations of CH_3I and Br^- increase, the rate of the reverse reaction increases.
>
> As long as the rate of the forward reaction is greater than the rate of the reverse reaction, the concentrations of the reactants (I^- and CH_3Br) will steadily decrease, and the concentrations of products (CH_3I and Br^-) will constantly increase. This leads to a decrease in the forward rate of the reaction and an increase in the rate of the reverse reaction. This continues until the two rates become equal. At this point, our system has reached a dynamic equilibrium.

b. Describe the changes that take place once the reaction reaches an equilibrium state. Are there changes in the concentrations of reactants and products at equilibrium? Explain your answer. *(Obj #21)*

> In a dynamic equilibrium for reversible chemical reactions, the forward and reverse reaction rates are equal, so although there are constant changes between reactants and products, there is no *net* change in the amounts of each. I^- and CH_3Br are constantly reacting to form CH_3I and Br^-, but CH_3I and Br^- are reacting to reform CH_3Br and I^- at the same rate. Thus, there is no net change in the amounts of I^-, CH_3Br, CH_3I, or Br^-.

50. Write K_C and K_P expressions for each of the following equations. *(Objs #22 & 23)*

a. $2CH_4(g) \rightleftharpoons C_2H_2(g) + 3H_2(g)$

$$K_C = \frac{[C_2H_2][H_2]^3}{[CH_4]^2} \qquad K_P = \frac{P_{C_2H_2} \, P_{H_2}^3}{P_{CH_4}^2}$$

b. $2N_2O(g) + O_2(g) \rightleftharpoons 4NO(g)$

$$K_C = \frac{[NO]^4}{[N_2O]^2[O_2]} \qquad K_P = \frac{P_{NO}^4}{P_{N_2O}^2 \, P_{O_2}}$$

c. $Sb_2S_3(s) + 3H_2(g) \rightleftharpoons 2Sb(s) + 3H_2S(g)$

$$K_C = \frac{[H_2S]^3}{[H_2]^3} \qquad K_P = \frac{P_{H_2S}^3}{P_{H_2}^3}$$

52. A mixture of nitrogen dioxide and dinitrogen tetroxide is allowed to come to equilibrium at 30 °C, and the gases partial pressures are found to be 1.69 atm N_2O_4 and 0.60 atm NO_2. *(Obj #24)*

   a. On the basis of these data, what is K_P for the following equation?

   $$NO_2(g) \rightleftharpoons \tfrac{1}{2}N_2O_4(g)$$

   $$K_P = \frac{P_{N_2O_4}^{1/2}}{P_{NO_2}} = \frac{(1.69)^{1/2}}{0.60} = \mathbf{2.2}$$

   b. On the basis of these data, what is K_P for the following equation?

   $$2NO_2(g) \rightleftharpoons N_2O_4(g)$$

   $$K_P = \frac{P_{N_2O_4}}{P_{NO_2}^2} = \frac{1.69}{(0.60)^2} = \mathbf{4.7}$$

   c. Table 16.1 lists the K_P for the equation in Part (b) as 6.7 at 25 °C. Explain why your answer to Part (b) is not 6.7.

   > Changing temperature leads to a change in the value for an equilibrium constant. (Because K_P for this reaction decreases with increasing temperature, the reaction must be exothermic.)

54. Predict whether each of the following reactions favors reactants, products, or neither at the temperature for which the equilibrium constant is given. *(Obj #25)*

   a. $CH_3OH(g) + CO(g) \rightleftharpoons CH_3CO_2H(g)$ $K_P = 1.2 \times 10^{-22}$ at 25 °C
   $K_P < 10^{-2}$ so reactants favored

   b. $CH_4(g) + 4Cl_2(g) \rightleftharpoons CCl_4(g) + 4HCl(g)$ $K_P = 3.3 \times 10^{68}$ at 25 °C
   $K_P > 10^2$ so products favored

   c. $CO(g) + Cl_2(g) \rightleftharpoons COCl_2(g)$ $K_P = 0.20$ at 600 °C
   $10^{-2} < K_P < 10^2$ so neither favored

56. Write the K_C expression for the following equation. Explain why the concentration of CH_3OH is left out of the expression. *(Objs #22 & 26)*

   $$CO(g) + 2H_2(g) \rightleftharpoons CH_3OH(l)$$

   $$K_C = \frac{1}{[CO][H_2]^2}$$

   > If the number of moles of $CH_3OH(l)$ in the container is doubled, its volume doubles too, leaving the concentration (mol/L) of the methanol constant. Increasing or decreasing the total volume of the container will not change the volume occupied by the liquid methanol,

so the concentration (mol/L) of the $CH_3OH(l)$ also remains constant with changes in the volume of the container. The constant concentration of methanol can be incorporated into the equilibrium constant itself and left out of the equilibrium constant expression.

$$K' = \frac{[CH_3OH]}{[CO][H_2]^2} \qquad \frac{K'}{[CH_3OH]} = \frac{1}{[CO][H_2]^2} = K_c$$

58. Ethylene, C_2H_4, is one of the organic substances found in the air we breathe. It reacts with ozone in an endothermic reaction to form formaldehyde, CH_2O, which is one of the substances in smoggy air that cause eye irritation.

$$2C_2H_4\,(g) + 2O_3(g) + \text{energy} \rightleftharpoons 4CH_2O(g) + O_2(g)$$

a. Why does the forward reaction take place more rapidly in Los Angeles than in a wilderness area of Montana with the same air temperature?

Los Angeles has a much higher ozone concentration than in the Montana wilderness.

b. For a variety of reasons, natural systems rarely reach equilibrium, but if this reaction was run in the laboratory, would increased temperature for the reaction at equilibrium shift the reaction to more reactants or more products?

Toward more products (Increased temperature favors the endothermic direction of reversible reactions.)

60. When the temperature of an equilibrium system for the following reaction is increased, the reaction shifts toward more reactants. Is the reaction endothermic or exothermic?

$$H_2(g) + Br_2(g) \rightleftharpoons 2HBr(g)$$

Increased temperature favors the endothermic direction of reversible reactions, so this reaction is endothermic in the reverse direction and **exothermic** in the forward direction.

62. Assume that you are picking up a few extra dollars to pay for textbooks by acting as a trainer's assistant for a heavyweight boxer. One of your jobs is to wave smelling salts under the nose of the fighter to clear his head between rounds. The smelling salts are ammonium carbonate, which decomposes in the following reaction to yield ammonia. The ammonia does the wakeup job. Suppose the fighter gets a particularly nasty punch to the head and needs an extra jolt to be brought back to his senses. How could you shift the following equilibrium to the right to yield more ammonia?

$$(NH_4)_2CO_3(s) + \text{energy} \rightleftharpoons 2NH_3(g) + CO_2(g) + H_2O(g)$$

Increased temperature will drive this endothermic reaction toward products, so warming the smelling salt container in your hands will increase the amount of ammonia released.

64. Formaldehyde, CH_2O, is one of the components of embalming fluids and has been used to make foam insulation and plywood. It can be made from methanol, CH_3OH (often called wood alcohol). The heat of reaction for the combination of gaseous methanol and oxygen gas to form gaseous formaldehyde and water vapor is −199.32 kJ per mole of CH_2O formed, so the reaction is exothermic.

$$2CH_3OH(g) + O_2(g) \rightleftharpoons CH_2O(g) + 2H_2O(g)$$

a. Increased temperature drives the reaction toward reactants and lowers the value for the equilibrium constant. Explain why this is true. *(Objs #27 & 28)*

Increased temperature increases the rate of both the forward and the reverse reactions, but it increases the rate of the endergonic reaction more than it increases the rate of the exergonic reaction. Therefore, changing the temperature of a chemical system at equilibrium will disrupt the balance of the forward and reverse rates of reaction and shift the system in the direction of the endergonic reaction. Because this reaction is exothermic in the forward direction, it must be endothermic in the reverse direction. Increased temperature shifts the system toward more reactants, decreasing the ratio of products to reactants and, therefore, decreasing the equilibrium constant.

b. This reaction is run by the chemical industry at 450-900 °C, even though the equilibrium ratio of product to reactant concentrations is lower than at room temperature. Explain why this exothermic chemical reaction is run at high temperature despite this fact. *(Obj #29)*

To maximize the percentage yield at equilibrium, the reaction should be run at as low a temperature as possible, but at low temperature, the rates of the forward and reverse reactions are both very low, so it takes a long time for the system to come to equilibrium. In this case, it is best to run the reaction at high temperature to get to equilibrium quickly. (The unreacted methanol can be recycled back into the original reaction vessel after the formaldehyde has been removed from the product mixture.)

Section 16.4 Disruption of Equilibrium

66. Urea, NH_2CONH_2, is an important substance in the production of fertilizers. The equation shown below describes an industrial reaction that produces urea. The heat of reaction is – 135.7 kJ per mole of urea formed. Predict whether each of the following changes in the equilibrium system will shift the system to more products, to more reactants, or to neither. Explain each answer in two ways: (1) by applying Le Châtelier's principle and (2) by describing the effect of the change on the forward and reverse reaction rates. *(Obj. #30-33)*

$$2NH_3(g) + CO_2(g) \rightleftharpoons NH_2CONH_2(s) + H_2O(g) + 135.7 \text{ kJ}$$

a. The concentration of NH_3 is increased by the addition of more NH_3. (In the industrial production of urea, an excess of ammonia is added so that the ratio of NH_3 to CO_2 is 3:1.)

- Using Le Châtelier's Principle, we predict that the system will **shift to more products** to partially counteract the increase in NH_3.

- The increase in the concentration of ammonia speeds the forward reaction without initially affecting the rate of the reverse reaction. The equilibrium is disrupted, and the system **shifts to more products** because the forward rate is greater than the reverse rate.

b. The concentration of $H_2O(g)$ is decreased by removing water vapor.

- Using Le Châtelier's Principle, we predict that the system will **shift to more products** to partially counteract the decrease in H_2O.

- The decrease in the concentration of $H_2O(g)$ slows the reverse reaction without initially affecting the rate of the forward reaction. The equilibrium is disrupted, and the system **shifts toward more products** because the forward rate is greater than the reverse rate.

c. The temperature is increased from 25 °C to 190 °C. (In the industrial production of urea, ammonia and carbon dioxide are heated to 190 °C.)

- Using Le Châtelier's Principle, we predict that the system shifts in the endothermic direction to partially counteract the increase in temperature. Because the forward reaction is exothermic, the reverse reaction must be endothermic. As the system **shifts toward more reactants**, energy is absorbed, and the temperature decreases.

- The increased temperature increases the rates of both the forward and reverse reactions, but it has a greater effect on the endothermic reaction. Thus, the system **shifts toward more reactants** because the reverse rate becomes greater than the forward rate.

68. Hydriodic acid, which is used to make pharmaceuticals, is made from hydrogen iodide. The hydrogen iodide is made from hydrogen gas and iodine gas in the following exothermic reaction.

$$H_2(g) + I_2(g) \rightleftharpoons 2HI(g) + 9.4\,kJ$$

What changes could you make for this reaction at equilibrium to shift the reaction to the right and maximize the concentration of hydrogen iodide in the final product mixture?

The **addition of either H$_2$ or I$_2$ (or both)** would increase the concentrations of reactants, increasing the rate of collision between them, increasing the forward rate, and shifting the system toward more product.

Lower temperature favors the exothermic direction of the reaction, so **lower temperature** would shift this reaction to a higher percentage of products at equilibrium.

70. Phosgene gas, $COCl_2$, which is a very toxic substance used to make pesticides and herbicides, is made by passing carbon monoxide gas and chlorine gas over solid carbon, which acts as a catalyst.

$$CO(g) + Cl_2(g) \rightleftharpoons COCl_2(g)$$

If the carbon monoxide concentration is increased by adding CO to an equilibrium system of this reaction, what effect, if any, does it have on the following? (Assume constant temperature.)

   a. The concentration of $COCl_2$ after the system has shifted to come to a new equilibrium.

   The system will shift toward products, which leads to **increased COCl$_2$**.

   b. The concentration of Cl_2 after the system has shifted to come to a new equilibrium.

   The system will shift toward products, which leads to **decreased Cl$_2$**.

   c. The equilibrium constant for the reaction.

   Equilibrium constants are unaffected by reactant and product concentrations, so the **equilibrium constant remains the same**.

Chapter 17
An Introduction to Organic Chemistry, Biochemistry, and Synthetic Polymers

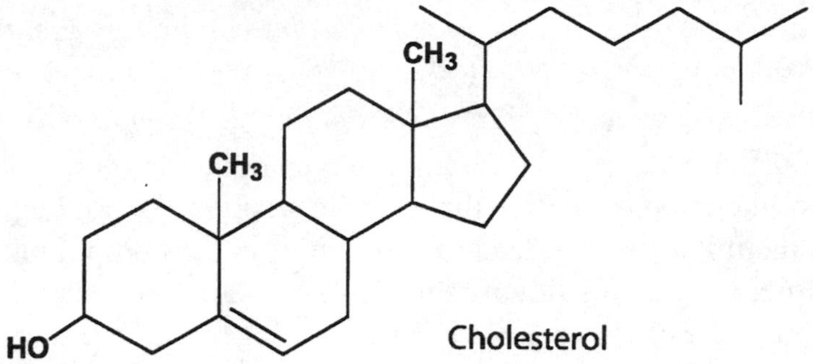

CH₃

CH₃

HO

Cholesterol

♦ Review Skills

17.1 Organic Compounds
- Formulas for Organic Compounds
- Alkanes
- Alkenes
- Alkynes
- Arenes (Aromatics)
- Alcohols
- Carboxylic Acids
- Ethers
- Aldehydes
- Ketones
- Esters
- Amines
- Amides
- Organic Compounds with More Than One Functional Group

  Special Topic 17.1: Rehabilitation of Old Drugs and Development of New Ones

17.2 Important Substances in Food
- Carbohydrates
- Amino Acids and Protein
- Fat

  Special Topic 17.2: Olestra
  Special Topic 17.3: Harmless Dietary Supplements or Dangerous Drugs?
- Steroids

17.3 Digestion
- Digestive Enzymes
- Digestion of Protein

  Internet: Chymotrypsin Protein Hydrolysis

17.4 Synthetic Polymers
- Nylon, a Synthetic Polypeptide
- Polyesters
- Addition Polymers

  Internet: Addition (Chain-growth) Polymers

  Special Topic 17.4: Recycling Synthetic Polymers

♦ Chapter Glossary

  Internet: Glossary Quiz

♦ Chapter Objectives

Review Questions

Key Ideas

Chapter Problems

Section Goals and Introductions

Section 17.1 Organic Compounds
Goals

- *To describe carbon-based compounds, called organic compounds.*
- *To describe the different ways that organic molecules can be represented and show you how to convert from one way to the others.*
- *To show how you can recognize different types of organic compounds.*

There are millions of different organic (carbon-based) compounds. The task of studying them becomes much easier when you recognize that organic compounds can be categorized according to structural similarities that lead to similarities in the compounds' important properties. For example, instead of studying the alcohols methanol, ethanol, and 2-propanol separately, you can study the characteristics of alcohols in general, because all alcohols have very similar characteristics. This section introduces you to some of the different types of organic compounds, shows you how your can recognize substances in each category, and shows you several ways of describing the structures of organic compounds.

Section 17.2 Important Substances in Food
Goal: To describe the different types of chemicals found in our food: carbohydrates, amino acids and proteins, fats and oils (triglycerides), and steroids.

Your understanding of organic compounds can be applied to understanding biomolecules, which are organic compounds that are important in biological systems. Like the organic compounds described in Section 17.1, recognizing that biomolecules can be placed in categories facilitates learning about them. You will learn about the structures of biomolecules in the categories of carbohydrates, amino acids, proteins, triglycerides, and steroids.

Section 17.3 Digestion
Goal: To describe the chemical changes that take place in digestion.

This section gives you a glimpse at the subject of biochemistry by describing some of the chemical changes of digestion. This includes a brief description of how enzymes facilitate this process. See the section on our Web site, called *Chymotrypsin Protein Hydrolysis*, that describes a proposed mechanism for an enzyme reaction.

www.chemplace.com/college/

Section 17.4 Synthetic Polymers
Goals

- *To describe synthetic polymers, including nylon, polyester, polyethylene, polypropylene, poly(vinyl chloride), and polystyrene.*
- *To describe the recycling of synthetic polymers.*

Scientists have developed ways of making many synthetic polymers that are similar to natural biomolecules. This section shows you how some of these polymers are made and describes their many different uses.

See the section on our Web site, called *Addition (Chain-growth) Polymers*, that provides more information on one type of polymer.

www.chemplace.com/college/

Chapter 17 Maps

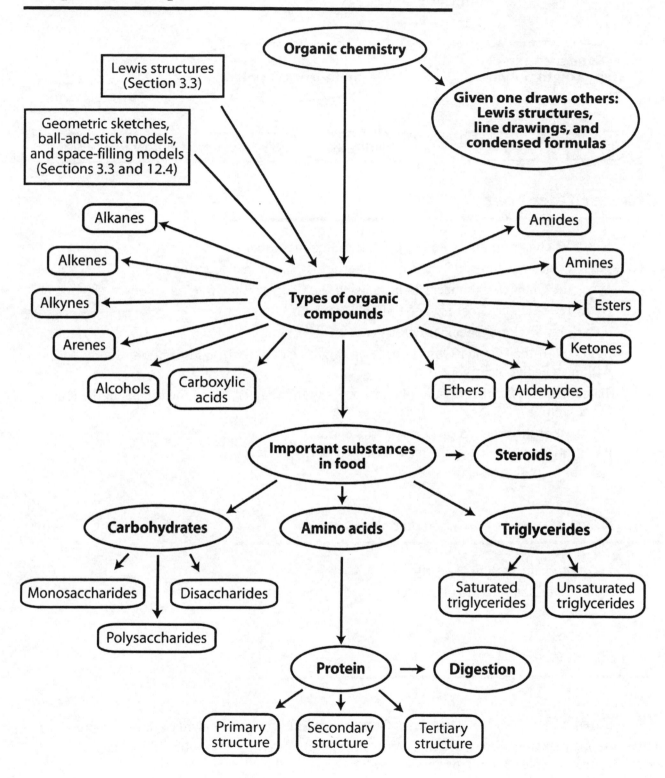

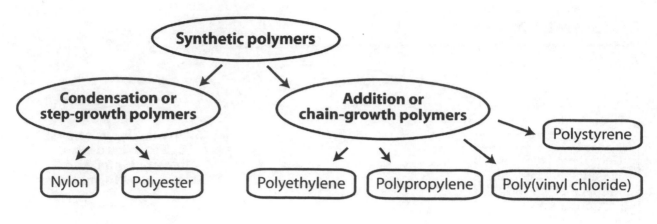

Chapter Checklist

☐ Read the chapter quickly before the lecture that describes it.

☐ Attend class meetings, take notes, and participate in class discussions.

☐ Work the Chapter Exercises, perhaps using the Chapter Examples as guides.

☐ Study the Chapter Glossary and test yourself on our Web site:

www.chemplace.com/college/

☐ Study all of the Chapter Objectives. You might want to write a description of how you will meet each objective.

☐ To get a review of the most important topics in the chapter, fill in the blanks in the Key Ideas section.

☐ Work all of the selected problems at the end of the chapter, and check your answers with the solutions provided in this chapter of the study guide.

☐ Ask for help if you need it.

Web Resources   www.chemplace.com/college/

Chymotrypsin Protein Hydrolysis
Addition (Chain-growth) Polymers
Glossary Quiz

Exercises Key

✍ Exercise 17.1 - Organic Compounds: Identify each of these structures as representing an alkane, alkene, alkyne, arene (aromatic), alcohol, carboxylic acid, ether, aldehyde, ketone, ester, amine, or amide. *(Obj #3)*

a. 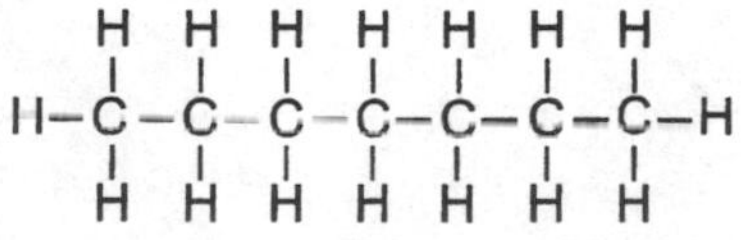
$$H-C-C-C-C-C-C-C-H$$

alkane

b.

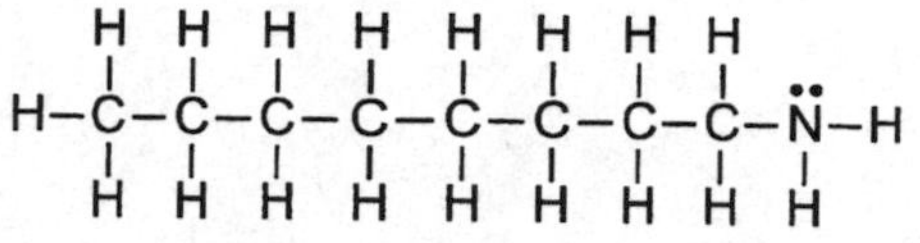

$$H-\overset{\overset{\displaystyle H}{|}}{C}-\overset{\overset{\displaystyle H}{|}}{C}-\overset{\overset{\displaystyle H}{|}}{C}-\overset{\overset{\displaystyle H}{|}}{C}-\overset{\overset{\displaystyle H}{|}}{C}-\overset{\overset{\displaystyle H}{|}}{C}-\overset{\overset{\displaystyle H}{|}}{C}-\overset{\overset{\displaystyle H}{|}}{C}-\overset{\overset{\displaystyle \cdot\cdot}{|}}{N}-H$$

amine

c.

$$H-\overset{\overset{\displaystyle H}{|}}{C}-\overset{\overset{\displaystyle H}{|}}{C}-\overset{\overset{\displaystyle H}{|}}{C}-\overset{\overset{\displaystyle H}{|}}{C}-\overset{\cdot\cdot}{\underset{\cdot\cdot}{O}}-\overset{\overset{\displaystyle H}{|}}{C}-\overset{\overset{\displaystyle H}{|}}{C}-\overset{\overset{\displaystyle H}{|}}{C}-\overset{\overset{\displaystyle H}{|}}{C}-H$$

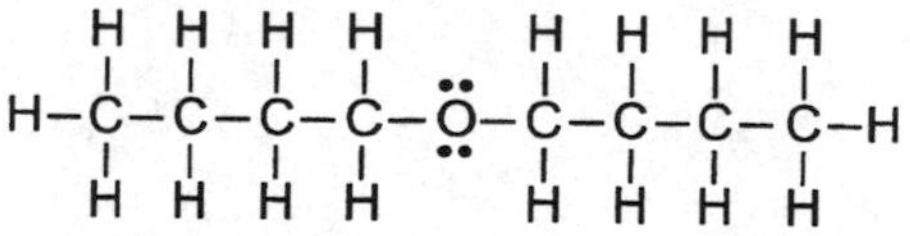

ether

d.

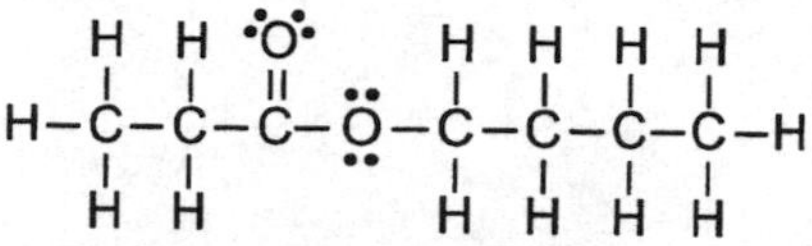

ester

e.

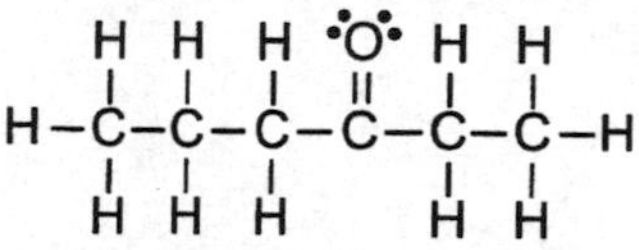

ketone

f.

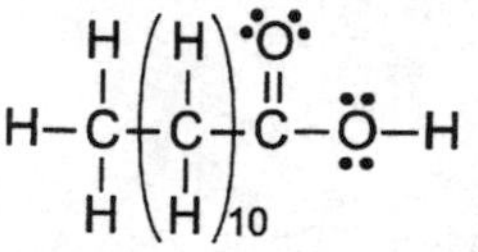

carboxylic acid

g.

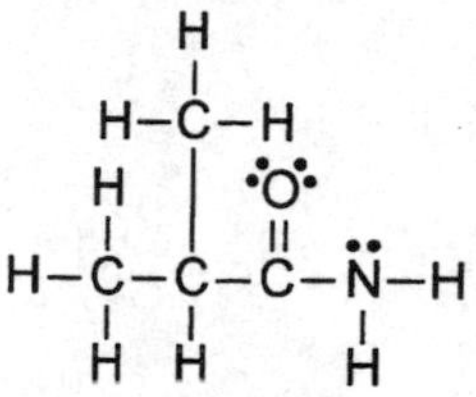

amide

h.

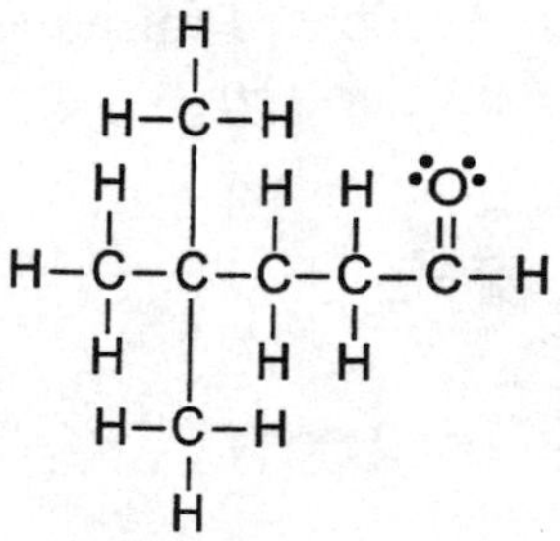

aldehyde

i.

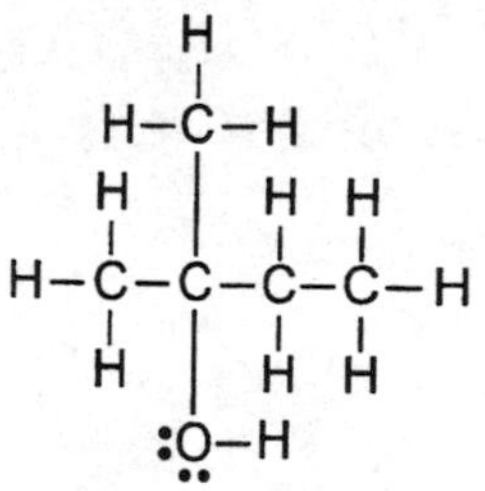

alcohol

j.

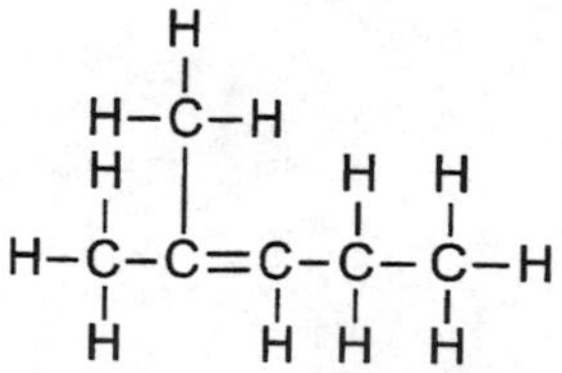

alkene

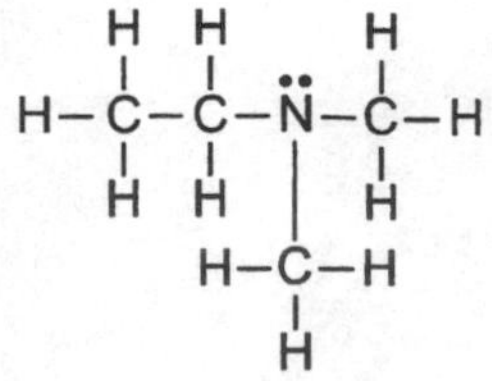

k.   **amine**

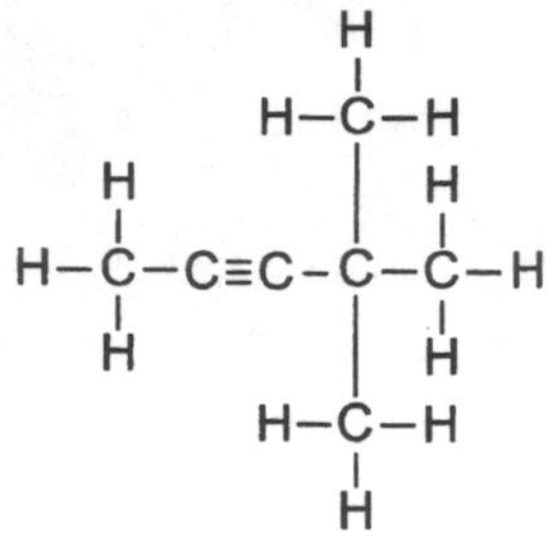

l.   **alkyne**

m.   **arene**

✍ **Exercise 17.2 - Condensed Formulas:**   Write condensed formulas to represent the Lewis structures in parts (a) through (l) of Exercise 17.1.   *(Obj #2)*

    a. $CH_3CH_2CH_2CH_2CH_2CH_2CH_3$ or $CH_3(CH_2)_5CH_3$

    b. $CH_3CH_2CH_2CH_2CH_2CH_2CH_2CH_2NH_2$ or $CH_3(CH_2)_7NH_2$

    c. $CH_3CH_2CH_2CH_2OCH_2CH_2CH_2CH_3$

    d. $CH_3CH_2CO_2CH_2CH_2CH_2CH_3$ or $CH_3CH_2COOCH_2CH_2CH_2CH_3$

    e. $CH_3CH_2CH_2COCH_2CH_3$

    f. $CH_3CH_2CH_2CH_2CH_2CH_2CH_2CH_2CH_2CH_2CH_2CO_2H$ or $CH_3(CH_2)_{10}CO_2H$

    g. or $CH_3CH_2CH_2CH_2CH_2CH_2CH_2CH_2CH_2CH_2CH_2COOH$ or $CH_3(CH_2)_{10}COOH$

    h. $(CH_3)_2CHCONH_2$

    i. $(CH_3)_3CCH_2CH_2CHO$

    j. $(CH_3)_2C(OH)CH_2CH_3$

    k. $(CH_3)_2CCHCH_2CH_3$

    l. $CH_3CH_2N(CH_3)_2$

    m. $CH_3CCC(CH_3)_3$

Exercise 17.3 - Line Drawings: Make line drawings that represent the Lewis structures in parts (a) through (j) of Exercise 17.1. *(Obj #2)*

a.

b.

c.

d.

e.

f.

g.

h.

i.

j.

Review Questions Key

1. Draw a Lewis structure, a geometric sketch, a ball-and-stick model, and a space-filling model for methane, CH₄.

    See Figure 12.4 on page 514 of the textbook.

2. Draw a Lewis structure, a geometric sketch, a ball-and-stick model, and a space-filling model for ammonia, NH₃.

    H–N̈–H
      |
      H   *See Figure 3.11 on page 113 of the textbook.*

3. Draw a Lewis structure, a geometric sketch, a ball-and-stick model, and a space-filling model for water, H₂O.

    H–Ö–H   *See Figure 3.12 on page 114 of the textbook.*

4. Draw a Lewis structure, a geometric sketch, a ball-and-stick model, and a space-filling model for methanol, CH₃OH.

    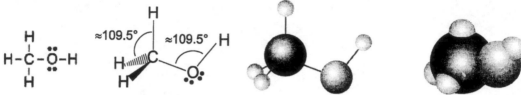

    Lewis structure   geometric sketch   ball-and-stick model   space-filling model

5. The following Lewis structure represents a molecule of formaldehyde, CH₂O. Draw a geometric sketch, a ball-and-stick model, and a space-filling model for this molecule.

    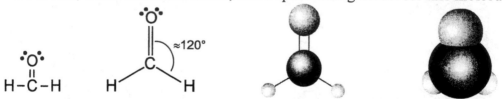

    Lewis structure   geometric sketch   ball-and-stick model   space-filling model

6. The following Lewis structure represents a molecule of hydrogen cyanide, HCN. Draw a geometric sketch, a ball-and-stick model, and a space-filling model for this molecule.

    Lewis structure   geometric sketch   ball-and-stick model   space-filling model

7. The following Lewis structure represents a molecule of ethanamide, CH₃CONH₂. Draw a geometric sketch for this molecule.

    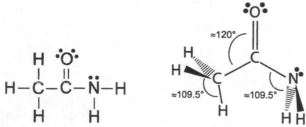

Key Ideas Answers

8. Hydrocarbons (compounds composed of carbon and hydrogen) in which all of the carbon-carbon bonds are **single** bonds are called alkanes.

10. When a(n) **small section** of an organic molecule is largely responsible for the molecule's chemical and physical characteristics, that section is called a functional group.

12. Compounds that contain the **benzene ring** are called arenes or aromatics.

14. Ethers consist of two **hydrocarbon groups** surrounding an oxygen atom.

16. Sugars are monosaccharides and **disaccharides**. Starches and cellulose are **polysaccharides**.

18. Maltose, a disaccharide consisting of two **glucose** units.

20. Sucrose is a disaccharide that contains glucose and **fructose**.

22. Almost every kind of plant cell has **energy** stored in the form of starch. Starch itself has two general forms, **amylose** and **amylopectin**.

24. All the polysaccharides are polymers, a general name for large molecules composed of **repeating units**, called monomers.

26. Protein molecules are polymers composed of **monomers** called **amino acids**.

28. A condensation reaction is a chemical change in which **water** or other small molecules are released.

30. The arrangement of atoms that are **close to each other** in the polypeptide chain is called the secondary structure of the protein.

32. The fat stored in our bodies is our primary **long-term** energy source.

34. A process called **hydrogenation** converts liquid triglycerides to solid triglycerides by adding hydrogen atoms to the double bonds and so converting them to single bonds.

36. A triglyceride that still has one or more carbon-carbon **double bonds** is an unsaturated triglyceride.

38. As the starting material for the production of many important body chemicals, including hormones (compounds that help regulate chemical changes in the body), the steroid **cholesterol** is necessary for normal, healthy functioning of our bodies.

40. In digestion, disaccharides are broken down into **monosaccharides** (glucose, galactose, and fructose), polysaccharides into glucose, **protein** into amino acids, and fat into **glycerol** and **fatty acids**.

42. The digestion of proteins begins in the stomach. The **acidic** conditions there weaken the links that maintain the protein molecules' tertiary structure. This process is called **denaturation**, because the loss of tertiary structure causes a corresponding loss of the protein's "natural" function.

44. For an enzyme-mediated reaction to take place, the reacting molecule or molecules, which are called **substrates**, must fit into a specific section of the enzyme's structure called the **active site**. A frequently used analogy for the relationship of substrate to active site is the way a key must fit into a lock in order to do its job. Each active site has (1) a(n) **shape** that fits a specific substrate or substrates only, (2) side chains that attract the enzyme's particular substrate(s), and (3) side chains specifically positioned to speed the reaction.

46. When small molecules, such as water, are released in the formation of a polymer, the polymer is called a condensation (or sometimes **step-growth**) polymer.

48. Polyesters are made from the reaction of a(n) **diol** (a compound with two alcohol functional groups) with a di-carboxylic acid.

Problems Key

Section 17.1 Organic Compounds

50. Classify each of the following as organic or inorganic (not organic) compounds.

    a. sodium chloride, NaCl, in table salt   **inorganic**

    b. hexane, C_6H_{14}, in gasoline   **organic**

    c. ethyl butanoate, $CH_3CH_2CH_2CO_2CH_2CH_3$, in a pineapple   **organic**

    d. water, H_2O, in your body   **inorganic**

52. Identify each of these Lewis structures as representing an alkane, alkene, alkyne, arene (aromatic), alcohol, carboxylic acid, aldehyde, ketone, ether, ester, amine, or amide. *(Obj #3)*

a. **ketone**   b. **alkane**

c. **carboxylic acid**   d. **amide**

e. **ether**   f. **aldehyde**

g. **alkene**   h. **ester**

i. **alcohol**   j. **alkyne**

k.   arene

54. Write condensed chemical formulas to represent the Lewis structures in parts (a) through (j) of Problem 52. (For example, 2-propanol can be described as $CH_3CH(OH)CH_3$.) *(Obj #2)*

  a. $CH_3(CH_2)_4COCH_3$ or $CH_3CH_2CH_2CH_2CH_2COCH_3$
  b. $CH_3CH_2CH(CH_3)CH_2CH(CH_3)_2$
      or $CH_3CH_2CH(CH_3)CH_2CH(CH_3)CH_3$
  c. $CH_3(CH_2)_{12}COOH$ or $CH_3(CH_2)_{12}CO_2H$
  d. $CH_3CH_2CH_2CONH_2$
  e. $CH_3CH_2OCH(CH_3)_2$ or $CH_3CH_2OCH(CH_3)CH_3$
  f. $(CH_3)_2CHCHO$ or $CH_3CH(CH_3)CHO$
  g. $CH_2C(CH_3)CHCH_2$
  h. $CH_3CH_2COOCH_3$ or $CH_3CH_2CO_2CH_3$
  i. $CH_3CH_2CH_2CH(OH) CH(CH_2OH)CH_2CH_3$
  j. CH_3CCCH_3

56. Write line drawings to represent the Lewis structures in Parts (a) through (i) of Problem 52. *(Obj #2)*

a.

b.

c.

d.

e.

f.

g.

h.

i.

58. The chemical structure of the artificial sweetener aspartame is below. Identify all of the organic functional groups that it contains.

60. Draw geometric sketches, including bond angles, for each of the following organic molecules.

a.

b.

c. $H-C\equiv N\colon$ $H-C\equiv N\colon$ 180°

63. Because the structure for a particular alkane can be drawn in different ways, two drawings of the same substance can look like isomers. Are each of the following pairs isomers or different representations of the same thing?

a. 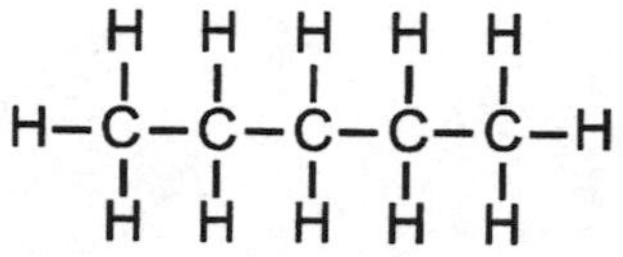
$$H-\overset{\overset{H}{|}}{\underset{\underset{H}{|}}{C}}-\overset{\overset{H}{|}}{\underset{\underset{H}{|}}{C}}-\overset{\overset{H}{|}}{\underset{\underset{H}{|}}{C}}-\overset{\overset{H}{|}}{\underset{\underset{H}{|}}{C}}-\overset{\overset{H}{|}}{\underset{\underset{H}{|}}{C}}-H$$
and
$$H-\overset{\overset{H}{|}}{\underset{\underset{H}{|}}{C}}-\overset{\overset{H-C-H}{|}}{\underset{\underset{H}{|}}{C}}-\overset{\overset{H}{|}}{\underset{\underset{H}{|}}{C}}-\overset{\overset{H}{|}}{\underset{\underset{H}{|}}{C}}-H$$
isomers

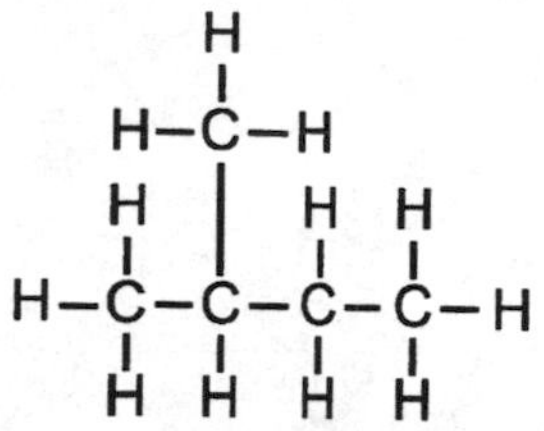

b. 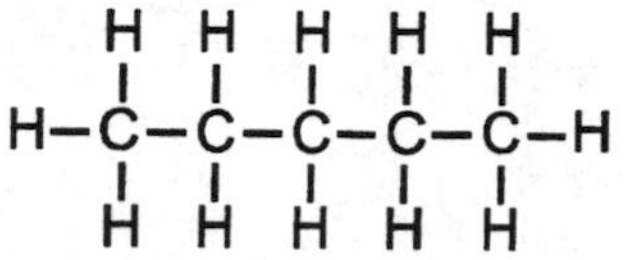
$$H-\overset{\overset{H}{|}}{\underset{\underset{H}{|}}{C}}-\overset{\overset{H}{|}}{\underset{\underset{H}{|}}{C}}-\overset{\overset{H}{|}}{\underset{\underset{H}{|}}{C}}-\overset{\overset{H}{|}}{\underset{\underset{H}{|}}{C}}-\overset{\overset{H}{|}}{\underset{\underset{H}{|}}{C}}-H$$
and
$$H-\overset{\overset{H}{|}}{\underset{\underset{H}{|}}{C}}-\overset{\overset{H-C-H}{|}}{\underset{\underset{H}{|}}{C}}-\overset{\overset{H}{|}}{\underset{\underset{H}{|}}{C}}-\overset{\overset{H}{|}}{\underset{\underset{H}{|}}{C}}-H$$
same

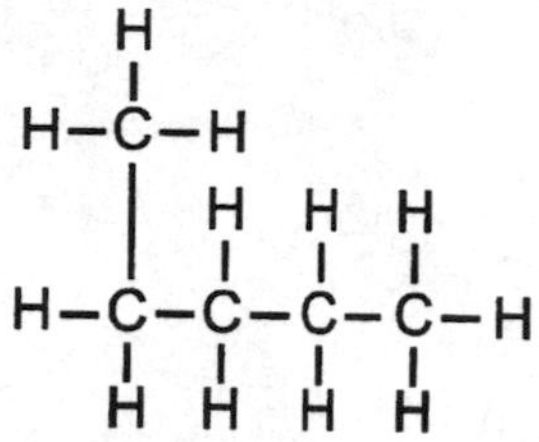

c.  and 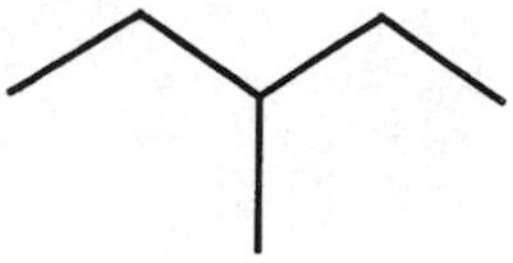 **isomers**

d. 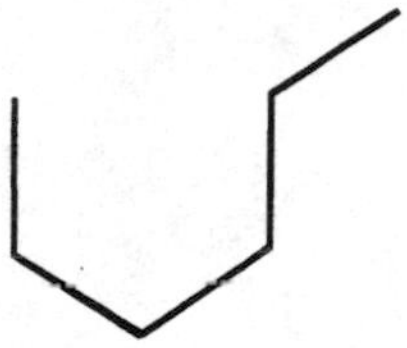 and  **same**

65. Draw line drawings for three isomers of C_5H_{12}.

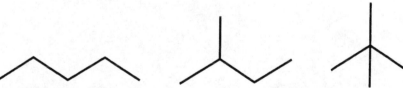

67. Two of the three isomers of C_3H_8O are alcohols and one is an ether. Draw condensed structures for these three isomers.

$CH_3CH_2CH_2OH$ $CH_3CH(OH)CH_3$ $CH_3OCH_2CH_3$

69. Draw a Lewis structure for an isomer of C_2H_5NO that is an amide, and draw a second Lewis structure for a second isomer of C_2H_5NO that has both an amine functional group and an aldehyde functional group.

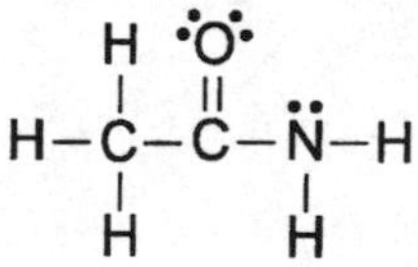

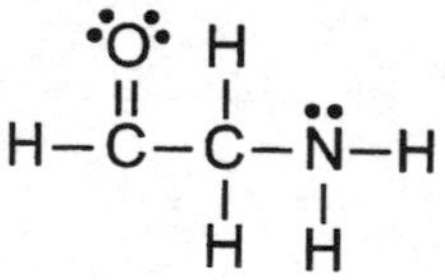

Section 17.2 Important Substances in Food

72. Identify each of the following structures as representing a carbohydrate, amino acid, peptide, triglyceride, or steroid. *(Obj #4)*

a. 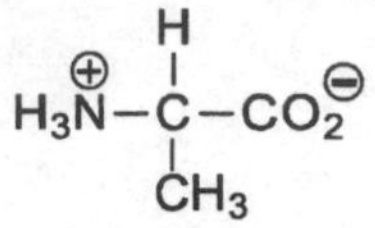 **amino acid**

b. 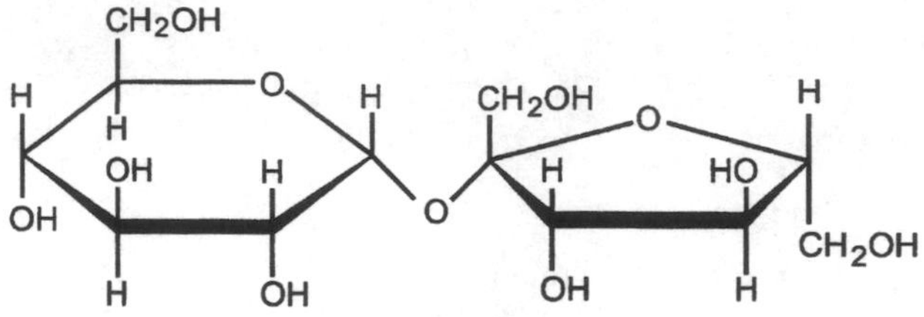 **carbohydrate**

c. 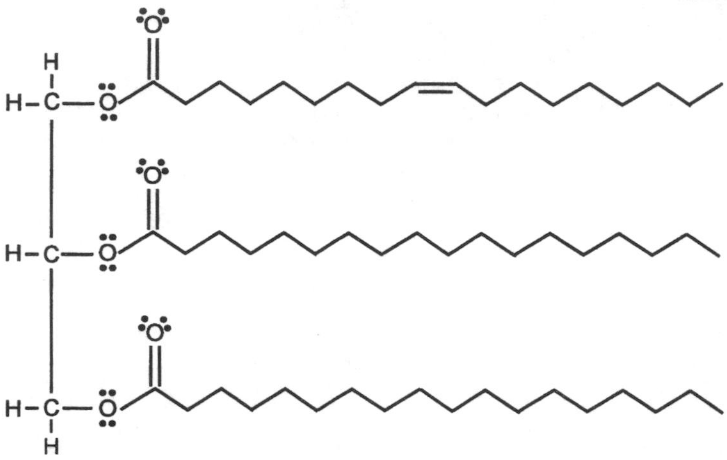 **triglyceride**

d. 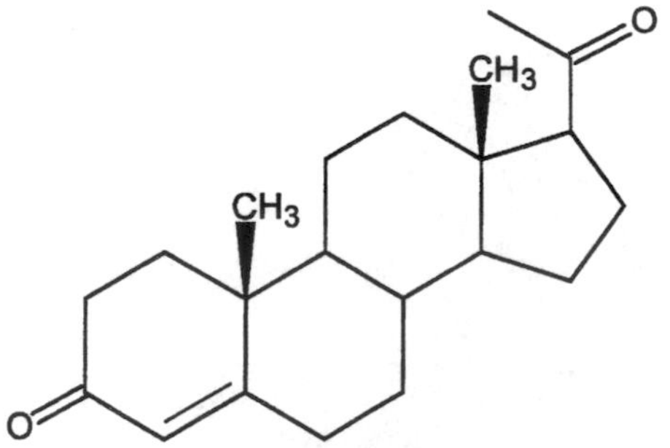 **steroid**

e.  **peptide**

74. Identify each of the following structures as representing a monosaccharide, disaccharide, or polysaccharide. *(Obj #5)*

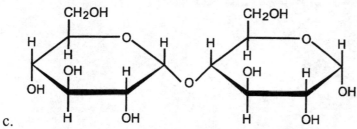

a.

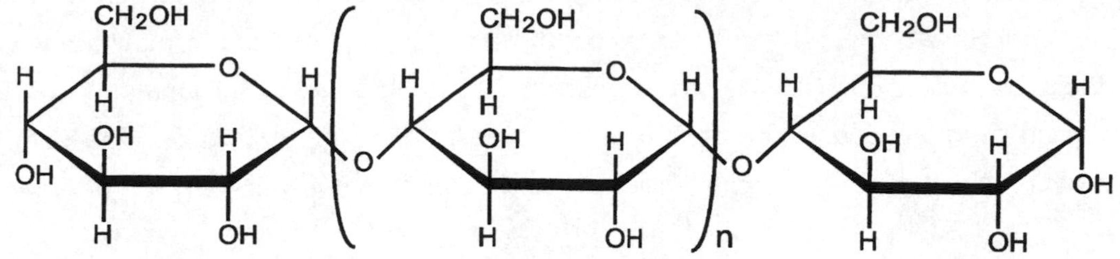

polysaccharide

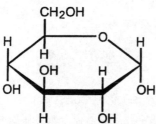

b. 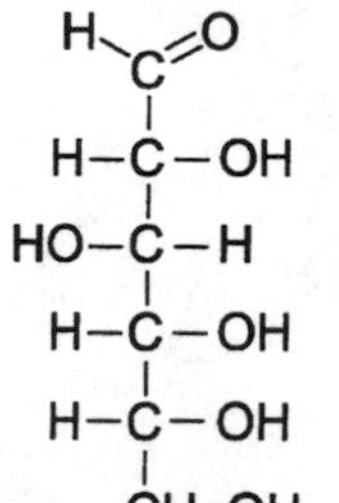 **monosaccharide**

c. **disaccharide**

d. **monosaccharide**

76. Identify each of the following as a monosaccharide, disaccharide, or polysaccharide.

a. maltose   **disaccharide**

b. fructose   **monosaccharide**

c. amylose   **polysaccharide**

d. cellulose   **polysaccharide**

78. Describe the general difference between glucose and galactose. *(Obj #6)*

Glucose and galactose differ in the relative positions of an –H and an –OH on one of their carbon atoms. In the standard notation for the open-chain form, glucose and galactose differ only in the relative position of the –H and –OH groups on the fourth carbon from the top. In the standard notation for the ring structures, the –OH group is down on the number 4 carbon of glucose and up on the number 4 carbon of galactose.

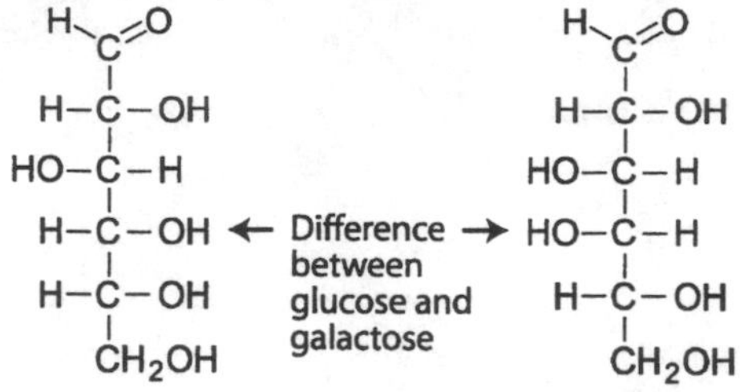

glucose galactose

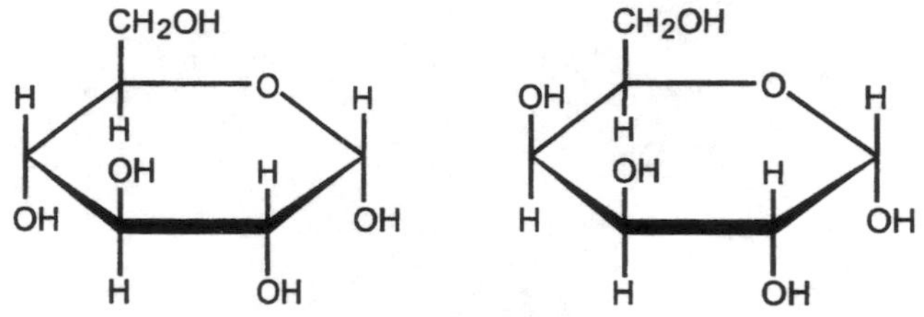
glucose galactose

80. What saccharide units form maltose, lactose, and sucrose? *(Obj #7)*

maltose – 2 glucose units
lactose – glucose and galactose
sucrose – glucose and fructose

82. Describe the similarities and differences between starches (such as amylose, amylopectin, and glycogen) and cellulose. *(Obj #8)*

Starch and cellulose molecules are composed of many glucose molecules linked together, but cellulose has different linkages between the molecules than starch. See Figure 17.21 on page 741 in the textbook.

84. Explain why glycine amino acid molecules in our bodies are usually found in the second form shown below rather than in the first. *(Obj #11)*

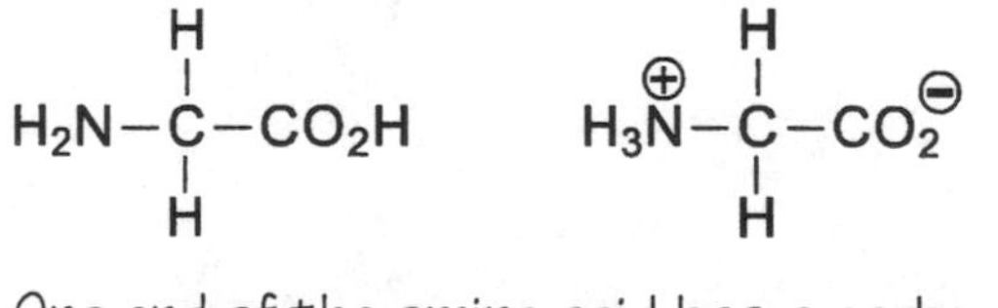

One end of the amino acid has a carboxylic acid group that tends to lose an H^+ ion, and the other end has a basic amine group that attracts H^+ ions. Therefore, in the conditions found in our bodies, amino acids are likely to be in the second form.

85. Using Figure 17.22 on pages 743 and 744 of the textbook, draw the Lewis structure of the dipeptide that has alanine combined with serine. Circle the peptide bond in your structure.

87. Show how the amino acids leucine, phenylalanine, and threonine can be linked together to form the tripeptide leu-phe-thr. *(Obj #12)*

89. When the artificial sweetener aspartame is digested, it yields methanol as well as the amino acids aspartic acid and phenylalanine. Although methanol is toxic, the extremely low levels introduced into the body by eating aspartame are not considered dangerous, but for people who suffer from phenylketonuria (PKU), the phenylalanine can cause severe mental retardation. Babies are tested for this disorder at birth, and when it is detected, they are placed on diets that are low in phenylalanine. Using Figure 17.22 on pages 743 and 744 of the textbook, identify the portions of aspartame's structure that yield aspartic acid, phenylalanine, and methanol.

91. Describe how disulfide bonds, hydrogen bonds, and salt bridges help hold protein molecules together in specific tertiary structures. *(Obj #14)*

Each of these interactions draw specific amino acids in a protein chain close together, leading to a specific shape of the protein molecule. Disulfide bonds are covalent bonds between sulfur atoms from two cysteine amino acids (Figure 17.28 on page 747 in the textbook). Hydrogen bonding forms between −OH groups in two amino acids, like serine or threonine, in a protein chain (Figure 17.29 on page 747 in the textbook). Salt bridges are attractions between negatively charged side chains and positively charged side chains.

For example, the carboxylic acid group of an aspartic acid side chain can lose its H⁺, leaving the side chain with a negative charge. The basic side chain of a lysine amino acid can gain an H⁺ and a positive charge. When these two charges form, the negatively charged aspartic acid is attracted to the positively charged lysine by a salt bridge (Figure 17.30 on page 747 in the textbook).

94. Draw the structure of the triglyceride that would form from the complete hydrogenation of the following triglyceride. *(Obj #19)*

Section 17.3 Digestion

96. When you wash some fried potatoes down with a glass of milk, you deliver a lot of different nutritive substances to your digestive tract, including lactose (a disaccharide), protein, and fat from the milk and starch from the potatoes. What are the digestion products of disaccharides, polysaccharides, protein, and fat? *(Obj #20)*

   Disaccharides – monosaccharides (glucose and galactose from lactose)
   Polysaccharides – glucose
   Protein – amino acids
   Fat – glycerol and fatty acids

98. Explain why each enzyme acts only on a specific molecule or a specific type of molecule. *(Obj #22)*

   Before an enzyme reaction takes place, the molecule or molecules that are going to react (called substrates) must fit into a specific section of the protein structure called the active site. Because the active site has a shape that fits specific substrates, because it has side chains that attract particular substrates, and because it has side chains in distinct positions that speed the reaction, each enzyme will only act on a specific molecule or a specific type of molecule.

Section 17.4 Synthetic Polymers

100. Explain why Nylon 66 is stronger than Nylon 610. *(Obj #24)*

One of the reasons for the exceptional strength of nylon is the hydrogen bonding between amide functional groups. A higher percentage of amide functional groups in nylon molecules' structures leads to stronger hydrogen bonds between them. Thus, changing the number of carbon atoms in the diamine and in the di-carboxylic acid changes the properties of nylon. Nylon 610, which has four more carbon atoms in the di-carboxylic acid molecules that form it than for Nylon 66, is somewhat weaker than Nylon 66 and has a lower melting point.

102. Describe the similarities and differences between the molecular structures of low-density polyethylene (LDPE) and high-density polyethylene (HDPE). *(Obj #26)*

Polyethylene molecules can be made using different techniques. One process leads to branches that keep the molecules from fitting closely together. Other techniques have been developed to make polyethylene molecules with very few branches. These straight-chain molecules fit together more efficiently, yielding a high-density polyethylene, HDPE, that is more opaque, harder, and stronger than the low-density polyethylene, LDPE.

104. Both ethylene and polyethylene are composed of nonpolar molecules. Explain why ethylene is a gas at room temperature while polyethylene is a solid at the same temperature.

Nonpolar molecules are attracted to each other by London forces, and increased size of molecules leads to stronger London forces. Polyethylene molecules are much larger than the ethylene molecules that are used to make polyethylene, so polyethylene molecules have much stronger attractions between them, making them solids at room temperature.

Chapter 18
Nuclear Chemistry

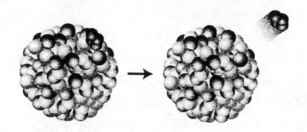

♦ Review Skills

18.1 The Nucleus and Radioactivity

- Nuclear Stability
- Types of Radioactive Emissions
- Nuclear Reactions and Nuclear Equations
- Rates of Radioactive Decay
- Radioactive Decay Series
- The Effect of Radiation on the Body

18.2 Uses of Radioactive Substances

- Medical Uses
- Carbon-14 Dating
- Other Uses for Radioactive Nuclides

18.3 Nuclear Energy

- Nuclear Fission and Electric Power Plants
- Nuclear Fusion and the Sun
  Special Topic 18.1: A New Treatment for Brain Cancer
  Special Topic 18.2: The Origin of the Elements

♦ Chapter Glossary
  Internet: Glossary Quiz

♦ Chapter Objectives

Review Questions

Key Ideas

Chapter Problems

Section Goals and Introductions

Section 18.1 The Nucleus and Radioactivity
Goals

- *To introduce the new terms nucleon, nucleon number, and nuclide.*
- *To show the symbolism used to represent nuclides.*
- *To explain why some nuclei are stable and others not.*
- *To provide you with a way of predicting nuclear stability.*
- *To describe the different types of radioactive decay.*
- *To show how nuclear reactions are different from chemical reactions.*
- *To show how nuclear equations are different from chemical equations.*
- *To show how the rates of radioactive decay can be described with half-life.*
- *To explain why short-lived radioactive atoms are in nature.*
- *To describe how radiation affects our bodies..*

This section provides the basic information that you need to understand radioactive decay. It will also help you understand the many uses of radioactive atoms, including how they are used in medicine and in electricity generation.

Section 18.2 Uses of Radioactive Substances

Goal: To describe many of the uses of radioactive atoms, including medical uses, archaeological dating, smoke detectors, and food irradiation.

Radiation and radioactive substances have often been viewed as dangerous and things that create problems rather than solving them, but there are actually many important and beneficial uses of radioactive atoms. Some of them are described in this section.

Section 18.3 Nuclear Energy
Goals

- *To describe nuclear fission (the splitting of larger atoms into smaller atoms) and nuclear fusion (the combination of smaller atoms into larger atoms) and to explain why energy is released by each.*
- *To describe briefly how nuclear power plants work.*
- *To show how the sun gets its energy.*

The potential energy stored in the nuclei of atoms can be converted into other forms of energy in the fission reactions in nuclear power plants and in the fusion reactions in the sun. This section tells you why this energy conversion takes place.

Nuclear power is a major source of energy for electrical generation worldwide. Nuclear power plants are found in over 30 countries and generate about 17% of the world's electricity. This section will help you to understand how these plants work.

Chapter 18 Map

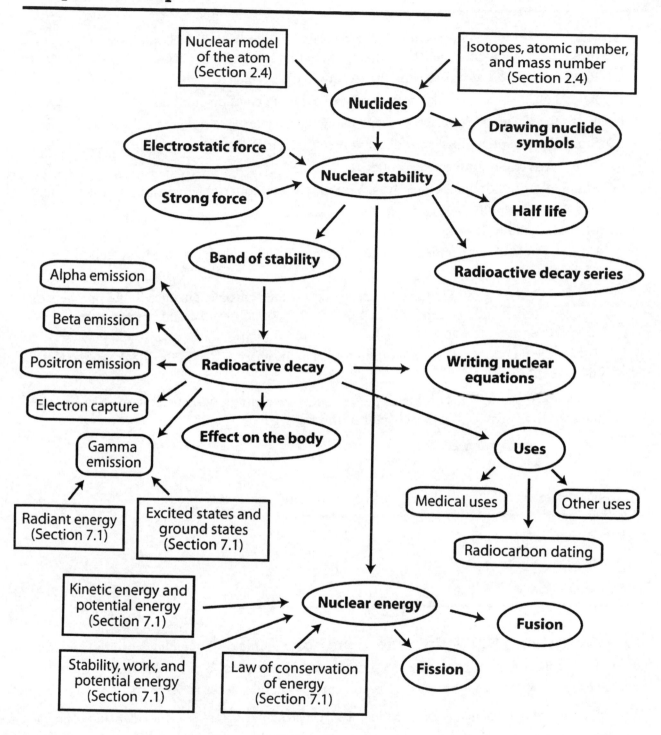

Chapter Checklist

☐ Read the Review Skills section. If there is any skill mentioned that you have not yet mastered, review the material on that topic before reading this chapter.

☐ Read the chapter quickly before the lecture that describes it.

☐ Attend class meetings, take notes, and participate in class discussions.

☐ Work the Chapter Exercises, perhaps using the Chapter Examples as guides.

☐ Study the Chapter Glossary and test yourself on our Web site:

www.chemplace.com/college/

☐ Study all of the Chapter Objectives. You might want to write a description of how you will meet each objective.

☐ Memorize the following.

Be sure to check with your instructor to determine how much you are expected to know of the following.

- For the types of radioactive decay, know the symbols, changes in the numbers of protons and neutrons, and changes in the atomic number and mass number described in Table 18.2.

☐ To get a review of the most important topics in the chapter, fill in the blanks in the Key Ideas section.

☐ Work all of the selected problems at the end of the chapter, and check your answers with the solutions provided in this chapter of the study guide.

☐ Ask for help if you need it.

Web Resources   www.chemplace.com/college/

Glossary Quiz

Exercises Key

Exercise 18.1 – Nuclide Symbols: One of the nuclides used in radiation therapy for the treatment of cancer has 39 protons and 51 neutrons. Write its nuclide symbol in the form of $_{Z}^{A}X$. Write two other ways to represent this nuclide. *(Objs #2, 3, & 4)*

Because this nuclide has 39 protons, its atomic number, Z, is 39. This identifies the element as yttrium. This nuclide of yttrium has 90 total nucleons (39 protons + 51 neutrons), so its nucleon number, A, is 90.

$_{39}^{90}Y$ ^{90}Y yttrium - 90

Exercise 18.2 – Nuclide Symbols: A nuclide with the symbol ^{201}Tl can be used to assess a patient's heart in a stress test. What is its atomic number and mass number? How many protons and how many neutrons are in the nucleus of each atom? Write two other ways to represent this nuclide. *(Objs #2, 3, & 4)*

The periodic table shows us that the atomic number for thallium is 81, so each thallium atom has 81 protons. The superscript in the symbol ^{201}Tl is this nuclide's mass number. The difference between the mass number (the sum of the numbers of protons and neutrons) and the atomic number (the number of protons) is equal to the number of neutrons, so this nuclide has 120 neutrons (201 – 81).

Atomic number = 81   mass number = 201   81 protons   120 neutrons

$$^{201}_{81}\text{Tl} \quad \text{thallium-201}$$

Exercise 18.3 - Nuclear Equations: Write nuclear equations for (a) alpha emission by plutonium-239, one of the substances formed in nuclear power plants, (b) beta emission by sodium-24, used to detect blood clots, (c) positron emission by oxygen-15, used to assess the efficiency of the lungs, and (d) electron capture by copper-64, used to diagnose lung disease. *(Obj #12)*

a. $^{239}_{94}\text{Pu} \quad \rightarrow \quad ^{235}_{92}\text{U} + ^{4}_{2}\text{He}$

b. $^{24}_{11}\text{Na} \quad \rightarrow \quad ^{24}_{12}\text{Mg} + ^{0}_{-1}e$

c. $^{15}_{8}\text{O} \quad \rightarrow \quad ^{15}_{7}\text{N} + ^{0}_{+1}e$

d. $^{64}_{29}\text{Cu} + ^{0}_{-1}e \quad \rightarrow \quad ^{64}_{28}\text{Ni}$

Exercise 18.4 - Nuclear Equations: Complete the following nuclear equations. *(Obj #13)*

a. $^{14}_{7}\text{N} + ^{4}_{2}\text{He} \quad \rightarrow \quad ^{17}_{8}\text{O} + ^{1}_{1}\text{H}$

b. $^{238}_{92}\text{U} + ^{14}_{7}\text{N} \quad \rightarrow \quad ^{247}_{99}\text{Es} + 5\,^{1}_{0}\text{n}$

c. $^{238}_{92}\text{U} + ^{2}_{1}\text{H} \quad \rightarrow \quad ^{239}_{93}\text{Np} + ^{1}_{0}\text{n}$

Exercise 18.5 - Half-life: One of the radioactive nuclides formed in nuclear power plants is hydrogen-3, called tritium, which has a half-life of 12.26 years. How long before a sample decreases to 1/8 of its original amount? *(Obj #14)*

In each half-life of a radioactive nuclide, the amount diminishes by one-half. The fraction 1/8 is ½ x ½ x ½, so it takes three half-lives to diminish to 1/8 remaining. Therefore, it will take **36.78 years** for tritium to decrease to 1/8 of what was originally there.

Exercise 18.6 - Half-life: Uranium-238 is one of the radioactive nuclides sometimes found in soil. It has a half-life of 4.51×10^9 years. What fraction of a sample is left after 9.02×10^9 years?   *(Obj #15)*

The length of time divided by the half-life yields the number of half-lives.

$$\frac{9.02 \times 10^9 \text{ years}}{4.51 \times 10^9 \text{ years}} = 2 \text{ half-lives}$$

Therefore, the fraction remaining would be **1/4** (½ × ½).

Review Questions Key

1. Describe the nuclear model of the atom, including the general location of the protons, neutrons, and electrons, the relative size of the nucleus compared to the size of the atom, and the modern description of the electron.

   Protons and neutrons are in a tiny core of the atom called the nucleus, which has a diameter about 1/100,000 the diameter of the atom. The position and motion of the electrons are uncertain, but they generate a negative charge that is felt in the space that surrounds the nucleus.

2. With reference to both their particle and their wave nature, describe the similarities and differences between gamma radiation and radio waves. Which has higher energy?

   In the particle view, radiant energy is a stream of tiny, massless packets of energy called photons. Different forms of radiant energy differ with respect to the energy of each of their photons. The energies of the photons of radio waves are much lower than for gamma radiation.

   In the wave view, as radiant energy moves away from the source, it has an effect on the space around it that can be described as a wave consisting of an oscillating electric field perpendicular to an oscillating magnetic field. Different forms of radiant energy differ with respect to the wavelengths and frequencies of these oscillating waves. The waves associated with radio waves have much longer wavelength than the waves associated with gamma radiation.

Complete the following statements by writing the word or phrase in each blank that best completes the thought.

3. Atoms that have the same number of protons but different numbers of neutrons are called **isotopes**. They have the same atomic number but different mass numbers.

4. The **atomic number** for an atom is equal to the number of protons in an atom's nucleus. It establishes the element's identity.

5. The **mass number** for an atom is equal to the sum of the numbers of protons and neutrons in an atom's nucleus.

6. **Energy** is the capacity to do work.

7. **Kinetic energy** is the capacity to do work due to the motion of an object.

8. The **Law of Conservation of Energy** states that energy can neither be created nor destroyed, but it can be transferred from one system to another and changed from one form to another.

9. **Potential energy** is a retrievable, stored form of energy an object possesses by virtue of its position or state.

10. The more stable a system is, the **lower** (higher or lower) its potential energy.

11. When a system shifts from a less stable to a more stable state, energy is **released** (absorbed or released).

12. The **ground state** of an atom is the condition in which its electrons are in the orbitals that give it the lowest possible potential energy.

13. The **excited state** of an atom is the condition in which one or more of its electrons are in orbitals that do not represent the lowest possible potential energy.

Key Ideas Answers

14. Because **protons** and **neutrons** reside in the nucleus of atoms, they are called nucleons.

16. There are two forces among the particles within the nucleus. The first, called the **electrostatic** force, is the force between electrically charged particles. The second force, called the **strong** force, holds nucleons (protons and neutrons) together.

18. Larger atoms with more protons in their nuclei require a greater **ratio** of neutrons to protons to balance the increased repulsion between protons.

20. One of the ways that heavy nuclides change to move back into the band of stability is to release two protons and two neutrons in the form of a helium nucleus, called an **alpha** particle.

22. When a radioactive nuclide has a neutron to proton ratio that is **too low**, it can move toward stability in one of two ways, positron emission or electron capture. In positron emission (β^+), a proton becomes a neutron and a positron. The neutron stays in the nucleus, and the positron speeds out of the nucleus at high velocity.

24. Because radioactive decay leads to more stable products, it always **releases** energy, some in the form of kinetic energy of the moving product particles, and some in the form of gamma rays. Gamma rays can be viewed as a stream of high-energy **photons**.

26. Different isotopes of the same element, which share the same chemical characteristics, often undergo very **different** nuclear reactions.

28. Nuclear reactions, in general, give off a lot more **energy** than chemical reactions.

30. Rates of radioactive decay are described in terms of half-life, the time it takes for **one-half** of a sample to disappear.

32. As alpha particles, which move at up to 10% the speed of light, move through the tissues of our bodies, they **pull** electrons away from the tissue's atoms.

34. Gamma photons are ionizing radiation, because they can **excite** electrons enough to actually remove them from atoms.

36. Because beta particles are smaller than alpha particles, and because they can move up to 90% the speed of light, they are about **100** times as penetrating as alpha particles.

38. Gamma photons that penetrate the body do more damage to **rapidly reproducing** cells than to others.

40. To date an artifact, a portion of it is analyzed to determine the $^{14}C/^{12}C$ **ratio**, which can be used to determine its age.

42. It takes about **10,000** times as much energy to remove a proton or a neutron from the nucleus of a hydrogen-2 atom as to remove its one electron.

44. There appears to be something stable about having 2, 8, 20, 28, 50, 82, or 126 protons or neutrons. The nuclides with **double magic numbers** have very high stability.

46. For atoms **larger than** iron-56, splitting larger atoms to form more stable, smaller atoms releases energy.

48. The nuclear reactor in a nuclear power plant is really just a big **furnace** that generates heat to convert liquid water to steam that turns a steam turbine generator to produce electricity.

Problems Key

Section 18.1: The Nucleus and Radioactivity

50. A radioactive nuclide that has an atomic number of 88 and a mass (nucleon) number of 226 is used in radiation therapy. Write its nuclide symbol in the form of $^A_Z X$. Write two other ways to represent this nuclide. *(Objs #2 & 4)*

   $^{226}_{88}Ra$ ^{226}Ra radium-226

52. A radioactive nuclide that has 6 protons and 5 neutrons is used to generate positron emission tomography (PET) brain scans. Write its nuclide symbol in the form of $^A_Z X$. Write two other ways to represent this nuclide. *(Objs #3 & 4)*

   $^{11}_{6}C$ ^{11}C carbon-11

54. A radioactive nuclide with the symbol $^{40}_{19}K$ is used for geologic dating. What is its atomic number and mass (nucleon) number? Write two other ways to represent this nuclide. *(Objs #2 & 4)*

   atomic number = 19   mass number = 40 ^{40}K potassium-40

56. A radioactive nuclide with the symbol $^{111}_{49}In$ is used to label blood platelets. How many protons and how many neutrons does each atom have? Write two other ways to represent this nuclide. *(Objs #3 & 4)*

   49 protons   62 neutrons ^{111}In indium-111

58. Barium-131 is used to detect bone tumors. What is its atomic number and mass number? How many protons and how many neutrons are in the nuclei of each atom? Write two other ways to represent this nuclide. *(Objs #2, 3, & 4)*

   atomic number = 56   mass number = 131   56 protons   75 neutrons $^{131}_{56}Ba$ ^{131}Ba

60. The radioactive nuclide with the symbol ^{75}Se is used to measure the shape of the pancreas. What is its atomic number and mass number? How many protons and how many neutrons are in the nuclei of each atom? Write two other ways to represent this nuclide. *(Objs #2, 3, & 4)*

   atomic number = 34   mass number = 75   34 protons   41 neutrons $^{75}_{34}Se$ selenium 75

62. Describe the two opposing forces between particles in the nucleus, and with reference to these forces, explain why the ratio of neutrons to protons required for a stable nuclide increases as the number of protons in a nucleus increases. *(Obj #5)*

> The first force among the particles in the nucleus is called the electrostatic force (or electromagnetic force). It is the force between electrically charged particles. Opposite charges attract each other, and like charges repel each other, so the positively charged protons in the nucleus of an atom have an electrostatic force pushing them apart. The second force, called the strong force, holds nucleons (protons and neutrons) together.

> You can think of neutrons as the nuclear glue that allows protons to stay together in the nucleus. Because neutrons are uncharged, there are no electrostatic repulsions among them and other particles, but each neutron in the nucleus of an atom is attracted to other neutrons and to protons by the strong force. Therefore, adding neutrons to a nucleus leads to more attractions holding the particles of the nucleus together without causing increased repulsion between those particles.

> Larger atoms with more protons in their nuclei require a greater ratio of neutrons to protons to balance the increased electrostatic repulsion between protons.

64. Write a general description of the changes that take place in alpha emission. Write the two symbols used for an alpha particle. Write the general equation for alpha emission, using X for the reactant element symbol, Y for the product element symbol, Z for atomic number, and A for mass number. *(Objs #6 & 7)*

> One of the ways that heavy nuclides change to move back into the band of stability is to release two protons and two neutrons in the form of a helium nuclei, called an alpha particle. In nuclear equations for alpha emission, the alpha particle is described as either α or $^{4}_{2}He$. In alpha emission, the radioactive nuclide changes into a different element that has an atomic number that is two lower and a mass number that is four lower.

$$^{A}_{Z}X \;\rightarrow\; ^{A-4}_{Z-2}Y + ^{4}_{2}He$$

66. Write a general description of the changes that take place in positron emission. Write the three symbols used for a positron. Write the general equation for positron emission, using X for the reactant element symbol, Y for the product element symbol, Z for atomic number, and A for mass number. *(Objs #6 & 7)*

> In positron emission (β^+), a proton becomes a neutron and an anti-electron. The neutron stays in the nucleus, and the positron speeds out of the nucleus at high velocity.

$$p \;\rightarrow\; n + e^+$$

> In nuclear equations for positron emission, the electron is described as either β^+, $^{0}_{+1}e$, or $^{0}_{1}e$. In positron emission, the radioactive nuclide changes into a different element that has an atomic number that is one lower but that has the same mass number.

$$^{A}_{Z}X \;\rightarrow\; ^{A}_{Z-1}Y + ^{0}_{+1}e$$

68. Consider three isotopes of bismuth, $^{202}_{83}\text{Bi}$, $^{209}_{83}\text{Bi}$, and $^{215}_{83}\text{Bi}$. Bismuth-209 is stable. One of the other nuclides undergoes beta emission, and the remaining nuclide undergoes electron capture. Identify the isotope that makes each of these changes, and explain your choices. *(Obj #8)*

> Bismuth-202, which has a lower neutron to proton ratio than the stable bismuth-209, undergoes electron capture, which increases the neutron to proton ratio. Bismuth-215, which has a higher neutron to proton ratio than the stable bismuth-209, undergoes beta emission, which decreases the neutron to proton ratio.

71. What nuclear process or processes lead to each of the results listed below? The possibilities are alpha emission, beta emission, positron emission, electron capture, and gamma emission. *(Obj #6)*

    a. Atomic number increases by 1.   **beta emission**
    b. Mass number decreases by 4.   **alpha emission**
    c. No change in atomic number or mass number.   **gamma emission**
    d. The number of protons decreases by 1.   **positron emission or electron capture**
    e. The number of neutrons decreases by 1.   **beta emission**
    f. The number of protons decreases by 2.   **alpha emission**

73. Describe the differences between nuclear reactions and chemical reactions. *(Obj #10)*

> - Nuclear reactions involve changes in the nucleus, as opposed to chemical reactions that involve the loss, gain, and sharing of electrons.
>
> - Different isotopes of the same element often undergo very different nuclear reactions, whereas they all share the same chemical characteristics.
>
> - Unlike chemical reactions, the rates of nuclear reactions are unaffected by temperature, pressure, and the other atoms to which the radioactive atom is bonded.
>
> - Nuclear reactions, in general, give off a lot more energy than chemical reactions.

74. Explain why $^{38}_{17}\text{Cl}$ and $^{38}_{17}\text{Cl}^-$ are very different chemically and why they each undergo identical nuclear reactions.

> These two particles differ only in the number of their electrons. Because chemical reactions involve the loss, gain, and sharing of electrons, the number of electrons for an atom is very important for chemical changes. The attractions between charges are also very important for chemical changes, so the different charges of these particles change how they act chemically.
>
> Nuclear reactions are determined by the stability of nuclei, which is related to the number of protons and neutrons in the nuclei. They are unaffected by the number of electrons and unaffected by the overall charge of the particles. Therefore, these particles, which both have 17 protons and 21 neutrons in their nuclei, have the same nuclear stability and undergo the same nuclear changes.

76. Marie Curie won the Nobel Prize for physics in 1903 for her study of radioactive nuclides, including polonium-218 (which was named after her native country, Poland). Polonium-218 undergoes alpha emission. Write the nuclear equation for this change. *(Obj #12)*

$$^{218}_{84}Po \rightarrow \; ^{214}_{82}Pb + \; ^{4}_{2}He$$

78. Cobalt-60, which is the most common nuclide used in radiation therapy for cancer, undergoes beta emission. Write the nuclear equation for this reaction. *(Obj #12)*

$$^{60}_{27}Co \rightarrow \; ^{60}_{28}Ni + \; ^{0}_{-1}e$$

80. Carbon-11 is used in PET brain scans because it emits positrons. Write the nuclear equation for the positron emission of carbon-11. *(Obj #12)*

$$^{11}_{6}C \rightarrow \; ^{11}_{5}B + \; ^{0}_{+1}e$$

82. Mercury-197 was used in the past for brain scans. Its decay can be detected, because this nuclide undergoes electron capture, which forms an excited atom that then releases a gamma photon that escapes the body and strikes a detector. Write the nuclear equation for the electron capture by mercury-197. *(Obj #12)*

$$^{197}_{80}Hg + \; ^{0}_{-1}e \rightarrow \; ^{197}_{79}Au$$

84. Complete the following nuclear equations. *(Obj #13)*

   a. $^{90}_{38}Sr \rightarrow \; ^{90}_{39}Y + \; ^{0}_{-1}\mathbf{e}$

   b. $^{17}_{9}F \rightarrow \; ^{17}_{8}O + \; ^{0}_{+1}\mathbf{e}$

   c. $^{222}_{86}Rn \rightarrow \; ^{218}_{84}Po + \; ^{4}_{2}\mathbf{Hc}$

   d. $^{18}_{9}F + \; ^{0}_{-1}\mathbf{e} \rightarrow \; ^{18}_{8}O$

   e. $^{235}_{92}U \rightarrow \; ^{231}_{90}\mathbf{Th} + \; ^{4}_{2}He$

   f. $^{7}_{4}Be + \; ^{0}_{-1}e \rightarrow \; ^{7}_{3}\mathbf{Li}$

   g. $^{52}_{26}Fe \rightarrow \; ^{52}_{25}\mathbf{Mn} + \; ^{0}_{+1}e$

   h. $^{3}_{1}H \rightarrow \; ^{3}_{2}\mathbf{He} + \; ^{0}_{-1}e$

   i. $^{14}_{6}C \rightarrow \; ^{14}_{7}N + \; ^{0}_{-1}e$

   j. $^{118}_{54}\mathbf{Xe} \rightarrow \; ^{118}_{53}I + \; ^{0}_{+1}e$

   k. $^{204}_{84}\mathbf{Po} + \; ^{0}_{-1}e \rightarrow \; ^{204}_{83}Bi$

   l. $^{238}_{92}U \rightarrow \; ^{234}_{90}Th + \; ^{4}_{2}He$

86. Silver-117 atoms undergo three beta emissions before they reach a stable nuclide. What is the final product?

$$^{117}_{47}Ag \rightarrow \; ^{117}_{48}Cd \rightarrow \; ^{117}_{49}In \rightarrow \; ^{117}_{50}\mathbf{Sn}$$

88. Tellurium-116 atoms undergo two electron captures before they reach a stable nuclide. What is the final product?

$$^{116}_{52}Te \rightarrow \; ^{116}_{51}Sb \rightarrow \; ^{116}_{50}\mathbf{Sn}$$

90. Samarium-142 atoms undergo two positron emissions before they reach a stable nuclide. What is the final product?

$$^{142}_{62}\text{Sm} \quad \rightarrow \quad ^{142}_{61}\text{Pm} \quad \rightarrow \quad \mathbf{^{142}_{60}Nd}$$

92. Bismuth-211 atoms undergo an alpha emission and beta emission before they reach a stable nuclide. What is the final product?

$$^{211}_{83}\text{Bi} \quad \rightarrow \quad ^{207}_{81}\text{Tl} \quad \rightarrow \quad \mathbf{^{207}_{82}Pb}$$

94. Complete the following nuclear equations describing the changes that led to the formation of previously undiscovered nuclides. *(Obj #13)*

   a. $^{246}_{96}\text{Cm} + ^{12}_{6}\text{C} \rightarrow \mathbf{^{252}_{102}No} + 6\,^{1}_{0}\text{n}$

   b. $^{249}_{98}\mathbf{Cf} + ^{16}_{8}\text{O} \rightarrow ^{263}_{106}\text{Sg} + 2\,^{1}_{0}\text{n}$

   c. $^{240}_{95}\text{Am} + ^{4}_{2}\text{He} \rightarrow ^{243}_{97}\text{Bk} + \mathbf{^{1}_{0}n}$

   d. $^{252}_{98}\text{Cf} + ^{10}_{5}\mathbf{B} \rightarrow ^{257}_{103}\text{Lr} + 5\,^{1}_{0}\text{n}$

96. In February 1981, the first atoms of the element Bohrium-262, $^{262}_{107}\text{Bh}$, were made from the bombardment of bismuth-209 atoms by chromium-54 atoms. Write a nuclear equation for this reaction. (One or more neutrons may be released in this type of nuclear reaction.) *(Obj #13)*

$$^{209}_{83}\text{Bi} + ^{54}_{24}\text{Cr} \rightarrow ^{262}_{107}\text{Bh} + ^{1}_{0}\text{n}$$

98. Cesium-133, which is used in radiation therapy, has a half-life of 30 years. How long before a sample decreases to 1/4 of what was originally there? *(Obj #14)*

   It takes 2 half-lives for a radioactive nuclide to decay to ¼ of its original amount (½ × ½). Therefore, it will take **60 years** for cesium-133 to decrease to ¼ of what was originally there.

100. Phosphorus-32, which is used for leukemia therapy, has a half-life of 14.3 days. What fraction of a sample is left in 42.9 days? *(Obj #15)*

   The 42.9 days is 3 half-lives (42.9/14.3), so the fraction remaining would be **1/8** (½ × ½ × ½).

102. Explain why short-lived radioactive nuclides are found in nature. *(Obj #16)*

   Although short-lived radioactive nuclides disappear relatively quickly once they form, they are constantly being replenished because they are products of other radioactive decays. There are three long-lived radioactive nuclides (uranium-235, uranium-238, and thorium-232) that are responsible for many of the natural radioactive isotopes.

105. The first six steps of the decay series for uranium-235 consist of the changes alpha emission, beta emission, alpha emission, beta emission, alpha emission, and alpha emission. Write the products formed after each of these six steps.

$$^{235}_{92}\text{U} \rightarrow ^{231}_{90}\text{Th} \rightarrow ^{231}_{91}\text{Pa} \rightarrow ^{227}_{89}\text{Ac} \rightarrow ^{227}_{90}\text{Th} \rightarrow ^{223}_{88}\text{Ra} \rightarrow ^{219}_{86}\text{Rn}$$

107. In the first five steps of the decay series for thorium-232, the products are $^{228}_{88}Ra$, $^{228}_{89}Ac$, $^{228}_{90}Th$, $^{224}_{88}Ra$, and $^{220}_{86}Rn$. Identify each of these steps as alpha emissions or beta emissions.

alpha, beta, beta, alpha, alpha

109. Explain why alpha particles are considered ionizing radiation. *(Obj #19)*

As alpha particles, which move at up to 10% the speed of light, move through the tissues of our bodies, they drag electrons away from the tissue's atoms. Remember that alpha particles are helium nuclei, so they each have a +2 charge. Thus, as the alpha particle moves past an atom or molecule, it attracts the particle's electrons. One of the electrons might be pulled toward the passing alpha particle enough to escape, but it might not be able to catch up to the fast moving alpha particle. The electron lags behind the alpha particle and is quickly incorporated into another atom or molecule, forming an anion, and the particle that lost the electron becomes positively charged. The alpha particle continues on its way creating many ions before it is slowed enough for electrons to catch up to it and neutralize its charge.

111. Explain why gamma photons are considered ionizing radiation. *(Obj #19)*

Gamma photons can excite electrons enough to actually remove them from atoms.

113. What types of tissues are most sensitive to emission from radioactive nuclides? Why do radiation treatments do more damage to cancer cells than to regular cells? Why are children more affected by radiation than adults are? *(Obj #21)*

The greatest effect is on tissues with rapidly reproducing cells where there are more frequent chemical changes. This is why nuclear emissions have a greater effect on cancerous tumors with their rapidly reproducing cells and on children, who have more rapidly reproducing cells than adults.

115. Why do you think radium-226 concentrates in our bones?

Because both radium and calcium are alkaline earth metals in group 2 on the periodic table, they combine with other elements in similar ways. Therefore, if radioactive radium-226 is ingested, it concentrates in the bones in substances that would normally contain calcium.

117. Why are alpha particles more damaging to tissues when the source is ingested than gamma rays would be? *(Obj #24)*

Alpha and beta particles lose all of their energy over a very short distance, so they can do more damage to localized areas in the body than the same number of gamma photons would.

Section 18.2: Uses of Radioactive Substances

118. Describe how cobalt-60 is used to treat cancer. *(Obj #25)*

Cobalt-60 emits ionizing radiation in the form of beta particles and gamma photons. Gamma photons penetrate the body and do more damage to rapidly reproducing cells that others. Typically, a focused beam of gamma photons from cobalt-60 is directed at a cancerous tumor. The gamma photons enter the tumor and create ions and free radicals that damage the tumor cells to shrink the tumor.

120. Explain how PET can show dynamic processes in the body, such as brain activity and blood flow. *(Obj #27)*

To get a PET scan of a patient, a solution that contains a positron-emitting substance is introduced into the body. The positrons that the radioactive atoms emit collide with electrons, and they annihilate each other, creating two gamma photons that move out in opposite directions.

$$e^+ \rightarrow \quad \leftarrow e^-$$

positron-electron collision
followed by the creation of
two gamma ray photons

$$\leftarrow \gamma\text{-ray} \qquad \gamma\text{-ray} \rightarrow$$

These photons can be detected, and the data can be computer-analyzed to yield images that show where in the body the radioactive substances collected. Depending on the nuclide used and the substance into which it is incorporated, the radioactive substance will move to a specific part of the body.

122. Describe how carbon-14 (radiocarbon) dating of artifacts is done. *(Obj #29)*

Carbon-14 atoms are constantly being produced in our upper atmosphere through neutron bombardment of nitrogen atoms.

$$^{14}_{7}N + ^{1}_{0}n \rightarrow ^{14}_{6}C + ^{1}_{1}H$$

The carbon-14 formed is quickly oxidized to form carbon dioxide, CO_2, which is converted into many different substances in plants. When animals eat the plants, the carbon-14 becomes part of the animal too. For these reasons, carbon-14 is found in all living things. The carbon-14 is a beta-emitter with a half-life of 5730 years (±40 years), so as soon as it becomes part of a plant or animal, it begins to disappear.

$$^{14}_{6}C \rightarrow ^{14}_{7}N + ^{0}_{-1}e$$

As long as a plant or animal is alive, the intake of carbon-14 balances the decay so that the ratio of ^{14}C to ^{12}C remains constant at about 1 in 1,000,000,000,000. When the plant or animal dies, it stops taking in fresh carbon, but the carbon-14 it contains continues to decay. Thus the ratio of ^{14}C to ^{12}C drops steadily. Therefore, to date an artifact, a portion of it is analyzed to determine the $^{14}C/^{12}C$ ratio, which can be used to calculate its age.

Initially, it was thought that determination of the age of something using this technique would be simple. For example, if the $^{14}C/^{12}C$ ratio had dropped to one-half of the ratio found in the air today, it would be considered to be about 5730 years old. A $^{14}C/^{12}C$ ratio of one-fourth of the ratio found in the air today would date it as 11,460 years old (two half-lives). This only works if we can assume that the $^{14}C/^{12}C$ ratio in the air was the same when the object died as it is now, and scientists have discovered that this is not strictly true.

Study of very old trees, such as the bristlecone pines in California, have allowed researchers to develop calibration curves that adjust the results of radiocarbon dating experiments for the variation in the $^{14}C/^{12}C$ ratio that go back about 10,000 years. These calibration curves are now used to get more precise dates for objects.

124. Describe how iridium-192 can be used to find leaks in welded pipe joints. *(Obj #31)*

The radioactive Iridium-192 is introduced to the pipe, and the connection is wrapped on the outside with film. If there is a crack in the connection, radiation leaks out and exposes the film.

126. Explain how scientists use radioactive tracers. *(Obj #33)*

Unstable nuclides have been used as radioactive tracers that help researchers discover a wide range of things. For example, incorporating carbon-14 into molecules helped scientists to study many of the aspects of photosynthesis. Because the radiation emitted from the carbon-14 atoms can be detected outside of the system into which the molecules are placed, the changes that involve carbon can be traced. Phosphorus-32 atoms can be used to trace phosphorus-containing chemicals as they move from the soil into plants under various conditions. Carbon-14, hydrogen-3, and sulfur-35 have been used to trace the biochemical changes that take place in our bodies.

Section 18.3: Nuclear Energy

128. Explain how the binding energy per nucleon can be used to compare the stability of nuclides. *(Obj #35)*

The greater the strengths of the attractions between nucleons, the more stable the nucleus and the greater the difference in potential energy between the separate nucleons and the nucleus. This is reflected in a greater binding energy per nucleon.

130. Explain why $^{4}_{2}He$, $^{12}_{6}C$, $^{16}_{8}O$, and $^{20}_{10}Ne$ are especially stable. *(Obj #37)*

This can be explained by the fact that they have an even number of protons and an even number of the neutrons. Paired nucleons (like paired electrons) are more stable than unpaired ones.

132. Give two reasons why $^{16}_{8}O$ is more stable than $^{15}_{8}O$.

> Paired nucleons are more stable than unpaired ones, and oxygen-16 with 8 protons and 8 neutrons would have its nucleons paired. Oxygen-15 with an odd number of neutrons (7) would be less stable. Oxygen-16 also has double magic numbers, making it especially stable. Finally, oxygen-15 has too few neutrons to be stable.

134. Describe how heat is generated in a nuclear power plant. *(Obj #40)*

> Heat is generated by the chain reaction of uranium-235, which is initiated by the bombardment of uranium fuel that is about 3% uranium-235. The products are significantly more stable than the initial reactants, so the system shifts from higher potential energy to lower potential energy, releasing energy as increased kinetic energy of the product particles. Higher kinetic energy means higher temperature.

136. Describe the role of the moderator in a nuclear reactor. *(Obj #42)*

> Both uranium-235 and uranium-238 absorb fast neutrons, but if the neutrons are slowed down, they are much more likely to be absorbed by uranium-235 atoms than uranium-238 atoms. Therefore, in a nuclear reactor, the fuel rods are surrounded by a substance called a moderator, which slows the neutrons as they pass through it.
>
> Several substances have been used as moderators, but normal water is most common.

138. Nuclear wastes must be isolated from the environment for a very long time because they contain relatively long-lived radioactive nuclides, such as technetium-99 with a half-life of over 2.1×10^5 years. One proposed solution is to bombard the waste with neutrons so as to convert the long-lived nuclides into nuclides that decay more quickly. When technetium-99 absorbs a neutron, it forms technetium-100, which has a half-life of 16 seconds and forms stable ruthenium-100 by emitting a beta particle. Write the nuclear equations for these two changes.

$$^{99}_{43}Tc + ^{1}_{0}n \rightarrow ^{100}_{43}Tc$$

$$^{100}_{43}Tc \rightarrow ^{100}_{44}Ru + ^{0}_{-1}e$$

Additional Problems

141. A radioactive nuclide that has an atomic number of 53 and a mass (nucleon) number of 131 is used to measure thyroid function. Write its nuclide symbol in the form of $^{A}_{Z}X$. Write two other ways to symbolize this nuclide.

> $^{131}_{53}I$ ^{131}I iodine-131

143. A radioactive nuclide that has 11 protons and 13 neutrons is used to detect blood clots. Write its nuclide symbol in the form of $^{A}_{Z}X$. Write two other ways to symbolize this nuclide.

> $^{24}_{11}Na$ ^{24}Na sodium-24

145. A radioactive nuclide with the symbol $^{133}_{55}Cs$ is used in radiation therapy. What is its atomic number and mass (nucleon) number? Write two other ways to represent this nuclide.

   atomic number = 55   mass number = 133 ^{133}Cs cesium-133

147. A radioactive nuclide with the symbol $^{51}_{24}Cr$ is used to determine blood volume. How many protons and neutrons does each atom have? Write two other ways to represent this nuclide.

   24 protons   27 neutrons ^{51}Cr chromium-51

149. Gallium-67 is used to diagnose lymphoma. What is its atomic number and mass number? How many protons and how many neutrons are in the nuclei of each atom? Write two other ways to represent this nuclide.

   atomic number = 31   mass number = 67   31 protons   36 neutrons $^{67}_{31}Ga$ ^{67}Ga

151. The radioactive nuclide with the symbol ^{32}P is used to detect eye tumors. What is its atomic number and mass number? How many protons and how many neutrons are in the nuclei of each atom? Write two other ways to represent this nuclide.

   atomic number = 15   mass number = 32   15 protons   17 neutrons $^{32}_{15}P$ phosphorus-32

153. Consider three isotopes of neon, $^{18}_{10}Ne$, $^{20}_{10}Ne$, and $^{24}_{10}Ne$. Neon-20, which is the most abundant isotope of neon, is stable. One of the other nuclides undergoes beta emission, and the remaining nuclide undergoes positron emission. Identify the isotope that makes each of these changes, and explain your choices.

   Neon-18, which has a lower neutron to proton ratio than the stable neon-20, undergoes positron emission, which increases the neutron to proton ratio. Neon-24, which has a higher neutron to proton ratio than the stable neon-20, undergoes beta emission, which decreases the neutron to proton ratio.

155. Write the nuclear equation for the alpha emission of bismuth-189.

   $$^{189}_{83}Bi \rightarrow {}^{185}_{81}Tl + {}^{4}_{2}He$$

157. Phosphorus-32, which is used to detect breast cancer, undergoes beta emission. Write the nuclear equation for this reaction.

   $$^{32}_{15}P \rightarrow {}^{32}_{16}S + {}^{0}_{-1}e$$

159. Write the nuclear equation for the positron emission of potassium-40.

   $$^{40}_{19}K \rightarrow {}^{40}_{18}Ar + {}^{0}_{+1}e$$

161. Radioactive selenium-75, used to determine the shape of the pancreas, shifts to a more stable nuclide via electron capture. Write the nuclear equation for this change.

   $$^{75}_{34}Se + {}^{0}_{-1}e \rightarrow {}^{75}_{33}As$$

163. Germanium-78 atoms undergo two beta emissions before they reach a stable nuclide. What is the final product?

   $$^{78}_{32}Ge \rightarrow {}^{78}_{33}As \rightarrow {}^{78}_{34}\mathbf{Se}$$

165. Iron-52 atoms undergo one positron emission and one electron capture before they reach a stable nuclide. What is the final product?

   $$^{52}_{26}Fe \rightarrow {}^{52}_{25}Mn \rightarrow {}^{52}_{24}\mathbf{Cr}$$

167. Arsenic-69 atoms undergo one positron emission and one electron capture before they reach a stable nuclide. What is the final product?

$$^{69}_{33}As \quad \rightarrow \quad ^{69}_{32}Ge \quad \rightarrow \quad \mathbf{^{69}_{31}Ga}$$

170. Complete the following nuclear equations.

a. $^{249}_{98}Cf + ^{15}_{7}N \quad \rightarrow \quad \mathbf{^{259}_{105}Db} + 5\,^{1}_{0}n$

b. $\mathbf{^{249}_{98}Cf} + ^{10}_{5}B \quad \rightarrow \quad ^{257}_{103}Lr + 2\,^{1}_{0}n$

c. $^{121}_{51}Sb + ^{1}_{1}H \quad \rightarrow \quad ^{121}_{52}Te + \mathbf{^{1}_{0}n}$

172. Nitrogen-containing explosives carried by potential terrorists can be detected at airports by bombarding suspicious luggage with low-energy neutrons. The nitrogen-14 atoms absorb the neutrons, forming nitrogen-15 atoms. The nitrogen-15 atoms emit gamma photons of a characteristic wavelength that can be detected outside the luggage. Write a nuclear equation for the reaction that forms nitrogen-15 from nitrogen-14.

$$^{14}_{7}N + ^{1}_{0}n \quad \rightarrow \quad ^{15}_{7}N$$

174. In March 1984, the nuclide hassium-265, $^{265}_{108}Hs$, was made from the bombardment of lead-208 atoms with iron-58 atoms. Write a nuclear equation for this reaction. (One or more neutrons may be released in this type of nuclear reaction.)

$$^{208}_{82}Pb + ^{58}_{26}Fe \quad \rightarrow \quad ^{265}_{108}Hs + ^{1}_{0}n$$

176. Krypton-79, which is used to assess cardiovascular function, has a half-life of 34.5 hours. How long before a sample decreases to 1/8 of what was originally there?

It takes 3 half-lives for a radioactive nuclide to decay to 1/8 of its original amount ($\frac{1}{2}$ x $\frac{1}{2}$ x $\frac{1}{2}$). Therefore, it will take 103.5 hours (**104 hours** to three significant figures) for krypton-79 to decrease to 1/8 of what was originally there.

178. Iron-59, which is used to diagnose anemia, has a half-life of 45 days. What fraction of it is left in 90 days?

The 90 days is 2 half-lives (90/45), so the fraction remaining would be **1/4** ($\frac{1}{2}$ x $\frac{1}{2}$).